AF789490

# 346
# Topics in Current Chemistry

*Editorial Board:*

H. Bayley, Oxford, UK
K.N. Houk, Los Angeles, CA, USA
G. Hughes, CA, USA
C.A. Hunter, Sheffield, UK
K. Ishihara, Chikusa, Japan
M.J. Krische, Austin, TX, USA
J.-M. Lehn, Strasbourg Cedex, France
R. Luque, Córdoba, Spain
M. Olivucci, Siena, Italy
J.S. Siegel, Nankai District, China
J. Thiem, Hamburg, Germany
M. Venturi, Bologna, Italy
C.-H. Wong, Taipei, Taiwan
H.N.C. Wong, Shatin, Hong Kong

## Aims and Scope

The series Topics in Current Chemistry presents critical reviews of the present and future trends in modern chemical research. The scope of coverage includes all areas of chemical science including the interfaces with related disciplines such as biology, medicine and materials science.

The goal of each thematic volume is to give the non-specialist reader, whether at the university or in industry, a comprehensive overview of an area where new insights are emerging that are of interest to larger scientific audience.

Thus each review within the volume critically surveys one aspect of that topic and places it within the context of the volume as a whole. The most significant developments of the last 5 to 10 years should be presented. A description of the laboratory procedures involved is often useful to the reader. The coverage should not be exhaustive in data, but should rather be conceptual, concentrating on the methodological thinking that will allow the non-specialist reader to understand the information presented.

Discussion of possible future research directions in the area is welcome.

Review articles for the individual volumes are invited by the volume editors.

**Readership: research chemists at universities or in industry, graduate students.**

More information about this series at
http://www.springer.com/series/128

Guangbin Dong
Editor

# C-C Bond Activation

With contributions by

N. Cramer · A. Dermenci · G. Dong · C.J. Douglas ·
A.M. Dreis · X.-F. Fu · Y. Gao · W.D. Jones · C.-H. Jun ·
J.S. Kingsbury · D.C. Moebius · Y. Nakao · J.-W. Park ·
E. Parker · V.L. Rendina · L. Souillart · T. Xu · Z.-X. Yu

Springer

*Editor*
Guangbin Dong
The University of Texas at Austin
Department of Chemistry & Biochemistry
Austin
USA

ISSN 0340-1022 ISSN 1436-5049 (electronic)
ISBN 978-3-662-52397-1 ISBN 978-3-642-55055-3 (eBook)
DOI 10.1007/978-3-642-55055-3
Springer Heidelberg New York Dordrecht London

© Springer-Verlag Berlin Heidelberg 2014
Softcover reprint of the hardcover 1st edition 2014
This work is subject to copyright. All rights are reserved by the Publisher, whether the whole or part of the material is concerned, specifically the rights of translation, reprinting, reuse of illustrations, recitation, broadcasting, reproduction on microfilms or in any other physical way, and transmission or information storage and retrieval, electronic adaptation, computer software, or by similar or dissimilar methodology now known or hereafter developed. Exempted from this legal reservation are brief excerpts in connection with reviews or scholarly analysis or material supplied specifically for the purpose of being entered and executed on a computer system, for exclusive use by the purchaser of the work. Duplication of this publication or parts thereof is permitted only under the provisions of the Copyright Law of the Publisher's location, in its current version, and permission for use must always be obtained from Springer. Permissions for use may be obtained through RightsLink at the Copyright Clearance Center. Violations are liable to prosecution under the respective Copyright Law.
The use of general descriptive names, registered names, trademarks, service marks, etc. in this publication does not imply, even in the absence of a specific statement, that such names are exempt from the relevant protective laws and regulations and therefore free for general use.
While the advice and information in this book are believed to be true and accurate at the date of publication, neither the authors nor the editors nor the publisher can accept any legal responsibility for any errors or omissions that may be made. The publisher makes no warranty, express or implied, with respect to the material contained herein.

Printed on acid-free paper

Springer is part of Springer Science+Business Media (www.springer.com)

# Preface

Stimulated by the need for green and more efficient approaches to functionalize hydrocarbon feedstocks, the vast development of carbon–hydrogen (C–H) bond activation has been evident in the past decade. In contrast, the related chemical processes involving activating carbon–carbon (C–C) bonds, another equally abundant chemical bond, have received much less attention until recently. The reason why C–C bond activation has been overlooked is straightforward. It is widely accepted that C–C bond formation is central to organic synthesis, such as cross-coupling, cycloaddition, alkylation, and aldol reactions; this is a constructive process. The question is why would people want to cleave C–C bonds? Is it a destructive process? As the editor of this volume, I hope by reading these contributions you will realize the answer to this question is clearly "no".

From a historical viewpoint, C–C bond cleavage reactions, such as sigmatropic rearrangements (e.g., Cope and Claisen rearrangements), retro-aldol, ozonolysis of alkenes, Beckman rearrangement, etc., have found broad applications in complex molecule synthesis. With the expansion of organotransition metal chemistry in the last 30 years, the unique reactivity of organometallic species adds new possibilities to this field. The goal of modern C–C bond activation is far from just splitting one molecule into two; instead, it converts relatively simple compounds into more complex, often value-added, products. More frequently, breaking one C–C bond allows two or more new C–C (or C–X) bonds to form, as shown in many chapters of this volume. Overall, C–C bond activation can not just be highly constructive; it can also provide opportunities to develop novel transformations through employing such distinct modes of reactivity.

This volume contains eight chapters from leading experts in the field, and is expected to provide an overview of the latest developments in C–C bond activation. It begins with mechanistic studies of transition metal-mediated C–C bond action by Prof. W.D. Jones, which laid the foundation for further discovery of catalytic reactions. The following chapters from Profs. Y. Nakao, C.-H. Jun, and C. J. Douglas focus on different strategies to activate unstrained C–C bonds, namely, C–CN bond activation, metal-organic cooperative catalysis, and activation of 8-acylquinolines.

The volume then slightly shifts direction to a Lewis acid-catalyzed diazoalkane–carbonyl homologation reaction by Prof. J.S. Kingsbury, followed by a thorough discussion of asymmetric transformations via C–C bond cleavage contributed by Prof. N. Cramer. The volume closes with two chapters covering activation of strained-ring systems: cycloaddition of cyclopropanes from Prof. Z. Yu and activation of four-membered cyclic ketones from my own research group.

As the editor, I am grateful to all the contributors for providing such high quality and in-depth review chapters. I am also indebted to Prof. M.J. Krische's recommendation for taking on this task. Given limited space and time, unfortunately, not all the important contributions in the field could be included, e.g., oxidative C–C bond cleavage and decarboxylation reactions. I apologize for omitting these efforts in this short volume.

Austin, USA
February 2014

Guangbin Dong

# Contents

Contents

Top Curr Chem (2014) 346: 1–32
DOI: 10.1007/128_2013_491
© Springer-Verlag Berlin Heidelberg 2013
Published online: 17 November 2013

# Mechanistic Studies of Transition Metal-Mediated C–C Bond Activation

**William D. Jones**

**Abstract** Organometallic compounds have been found to be of use in cleaving C–C bonds, as strong metal–carbon bonds can be formed to replace the bond that is broken. Studies of the mechanism of C–C cleavage can provide insight into how these bonds can be cleaved, and can give valuable information that can be used to develop new strategies for breaking C–C bonds and using the products in catalysis. In this chapter, we will examine a number of systems where mechanistic information has been obtained in C–C cleavage.

**Keywords** Activation · Carbon–carbon · Cleavage · Mechanism · Thermodynamics

**Contents**



## 1 Introduction

Organometallic compounds have been found to be of use in cleaving C–C bonds, as strong metal–carbon bonds can be formed to replace the bond that is broken. Studies of the mechanism of C–C cleavage can provide insight into how these

---

W.D. Jones (✉)
Department of Chemistry, University of Rochester, Rochester, NY 14627, USA
e-mail: william.jones@rochester.edu

bonds can be cleaved, and can give valuable information that can be used to develop new strategies for breaking C–C bonds and using the products in catalysis. In this chapter we will examine a number of systems where mechanistic information has been obtained in C–C cleavage.

While chemists have devised many ways to make carbon–carbon bonds, only a few methods are known for cleaving carbon–carbon bonds. For example, the retro Diels–Alder reaction and the Cope rearrangement are common reactions that involve C–C cleavage in organic chemistry. The olefin metathesis reaction and the CO de-insertion reaction also represent common examples of C–C cleavage in organometallic chemistry. Despite these common examples, more general methods for cleaving C–C bonds are lacking.

Several general methods have been recognized as favoring C–C cleavage with transition metal complexes. Relief of ring strain has been used to open cyclopropanes. The attainment of aromaticity can also help C–C cleavage. Proximity has also been shown to assist C–C cleavage, such as the activation of 8-acylquinolines. Even aryl–methyl bonds can be cleaved if forced into proximity with the metal [1].

## 2 C–C Cleavage of Biphenylene

Biphenylene has served as a substrate for C–C cleavage as it provides a number of features that make this reaction possible. First, there is ring strain associated with the C–C ring, which makes the aryl–aryl C–C bond weak. The bond strength can be estimated using thermochemical data for biphenylene and biphenyl as indicated in Scheme 1. The first reaction represents the homolysis of the C–C bond in biphenylene, and while the heat of formation of biphenylene is known, the heat of formation of the diradical product is not. However, cleavage of two aryl C–H bonds in biphenyl will provide this same product, whose $\Delta H_f°$ can be obtained as 165.9 kcal mol$^{-1}$. This permits estimation for the biphenylene C–C bond as 65.4 kcal mol$^{-1}$, much weaker than the C–C bond in biphenyl (114.4 kcal mol$^{-1}$) (thermodynamic data from [2, 3]).

A second reason why biphenylene has been found to be a good substrate for C–C cleavage is that when a transition metal cleaves the bond by insertion via oxidative addition, *two* metal–aryl bonds are formed in the product. Metal–aryl bonds are among the strongest metal–carbon bonds (e.g., $D_{Ir-Ph}$ is 81 kcal mol$^{-1}$ in Cp*Ir(PMe$_3$)(Ph)H) [4], and this gives an added advantage to biphenylene as a substrate. A third advantage of biphenylene is that it has a $\pi$-system to which a metal can bind. This provides the metal with direct access to the carbon atoms whose bond will be cleaved, and hence new bonds can be formed before the C–C bond is broken.

**Scheme 1** Energetics of biphenylene cleavage

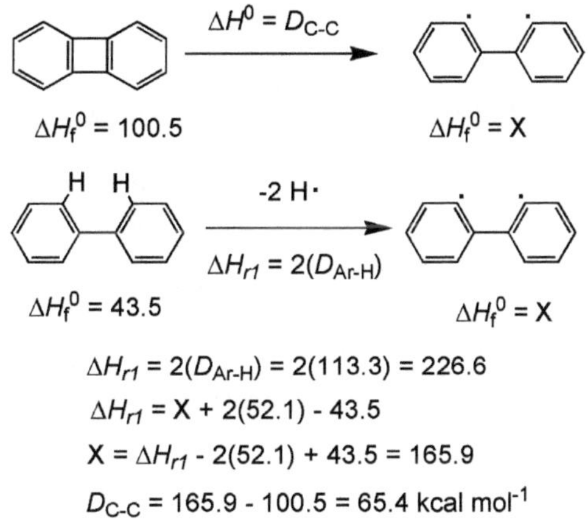


In one of the earliest reports, biphenylene was heated with $Cr(CO)_6$ to give 30–60% $\Delta^{9,9'}$-bifluorene and fluorenone (1). A suggested mechanism was that carbonyl insertion was involved, but few other details were reported [5].

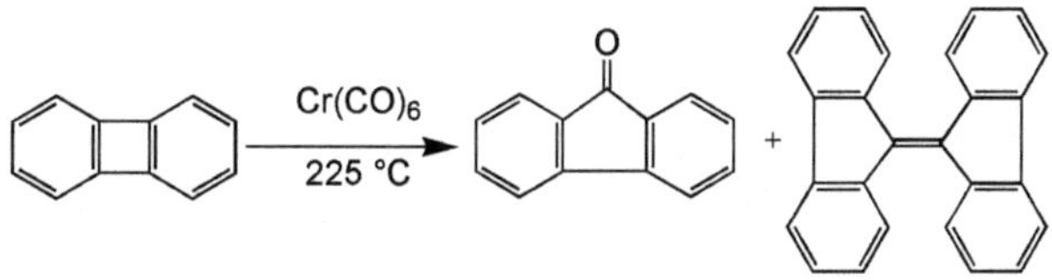


(1)

In another example, Eisch found that $Ni(cyclooctadiene)_2$ in the presence of bipyridine or phosphine ligands would insert into biphenylene, and the insertion adduct could be isolated when $PEt_3$ was employed. This species was unstable, however, and formed a dimer upon loss of one $PEt_3$ ligand. Heating this species to 146°C led to the formation of nickel metal and tetraphenylene (2), with the sequential isolation of these intermediates providing some evidence for the mechanism [6]. Vollhardt later reported that the catalytic dimerization of biphenylene to tetraphenylene could be carried out at 100°C using 10% Ni(cod) $(PMe_3)_2$ as catalyst [7].

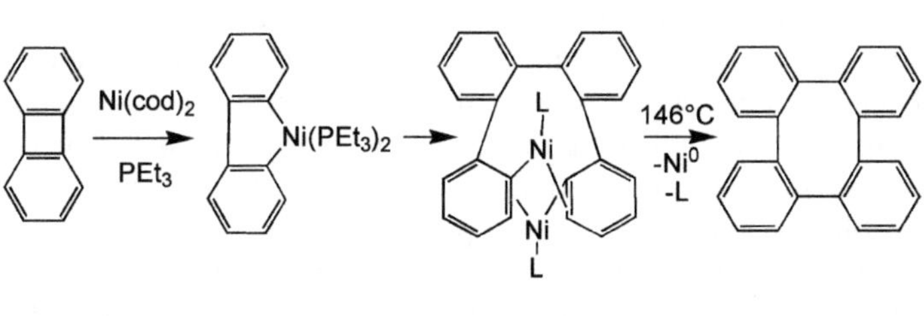


(2)

Crabtree also reported the insertion of Ir(I) into the C–C bond of biphenylene using $[Ir(cod)Cl]_2$. Here, an Ir(III) dimeric product was obtained that could be

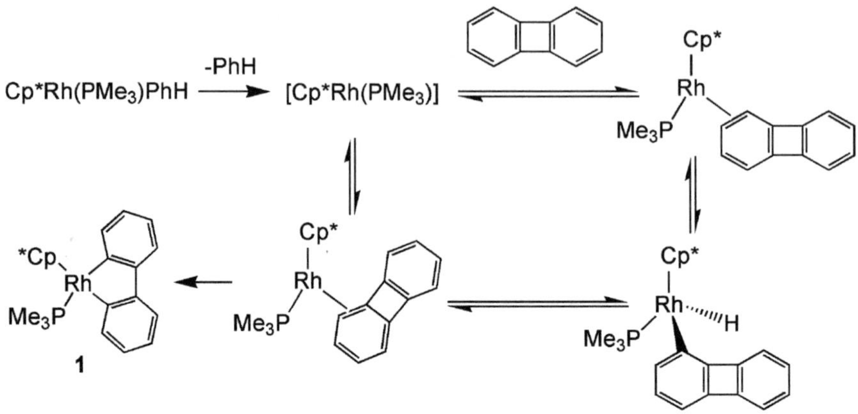


**Scheme 2** Cleavage of biphenylene by [Cp*Rh($PMe_3$)]

cleaved to monomers using phosphine ligands or CO, but no further insertion chemistry was observed (3) [8].

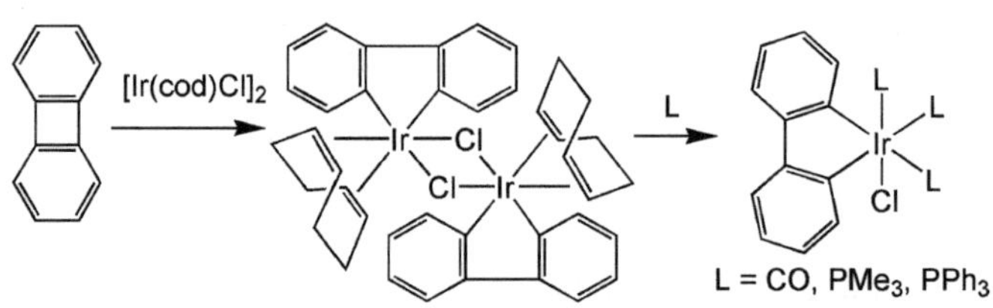


(3)

Our group first became interested in C–C cleavage while examining substrates for C–H bond activation. We had found that the reactive fragment [Cp*Rh($PMe_3$)] could be generated by photolysis of Cp*Rh($PMe_3$)$H_2$ or thermolysis of Cp*Rh($PMe_3$)PhH, and that this fragment could undergo oxidative addition with aliphatic and aromatic C–H bonds [9]. When biphenylene was examined as a substrate, activation of the α-C–H bond was observed at 65°C. Over the next few weeks, this C–H insertion product converted quantitatively to the C–C insertion product [10]. When the rearrangement was monitored in the presence of excess deuterated biphenylene, about 50% of the C–C insertion product contained the deuterated biphenylene, indicating that exchange was about as fast as C–C activation. A mechanism was proposed that was consistent with these observations involving $\eta^2$-biphenylene intermediates prior to C–H or C–C insertion, as similar $\eta^2$-arene complexes have been observed previously in this system prior to C–H oxidative addition (Scheme 2) [11]. It was also discovered that Cp*Rh($PMe_3$)$H_2$

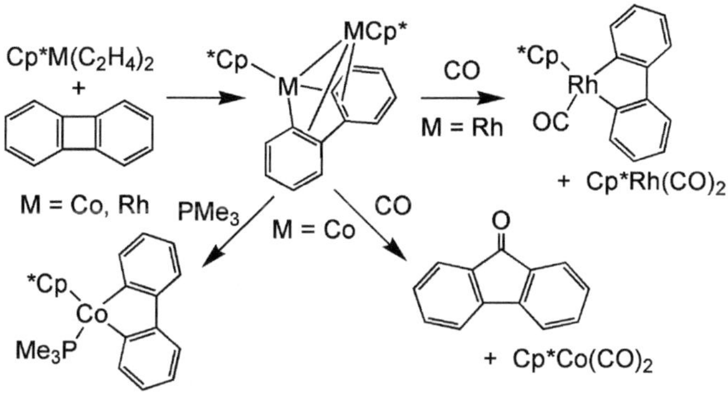


**Scheme 3** Cleavage of biphenylene by [Cp*M(CO)] (M = Co, Rh)

can act as a catalyst for the hydrogenation of biphenylene to biphenyl (i.e., C–C hydrogenolysis). The C–C insertion product **1** is not an intermediate in the catalysis, as **1** is resistant to hydrogenation. The related trisdimethylpyrazolylborate complex Tp′Rh($PMe_3$)$H_2$ also serves as a catalyst for biphenylene hydrogenolysis to biphenyl with a rate that is ~3× faster [12].

In related chemistry, both Cp*Rh($C_2H_4$)$_2$ and Cp*Co($C_2H_4$)$_2$ were found to react with biphenylene to give C–C insertion products that are dinuclear [13]. The rhodium dinuclear product could be cleaved with CO to give Cp*Rh(CO)$_2$ and Cp*Rh (CO)(2,2′-biphenylyl) whereas carbonylation of the cobalt complex gave Cp*Co (CO)$_2$ and fluorenone (Scheme 3). Cleavage of the cobalt dimer with $PMe_3$ led to Cp*Co($PMe_3$)(2,2′-biphenylyl), but the rhodium dimer did not form **1** even after reaction with excess $PMe_3$ at 160°C. As might be anticipated, the metal carbonyls Cp*Rh(CO)$_2$ and Cp*Co(CO)$_2$ serve as catalysts for the carbonylation of biphenylene to fluorenone at 160°C and 500 torr CO. The rhodium catalyst is stable but slow (1 t.o./day; t.o. = turnover) under these conditions, and the cobalt complex decomposes after a few turnovers.

Investigations of zerovalent group 10 metal complexes also revealed evidence for C–C cleavage of biphenylene. The platinum complex Pt($PEt_3$)$_3$ reacts with biphenylene at 120°C to give tetraphenylene [14]. Two platinum-containing intermediates are observed by $^{31}$P NMR spectroscopy. One was identified as the C–C insertion complex **2** and the other as the tetraphenylene insertion complex **3** (Scheme 4). Complex **2** forms if the reaction is carried out at 80°C with 1 equiv. of biphenylene. Complex **3** is formed cleanly if **2** is reacted with additional biphenylene at 80°C in the absence of $PEt_3$. Examination of the mechanism of the catalytic reaction was made by looking at the effects of added $PEt_3$ and biphenylene. It was found that the *ratio* of the concentrations [biphenylene]/[$PEt_3$] was the factor that controlled the resting state of the catalyst. The higher the ratio, the higher the **3/2** ratio is seen in the resting state. This observation was interpreted in terms of a mechanism involving reversible $PEt_3$ dissociation from **2** followed by reaction with biphenylene to give a Pt(IV) bis-biphenylyl

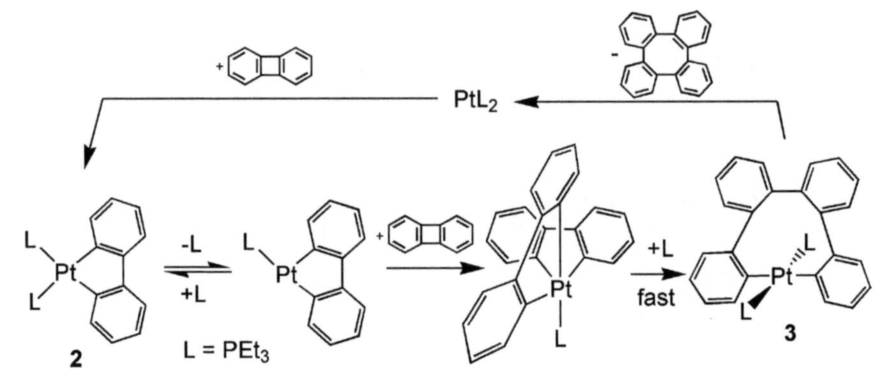


**Scheme 4** Catalytic dimerization of biphenylene by [$PtL_2$] (L = $PEt_3$)

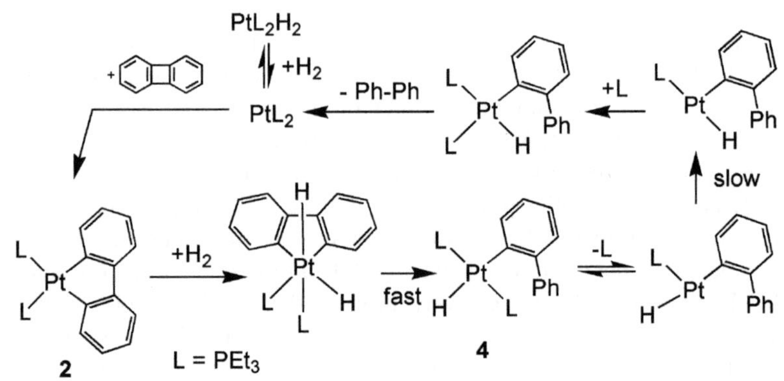


**Scheme 5** Hydrogenolysis of biphenylene by [$PtL_2$] (L = $PEt_3$)

intermediate that then reductively eliminates a C–C bond to give **3**. Complex **3** eliminates tetraphenylene at higher temperatures (120°C), and the $L_2Pt^0$ fragment then re-enters the cycle. It was determined that the back-reaction of the unsaturated intermediate with $PEt_3$ was 131 times faster than the forward reaction with biphenylene. Note that the mechanism for formation of tetraphenylene revealed here by the kinetics is different from that seen with nickel(0), where a binuclear intermediate was observed. The platinum catalysis was slow – about 1 t.o./week at 120°C and millimolar concentrations. The analogous palladium complex, however, showed rates of about 20 t.o./h under similar conditions. The only species observed during catalysis with palladium was the analog of **2**. With the platinum complex, the catalysis ultimately ends by competitive C–H activation which leads to inert $Pt(PEt_3)_2(aryl)_2$ complexes.

Addition of hydrogen during the catalysis with $Pt(PEt_3)_3$ leads to hydrogenolysis of the biphenylene C–C bond to form biphenyl [15]. The resting state during catalysis is the *trans*-hydrido biphenylyl complex **4** (Scheme 5). The rate of reaction of **2** with dihydrogen is not affected by added $PEt_3$, implying a direct reaction of **2** with dihydrogen to generate a platinum(IV) dihydride, which then rapidly reductively eliminates C–H to produce **4**. Reaction of **4** with dihydrogen produces biphenyl and $Pt(PEt_3)_2H_2$ and the reaction is strongly inhibited by added $PEt_3$, implying that the reaction proceeds by loss of phosphine followed by a slow *trans–cis* isomerization to allow reductive elimination of biphenyl.

Examination of other platinum phosphine derivatives showed variations on the above chemistry [16]. The use of bis-(di-*tert*-butylphosphino)methane allowed oxidative addition of the C–C bond of biphenylene, but no further reaction with biphenylene occurred (4). No reaction occurred with PhCCPh, or $H_2$, but $[Pt(PPh_2\text{-}t\text{-}Bu)_2]$ was seen to displace the phosphines. In view of the mechanistic studies with $Pt(PEt_3)_2$, the lack of reactivity with biphenylene is not unexpected due to the chelating ligand that would have to dissociate to permit addition of a second C–C bond. However, the lack of reaction with dihydrogen is surprising, as the corresponding reaction involving **2** was not inhibited by added $PEt_3$, implying that phosphine dissociation is not required.

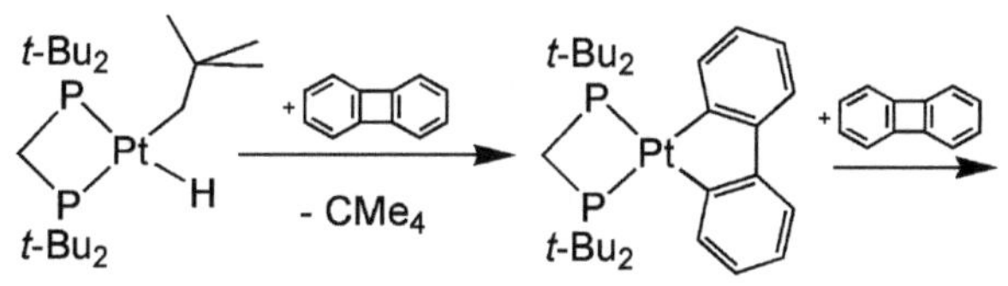


(4)

The fragment $[Pt(PPh_2\text{-}t\text{-}Bu)_2]$ was found to insert into the C–C bond of biphenylene, and could slowly catalyze the formation of tetraphenylene [16]. Here, however, intramolecular cyclometallation of the phosphine phenyl ring leads to an off-cycle dead-end intermediate, resulting in slow catalysis (5).

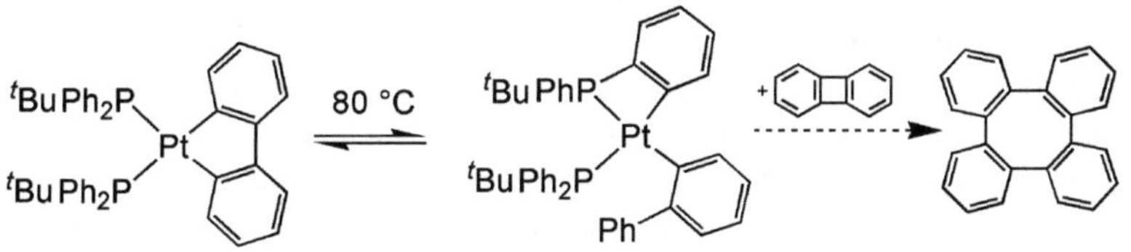


(5)

$Pt(PPh_3)_3$ was not found to react with biphenylene [16]. However, the insertion adduct (**5**) can be prepared from the reaction of $Pt(PPh_3)_2Cl_2$ with 2,2′-dilithiobiphenyl. This species then reacts with biphenylene, but phenylterphenylene is the major product (6). Terphenylene and tetraphenylene are formed as minor products, and some hexaphenylene (two distinct isomers are known [17]) is also observed. Use of deuterium labeled $PPh_3$ ligands revealed that one of the phosphine phenyl groups was incorporated into the terphenyl that is formed, indicating that P–C cleavage was involved in its formation.

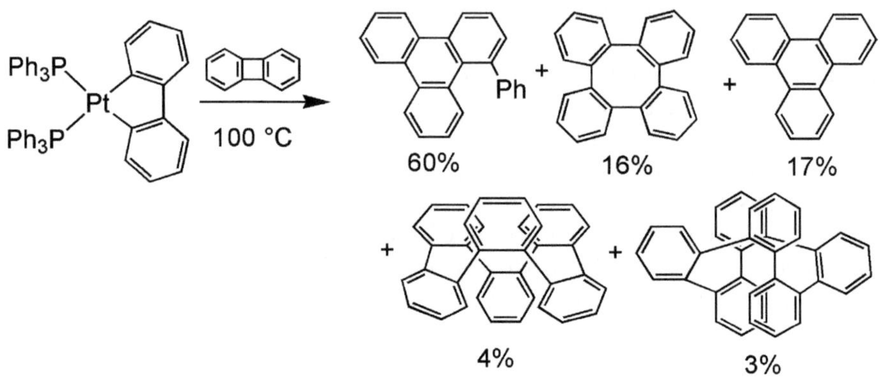


(6)

As mentioned above, early studies showed that nickel(0) also is capable of biphenylene C–C activation. The use of bis-(diisopropylphosphino)ethane as a ligand (dippe) allowed for the formation and study of several reactive complexes. In particular, the complexes (dippe)Ni(alkyne) were found to be catalysts for the formation of phenanthrenes from biphenylene and acetylenes [18]. Diphenylacetylene was the most active, providing 12 t.o./h at 70°C. Dimethylacetylene was slower, giving ~1 t.o./h. Acetylenes with electron-withdrawing groups tended to give alkyne cyclotrimerization products rather than phenanthrenes. Examination of the mechanism of reaction revealed that small quantities of $O_2$ were required to generate the active catalytic species. Titration of the reactants with oxygen shows a maximum in catalytic rate with 40 mol% $O_2$ added. The $^{31}$P NMR spectrum of the sample shows the formation of dippe phosphineoxide. This led to the proposal that the active catalyst was the alkyne complex of Ni(0), in which only reactive ligands were present (Scheme 6).

As the above chemistry appeared to occur by oxidizing the phosphine to remove the ligand from the coordination sphere, it was decided to look at hemi-labile ligands as a way to circumvent this problem. A P–N analog of the (dippe)Ni (alkyne) compounds was prepared using a dimethyl amino group in place of a diisopropylphosphine group. The labile $NMe_2$ group now rendered the nickel complex as a good catalyst for phenanthrene formation [19]. Both diphenylacetylene and *tert*-butylphenylacetylene showed catalytic product formation at 70°C (7). However, electron deficient acetylenes such as trifluoromethylphenylacetylene gave cyclotrimerization products instead, as seen with dippe.

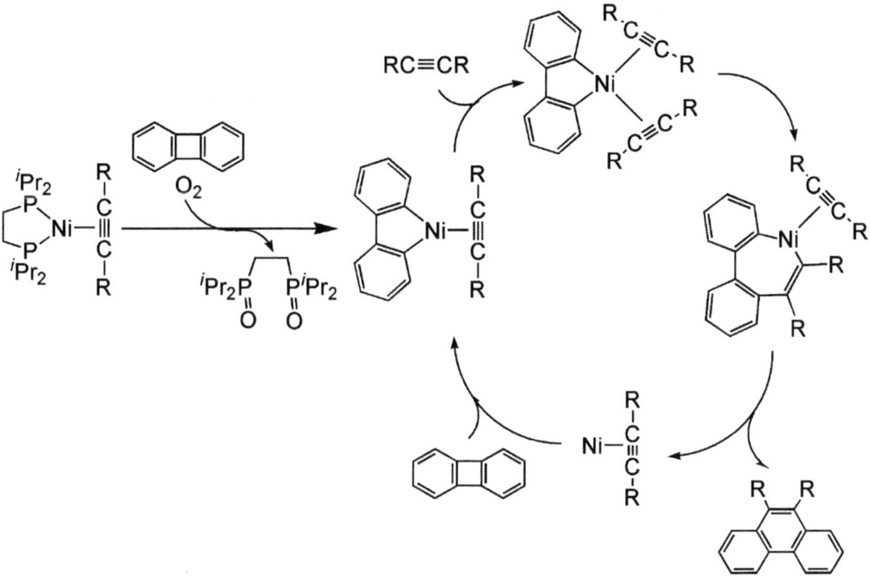


**Scheme 6** Nickel catalyzed insertion of alkynes into biphenylene

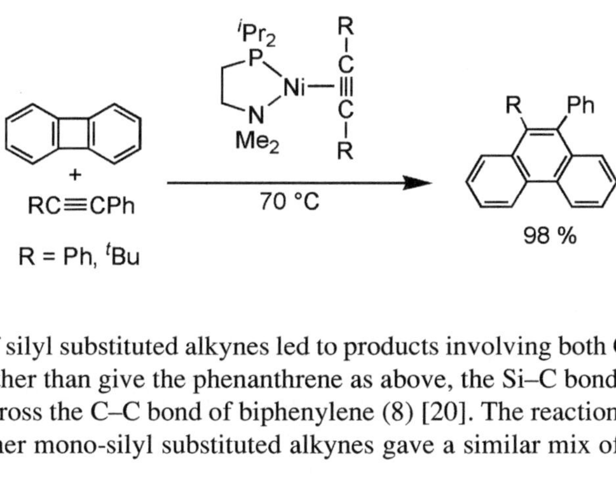


(7)

The use of silyl substituted alkynes led to products involving both C–C and C–Si cleavage. Rather than give the phenanthrene as above, the Si–C bond of the alkyne was added across the C–C bond of biphenylene (8) [20]. The reaction was catalytic in nickel. Other mono-silyl substituted alkynes gave a similar mix of products.

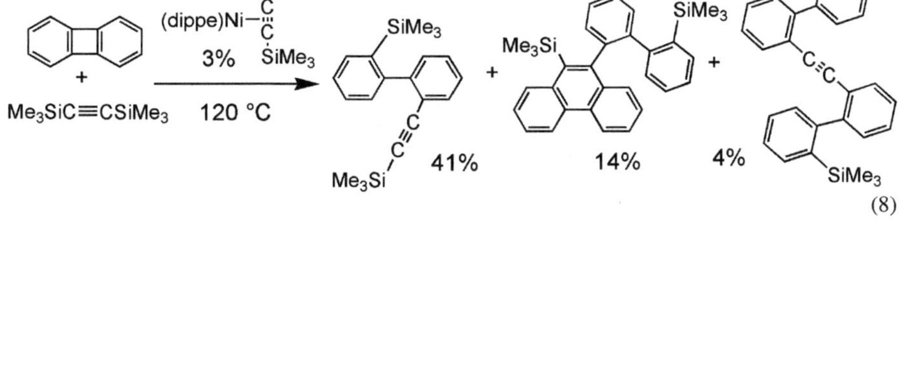


(8)

Palladium phosphines served as improved catalysts for additions across the C–C bond of biphenylene. Olefins could be incorporated by a Heck-type vinylation, presumably involving insertion into the aryl–Pd bond of a biphenylyl metallacycle followed by β-elimination and reductive elimination of product (9). Suzuki-type additions could be made using arylboronic acids (10). It was also found that weakly acidic C–H bonds could serve as addition partners, such as methyl ketones or benzylic nitriles (11, 12) [21]. In these reactions, Pd(0) was proposed to insert first into the biphenylene C–C bond generating $L_2Pd(2,2'\text{-biphenylyl})$, which was then protonated by *p*-cresol to leave a Pd–aryl bond that went on to couple with the conjugate base of the substrate to give the product.

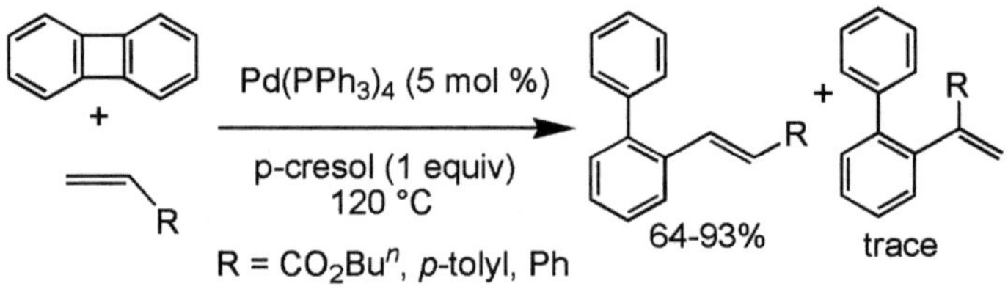


(9)

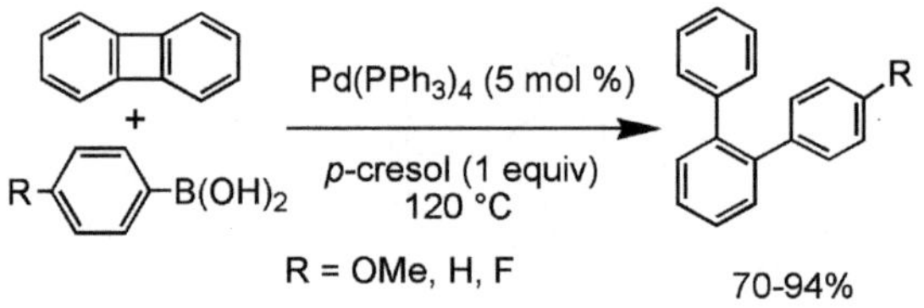


(10)

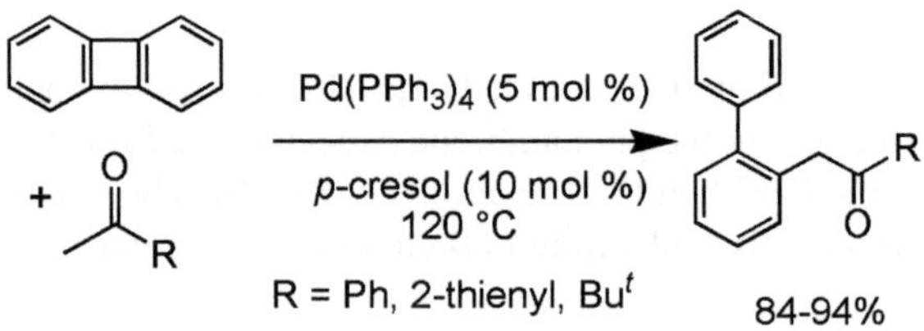


(11)

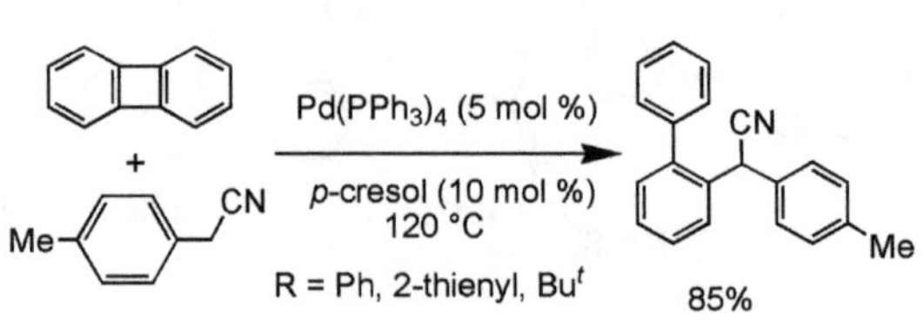


(12)

A rhodium complex similar to the platinum complex described in (4) also showed an ability to catalyze formation of phenanthrenes from biphenylene and alkynes [22]. Cyclotrimerization was observed as a side reaction, and with silyl substituted alkynes some [1,2]-silyl rearrangements were seen, leading to [1,1] addition of the alkyne across the C–C bond (13).

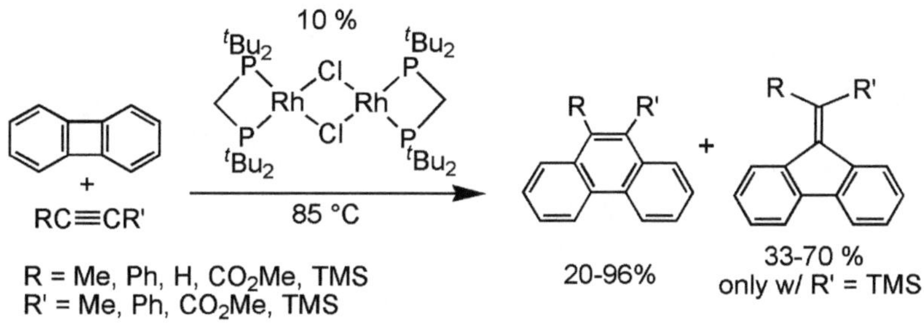


(13)

## 3 C–C Cleavage of C–CN Bonds

The oxidative addition of C–CN bonds at low valent transition metals was documented well over 40 years ago (for the cleavage of C–CN bonds via oxidative addition see [23–29]). The reverse reaction, reduction elimination to form a C–CN bond, has also been observed (for the formation of C–CN bonds via reductive elimination see [30–35]). We discovered in 2000 a case where C–CN cleavage was clean and reversible. Using $[Ni(dippe)H]_2$ as a source of [Ni(dippe)], reaction with benzonitrile leads first to the formation of Ni(dippe)($\eta^2$-NCPh), which was isolated and characterized by X-ray crystallography (Fig. 1a). $\pi$-Coordination of benzyl nitrile has been proposed previously in $Ni(PCy_3)(\pi\text{-}NCCH_2Ph)$ on the basis of IR data, but no structure was obtained [3]. If this nickel complex is allowed to stand in solution for several days at room temperature (or heated to 60°C for a few hours), conversion to the C–CN oxidative addition product Ni(dippe)(Ph)(CN) is observed. This product was also characterized by X-ray crystallography, proving the structure of this isomeric form (Fig. 1b). Furthermore, the reaction did not quite go to completion, and it was discovered that there is an equilibrium between these two forms of the compound (14) [36]. The equilibrium position can be controlled by variation of the para-substituent on the phenyl group, or by changing the polarity of the solvent [37]. A polar solvent, such as THF, drives the equilibrium towards the polar C–CN cleavage product, whereas a nonpolar solvent, such as toluene, drives

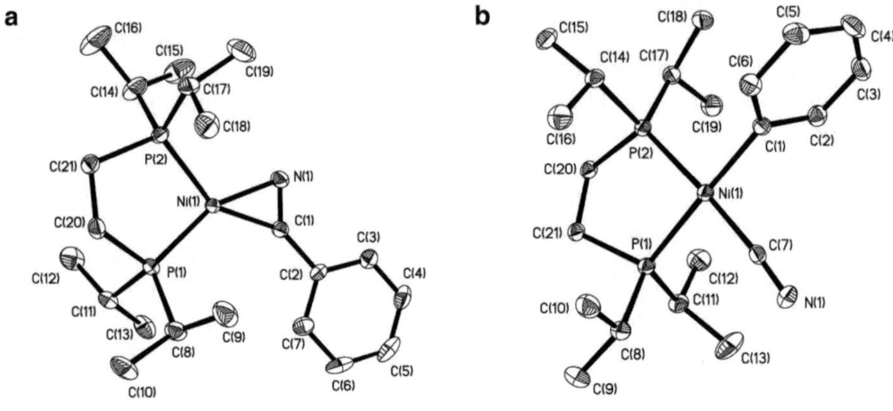


**Fig. 1** X-Ray structures of (**a**) Ni(dippe)($\eta^2$-NCPh) and (**b**) Ni(dippe)(Ph)(CN). Reprinted with permission from [36], Copyright (2000) American Chemical Society

the equilibrium towards the less polar $\pi$-nitrile complex. A Hammett plot for the equilibrium in (14) shows a slope of $\rho = +6.1$, indicating substantial negative charge at the ipso carbon in the oxidative addition product.

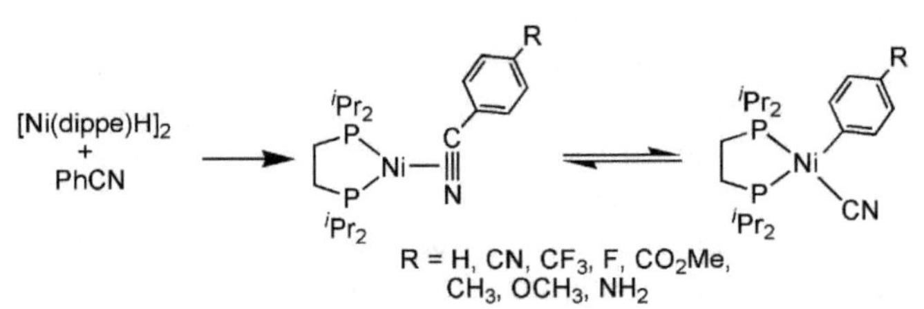


(14)

Soon thereafter, C–CN cleavage of alkyl nitriles was investigated with this nickel system. Alkyl nitriles also react to form $\pi$-complexes at room temperature (15). Heating results in oxidative addition to the C–CN bond, which in the case of acetonitrile gives the methyl cyanide complex [38]. Other alkyl derivatives, however, undergo $\beta$-elimination to give the olefin, Ni(dippe)($\eta^2$-olefin) and transient Ni(dippe)(H)(CN). The latter is unstable and decomposes to give Ni(dippe)$(CN)_2$. Unlike the case with aryl nitriles, the acetonitrile insertion goes to completion, and does not appear to be reversible.

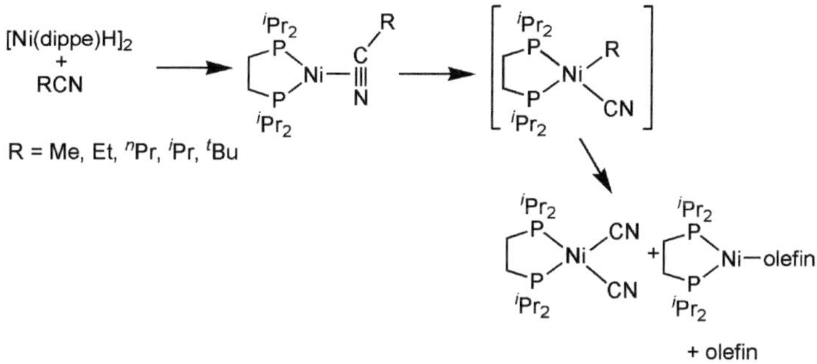

(15)

To gain a better understanding of the mechanism of C–CN cleavage, DFT calculations were undertaken. The first system chosen for study was the activation of acetonitrile, since it appeared to be the simplest. For the calculations the dippe ligand was replaced by the simpler dmpe (bisdimethylphosphinoethane) ligand. The ground state structures of both the π-complex and the C–CN oxidative addition product were obtained, and the insertion was found to be exothermic by 2.4 kcal $mol^{-1}$. In all calculations a polarizable continuum model (PCM) was used to correct the energies for solvation. This was an important contribution, as the metal–cyanide products are highly polar (dipole moment = 14.3 Debye). For comparison, the C–H oxidative addition product was also calculated, Ni(dmpe)($CH_2CN$)H, and found to lie 13.2 kcal $mol^{-1}$ above the π-complex, which explains why C–H addition products are not seen with this system. This was verified experimentally by synthesizing this product as shown in (16). The complex undergoes reductive elimination at −40°C to give the π-complex (the $BEt_3$ can be trapped with added pyridine) [39].

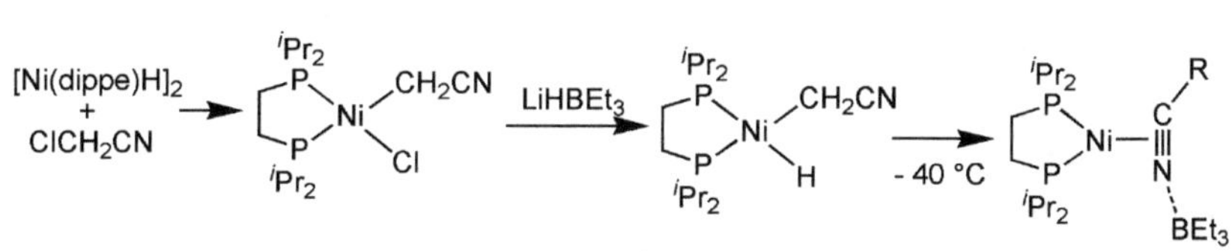

(16)

A search for a structure that could represent the transition state for C–CN cleavage led to a marginally stable structure in which the C–H bond of the acetonitrile is σ-bound to the nickel, and there is some interaction with the cyano group. This weakly bound species lies in a well only 3 kcal $mol^{-1}$ below the free fragments [Ni(dmpe)] and $CH_3CN$. Its geometry is tetrahedral, i.e., the C–CN bond is perpendicular to the $NiP_2$ plane. This key intermediate links all three of the other stable structures. Moving the nitrogen towards the nickel leads to a transition state to produce the π-complex with a barrier of ~1 kcal $mol^{-1}$. Moving the hydrogen towards the nickel leads to a transition state leading to the C–H cleavage product with a barrier of ~4 kcal $mol^{-1}$. Moving the methyl carbon towards the nickel leads

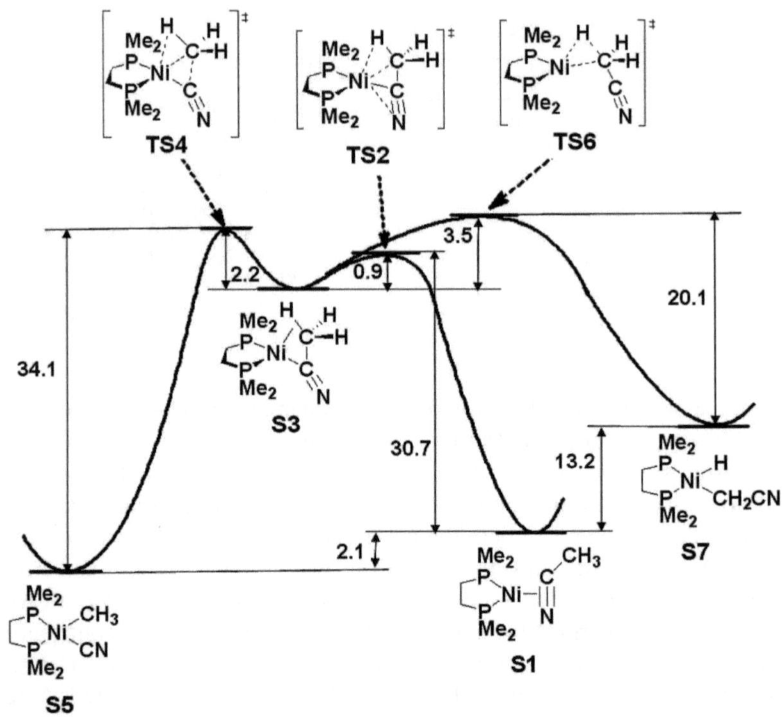


**Fig. 2** Free energy picture for the reaction of [Ni(dmpe)] with $CH_3CN$ (PCM in THF). Energies are in kcal $mol^{-1}$. Reprinted with permission from [37], Copyright (2007) American Chemical Society

to a transition state to produce the C–CN cleavage product with a barrier of ~2 kcal $mol^{-1}$. Therefore, the DFT calculations mimic the behavior of the actual system quite well (Fig. 2).

Examination of transition state TS4 for C–CN cleavage shows some revealing features. First, the C–CN bond is slightly lengthened to 1.68 Å (vs 1.49 Å in S1 and S3). The Ni–CN distance is 1.82 in TS4 vs 1.88 Å in product S5, and the Ni–$CH_3$ distance is 2.12 Å in TS4 vs 1.96 in product S5. Therefore, it appears necessary substantially to make the new Ni–C bonds before cleaving the C–CN bond. The Ni–CN distance in the transition state for C–C cleavage is actually *shorter* than in the final product! In addition, the C–CN bond is at an angle of 38° to the $NiP_2$ plane, not in the plane where there would be greater steric interference with the phosphorus atoms attached to nickel. This twisted geometry allows the closest approach of the C–C bond prior to its cleavage.

The benzonitrile C–CN cleavage has also been examined by DFT using [Ni(dmpe)] as a model for the dippe complexes [40]. Here, once again, both ground state structures were first calculated for Ni(dmpe)($\eta^2$-NCPh) and Ni(dmpe)(Ph)(CN). Excellent agreement was seen between the calculated structures and those found by X-ray diffraction, including the orientation of the phenyl ring relative to the $NiP_2$ plane (see Fig. 1). The energetics showed C–CN cleavage to be favored by only 0.9 kcal mol$^{-1}$, consistent with the observed equilibrium (with PCM correction in THF). It was thought initially that C–C cleavage would occur by first rotating the phenyl ring in Ni(dmpe)($\eta^2$-NCPh) to be perpendicular to the square plane, and then migrating the phenyl to the nickel to give directly the observed structure for the C–CN cleavage product. This turned out not to be the case.

In searching for this transition state, a stable $\eta^2$-arene complex was found in which the cyano group is attached to one of the two arene carbons that interacts with the metal. This species lies 12.1 kcal mol$^{-1}$ above the $\eta^2$-NCPh complex. Experimental evidence for such a species was obtained by addition of benzonitrile to a solution of $[Ni(dippe)H]_2$ at −60°C. At this temperature, the major product was assigned as the Ni(0) complex Ni(dippe)($\eta^2$-$C_6H_5CN$), with the metal bound to the ring in a π-fashion. The $^{31}$P NMR spectrum showed two distinct phosphorus ligands, but the $^1$H spectrum showed a symmetrical phenyl group ($2H_o$, $2H_m$, $1H_p$). The symmetrical phenyl group could be accommodated if the $NiP_2$ unit was fluxional, migrating around the arene ring, but why then would the two phosphorus atoms appear distinct? The answer was found by calculating the geometry of the transition state for migration around the ring. Initially, the $NiP_2$ unit is bound to the double bond adjacent to the cyano group in a fashion that renders the phosphines distinct – one is near the CN, the other is away from the CN. In the transition state the nickel is $\eta^3$-bound to three aromatic carbons as an allyl unit with the $NiP_2$ plane bisecting the allyl (17). It continues to the adjacent C=C double bond, but the phosphorus atoms maintain their distinct identities. If the metal continues this type of migration around the entire ring, the $NiP_2$ unit returns to its original position *without having interchanged the two phosphorus atoms*. Hence, the phosphine environments can remain distinct even though the fluxional process renders the phenyl group symmetric. Upon warming this solution to room temperature, it converts to the $\eta^2$-NCPh complex.

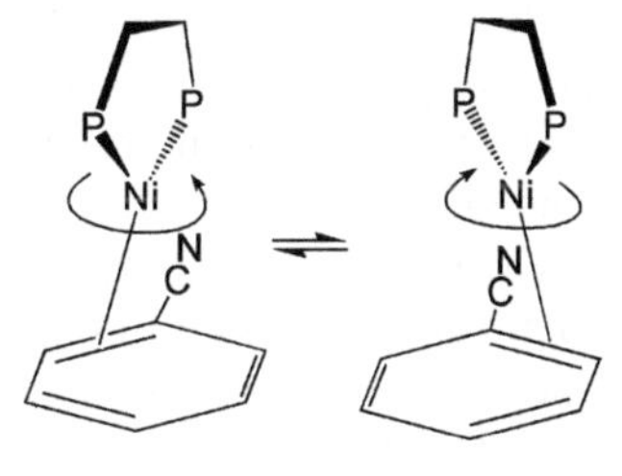


(17)

The transition state for C–CN cleavage was found, beginning with the stable $\eta^2$-$C_6H_5CN$ adduct in (17) and restricting the Ni–CN distance, optimizing the

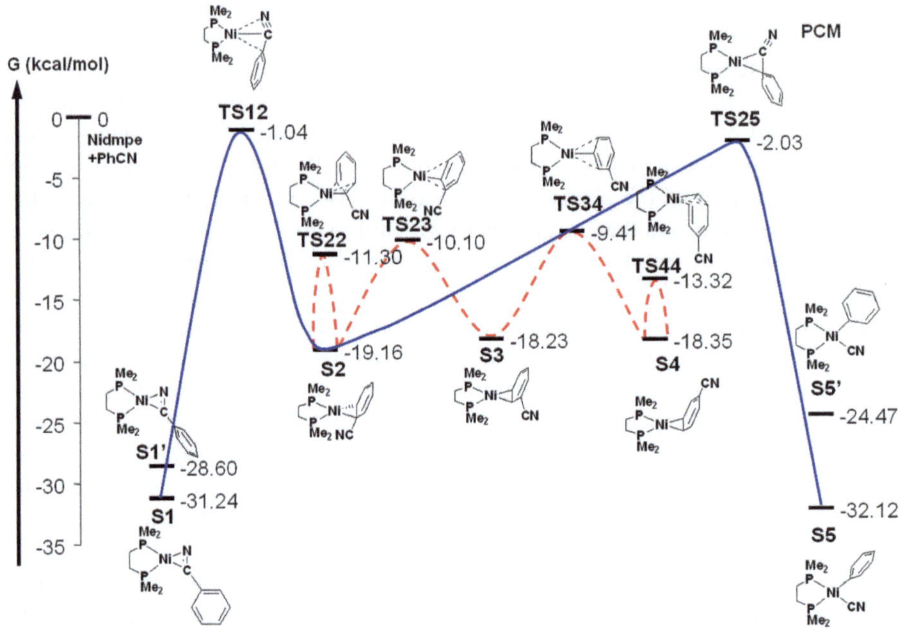


**Fig. 3** Free energy picture for the reaction of [Ni(dmpe)] with PhCN (PCM in THF). Energies are in kcal mol$^{-1}$. Reprinted with permission from [40], Copyright (2008) American Chemical Society

geometry at each step. Eventually, the C–CN bond breaks to give the oxidative addition product. The transition state was located starting with the structure just prior to the full bond cleavage. The optimized transition state shows features similar to that seen with acetonitrile. First, the C–CN bond is lengthened only slightly from 1.466 Å in π-NCPh adduct S1 to 1.590 Å in TS25. The Ni–CN distance is 1.874 Å in the transition state and 1.867 Å in the product S5. The Ni–Ph distance is 2.033 Å in the transition state and 1.926 Å in product S5. Here, once again, the new nickel–carbon bonds are made before the C–CN bond is substantially cleaved. The C–CN bond lies at an angle of 28° to the $NiP_2$ plane, somewhat smaller than in the acetonitrile case. A full energy picture is shown in Fig. 3 and includes the barriers and energies for the fluxional migration around the arene ring.

These concepts were extended to ortho-, meta-, and para-dicyanobenzenes to give $\eta^2$-NCaryl and C–CN cleavage products [41]. Some interesting differences were noted compared to the benzonitrile system. For example, the $\eta^2$-arene complex of ortho-dicyanobenzene can form a symmetrical $\eta^2$-complex when the Ni(dippe) fragment is bound to the double bond between the two cyano groups. Since this complex has mirror symmetry, the phosphorus atoms are equivalent and therefore the low temperature $^{31}$P NMR spectrum of this complex appears as a singlet at low temperature (18).

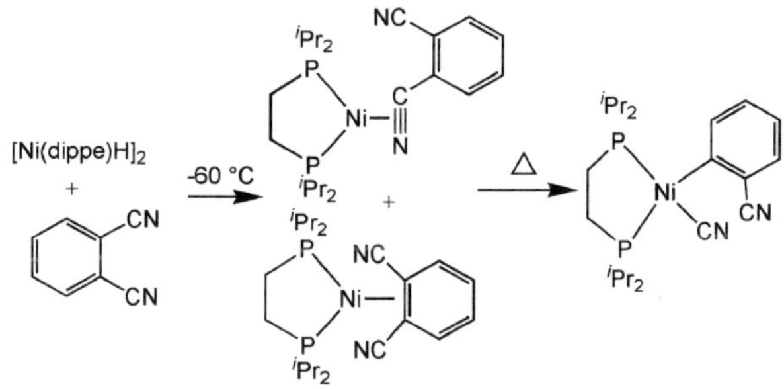


(18)

Examination of meta-dicyanobenzene also shows formation of both $\eta^2$-aryl-CN and $\eta^2$-$C_6H_4(CN)_2$ isomers at low temperature. Now, however, all of the $\eta^2$-arene complexes are asymmetric and therefore the low temperature $^{31}P$ spectrum shows two distinct doublets as seen with benzonitrile (19). Fluxional migration around the ring does not equilibrate the two phosphorus atoms [41].

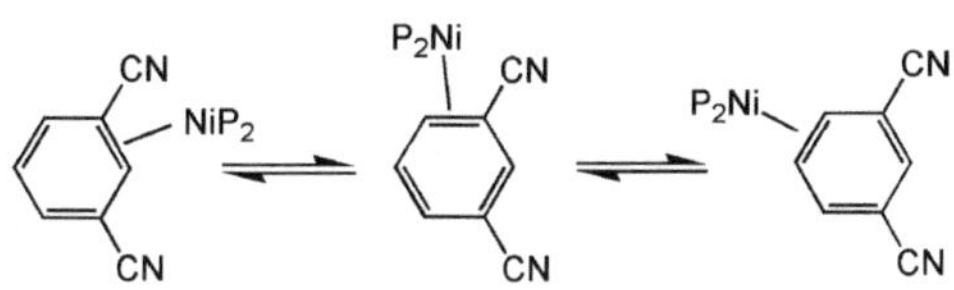


(19)

Examination of the analogous reaction with para-dicyanobenzene was logically expected to display a singlet in the low temperature $^{31}P$ NMR spectrum, as one of the $\eta^2$-arene complexes would be symmetric. Surprisingly, however, at low temperature two distinct phosphorus signals were observed. This was ultimately interpreted in terms of a [1,4]-shift of the $NiP_2$ fragment as opposed to a [1,2]-shift (20). DFT calculations showed that the transition state for the [1,4]-shift was lower in energy than the [1,2]-shift by ~1 kcal mol$^{-1}$, which can be attributed to the preference for binding to double bonds with cyano substitution [41].

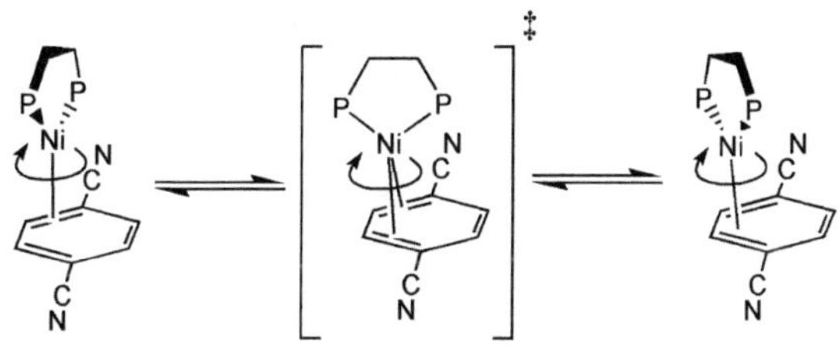


(20)

Cyanonaphthalene complexes of [Ni(dippe)] were also investigated. Reaction of 1-cyanonaphthalene with $[Ni(dippe)H]_2$ produced two stable Ni(0) complexes at ambient temperature. One was the expected $\eta^2$-(*C*,*N*)-naphthalene-CN complex and the other was assigned as the $\eta^2$-(*C*,*C*)-naphthalene-CN complex with the metal bound to the arene C–C double bond. Earlier studies have shown that nickel forms stable $\eta^2$-complexes with polycyclic aromatics [37, 42]. Upon heating, these two complexes convert to the C–CN insertion product (21). DFT calculations were in good agreement with the experimental observations, and the only two $\eta^2$-naphthaleneCN complexes of low energy were the 1,2-complex shown in (21) and the 3,4-complex (29).

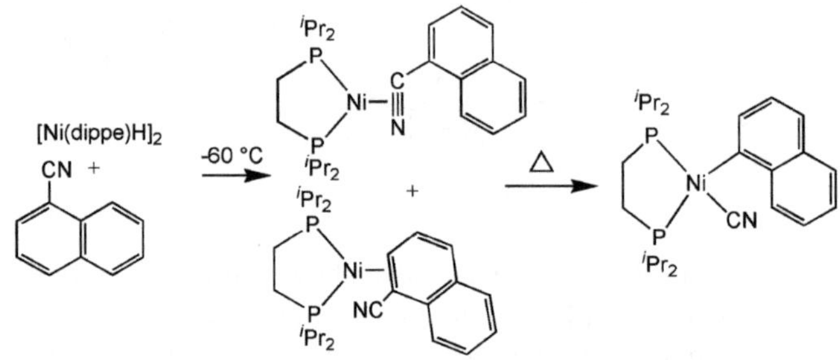


(21)

2-Cyanonaphthalene reacts with $[Ni(dippe)H]_2$ to give very similar intermediates and products. 1,4-Dicyanonaphthalene, however, reacts to give almost exclusively the $\eta^2$-naphthalene complex prior to going on to break the C–CN bond. X-Ray structures were obtained for both the $\eta^2$-arene complex and the C–CN cleavage product. 9-Cyanoanthracene reacts with $[Ni(dippe)H]_2$ to give a mixture of both the arene π-complex and the $\eta^2$-(*C*,*N*)-anthracene-CN complex (22). Upon heating, no cleavage of the C–CN bond was observed, despite the fact that DFT calculations indicate that the reaction should be exothermic by at least 3 kcal $mol^{-1}$. It was concluded that the requisite $\eta^2$-arene complex could not be accessed due to steric crowding. The DFT calculated transition state for C–CN cleavage was ~3 kcal $mol^{-1}$ above the energy of the free fragments, so it appears as if dissociation occurs prior to cleavage of the bond [41].

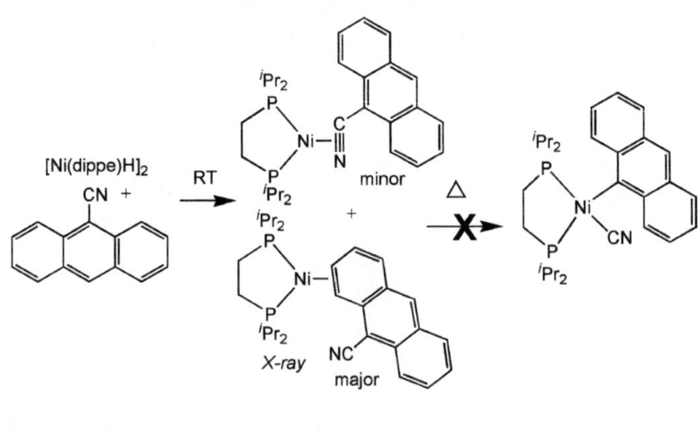


(22)

9-Cyanophenanthrene also showed interesting results when reacted with $[Ni(dippe)H]_2$. Both $\eta^2$-NC–aryl and $\eta^2$-phenanthrene–CN were formed at low temperature, but at ambient temperature only the $\eta^2$-phenanthrene–CN was observed. The X-ray structure of the latter showed it to be in the 9,10 position, i.e., the position necessary to approach the C–CN transition state. Yet heating this

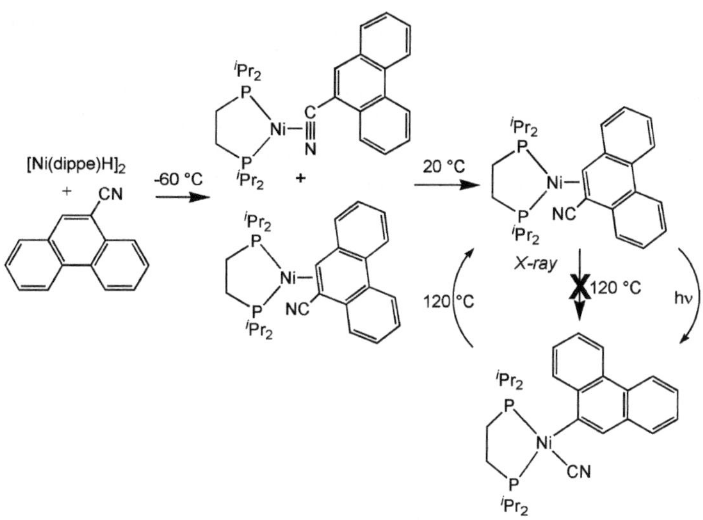


**Scheme 7** Reaction of cyanophenanthrene with [Ni(dippe)]

complex did not result in C–CN cleavage (Scheme 7). DFT calculations indicated the cleavage should be exothermic by ~2 kcal mol$^{-1}$, and the transition state was essentially the same as the dissociation energy. It was found that irradiation of the complex led to the C–CN cleavage product in ~30% yield. Upon heating this product to 120°C it reverts back to the $\eta^2$-phenanthrene–CN complex, indicating that, in this case, the C–CN cleavage is actually uphill thermodynamically [41].

An area of tremendous industrial importance that involves C–CN cleavage is the hydrocyanation of butadiene to produce adiponitrile. Hydrogenation of adiponitrile gives 1,6-diaminohexane, a coupling partner with adipic acid to give nylon-6,6. In the DuPont adiponitrile process, HCN is added twice across butadiene in anti-Markovnikov fashion to produce adiponitrile. The problem is that the first addition goes preferably to give the Markovnikov product, which is branched rather than linear. The DuPont process involves a nickel-based catalyst that can cleave C–CN bond reversibly, allowing for the branched isomer 3-methyl-2-butene nitrile (3M2BN) to be equilibrated with the linear isomer 3-pentene nitrile (3PN). Isomerization in the presence of Lewis acid gives 4-pentene nitrile (4PN) which is consumed in a second HCN addition to produce adiponitrile (ADN). The key reaction involved the isomerization of an allyl cyanide unit, moving the cyano group from one end of the allyl to the other (23).

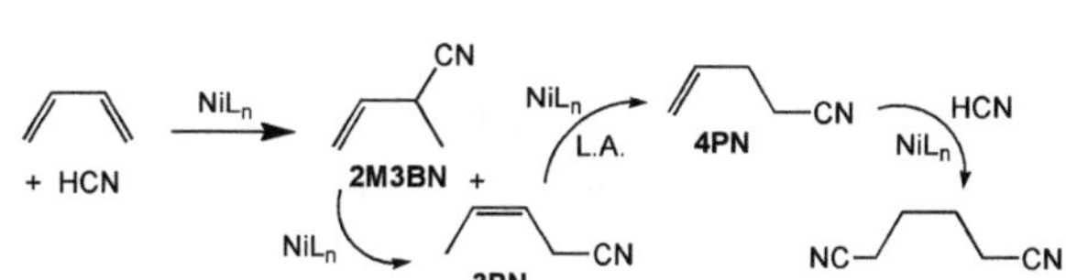


(23)

Early mechanistic studies on this catalysis with phosphite ligands on nickel was carried out by the DuPont group. They saw evidence for HCN addition and diene insertion to give π-allyl nickel cyanide complexes, which then underwent reductive elimination of both branched (2M3BN, 33%) and linear (3PN, 66%) nitrile [43–45]. Steric effects were believed to be critical in determining these product ratios. Vogt has shown that the Trypt phosphine ligand gives 98% 3-pentene (24) [46].

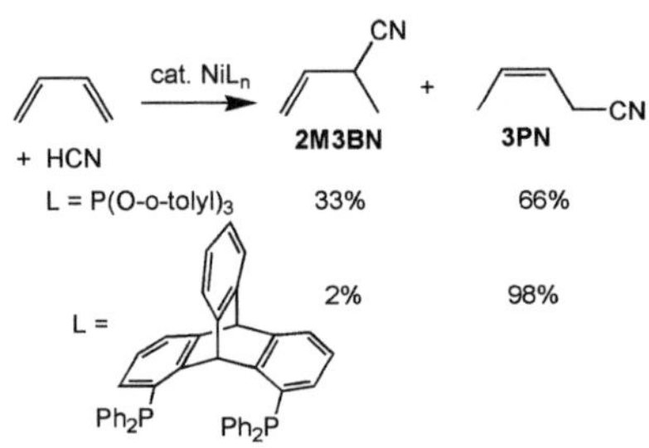


(24)

Our work with [Ni(dippe)] prompted the examination of allyl nitriles as substrates for C–CN activation. It was observed that $[Ni(dippe)H]_2$ reacts with allylcyanide to give first the $\eta^2$-olefin complex – no $\eta^2$-NC complex is observed. This species then rearranges to give a π-allyl cyanide complex, a 5-coordinate square pyramidal structure with apical cyanide [47]. The C–CN addition is reversible and over time C–H activation occurs to isomerize the double bond into conjugation with the nitrile. As Lewis acids are used in the adiponitrile process to rearrange 3PN to 4PN, the addition of $BPh_3$ to this system was also examined. C–CN cleavage is observed exclusively to give the 5-coordinate square pyramidal product with $BPh_3$ attached to the cyanide ligand (25) [48].

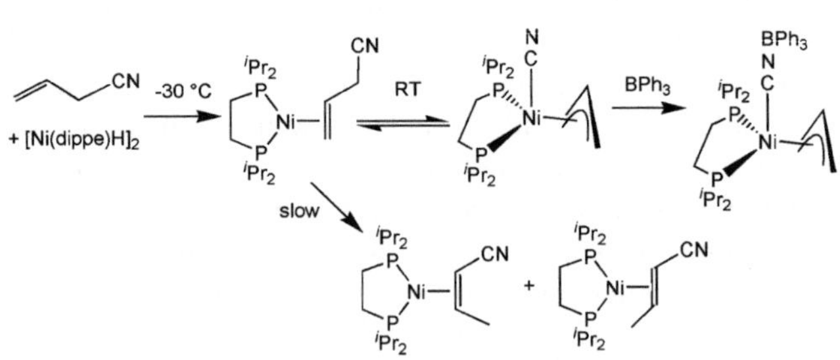


(25)

Sabo-Etienne used $[Ni(PPh_3)_2]$ to model the rearrangement of 2M3BN. This system produces 3PN in 81% yield. DFT calculations using $PH_3$ in place of $PPh_3$ suggested a mechanism for rearrangement where the cyano group was transferred directly to the metal, giving a metal cyanide and a σ-allyl group, which then coordinated to give a π-allyl ligand. Formation of a branched σ-allyl complex

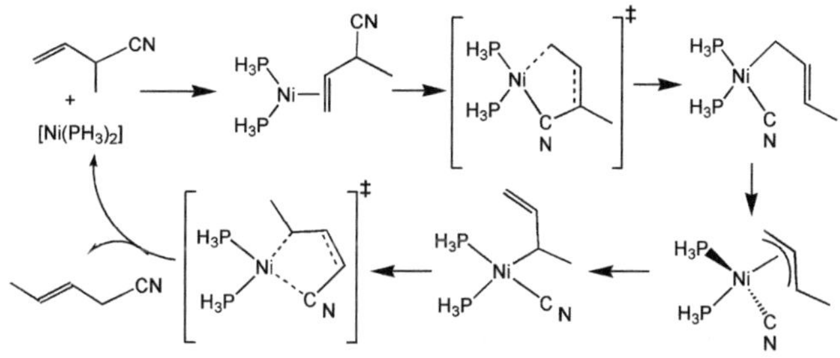


**Scheme 8** Nickel catalyzed isomerization of 2-methyl-3-butenenitrile to 3-pentenenitrile via 1,3-allyl shift

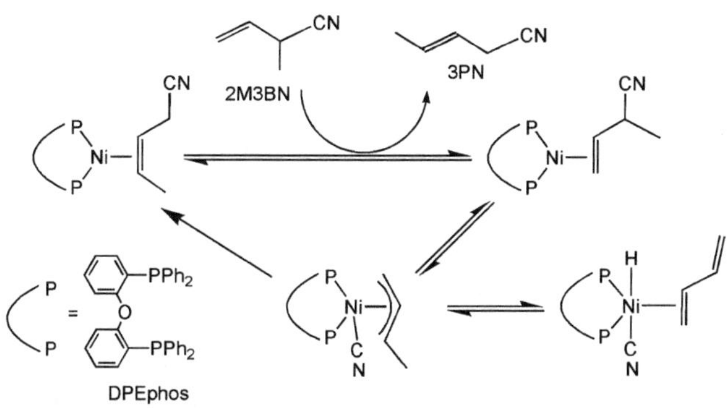


**Scheme 9** Nickel catalyzed isomerization of 2-methyl-3-butenenitrile to 3-pentenenitrile via π-allyl cyanide intermediate

followed by transfer of cyanide back to the allyl allowed isomerization to 3PN (Scheme 8) [49].

Vogt has also studied 2M3BN isomerization to 3PN using a nickel(0) DPEphos system. Here, the conversion to 3PN was found to be zero order in [2M3BN] and first order in nickel, either with or without $ZnCl_2$ Lewis acid. The activation parameters for the isomerization were found to be $\Delta H^{\ddagger} = 60.5(5.2)$ kJ mol$^{-1}$ and $\Delta S^{\ddagger} = -112(15)$ J mol$^{-1}$ K$^{-1}$. These data are consistent with a rate determining reductive elimination of 3PN, followed by an associative displacement of 3PN by 2M3BN (Scheme 9) [50].

The fragment [Ni(dippe)] was also examined for its reactivity with 2M3BN in stoichiometric experiments. Initial formation of the $\eta^2$-alkene adduct (two isomers) leads to competitive C–H and C–CN activation. The π-allyl cyanide is observed, but the π-allyl hydride is not. It could, however, be prepared by reaction of (dippe)Ni($\eta^3$-$CH_2CHC(Me)(CN))^+$ with $LiHBEt_3$. It was not directly observed, but converted rapidly to the isomerized, conjugated olefin complexes of 2M2BN (80%) plus some of the $\eta^2$-2M3BN complex, indicating that the C–H activation step is partially reversible. Ultimately, the 3PN products isomerize to conjugated 2PN products

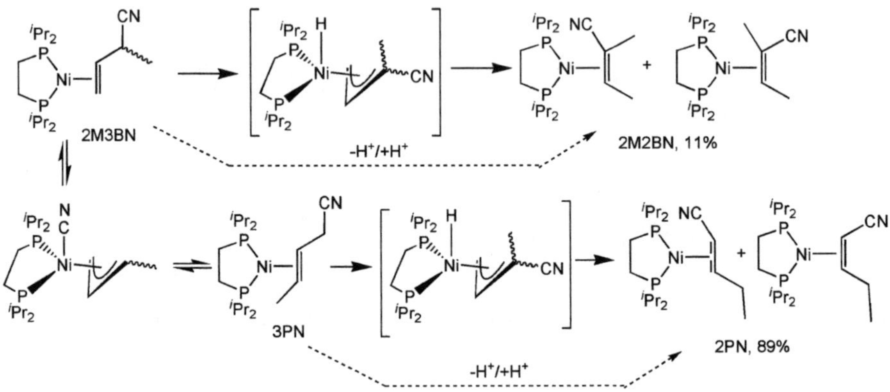


**Scheme 10** Possible proton catalyzed isomerizations of butenenitriles

irreversibly. A summary of the observed intermediates is shown in Scheme 10. Species in brackets were not directly observed, but all others could be characterized [51].

This system is catalytic if excess 2M3BN is added at 100°C, producing a mixture of linear and branched nitrile products. There was a pronounced effect of solvent polarity upon the selectivity of the reaction. Nonpolar solvents such as decane gave very high linear:branched product ratios, whereas very polar solvents such as acetonitrile gave high branched:linear ratios. Table 1 shows these product ratios as a function of solvent polarity. Comparison of acetone with di-*tert*-butyl ketone and acetonitrile with pivalonitrile shows that solvent steric factors also play a role in the selectivities, suggesting some type of intimate interactions being involved in changing the relative energies of the transition states [51]. The addition of piperidine as base was found to have a substantial shift in the isomer ratio towards branched products, but only in very polar solvents; no effect was seen in decane or THF. This observation suggests that a proton transfer mechanism for isomerization might be operating under these conditions, as indicated by the dotted lines in Scheme 10.

Temperature-dependent studies were also made with the [Ni(dippe)] system in both nonpolar (decane) and polar (DMF) solvents, as the former gives products resulting predominately from C–CN cleavage (93%) whereas the latter gives products predominately from C–H cleavage (90%). In general, the linear to branched product ratio increases by about a factor of 3 with increasing temperature (60–100°C) in both nonpolar decane and polar DMF [51]. Garcia investigated bis (dicyclohexylphosphino)ethane complexes of Ni(0) for isomerization of 2M3BN [52]. As with the dippe system, a number of similar intermediates in the isomerization could be observed and identified.

Extensive DFT calculations were made using bis(dimethylphosphino)ethane (dmpe) as a model for dippe. Thirty four ground state and transition state species were calculated using the species in Scheme 11 as models for the reaction pathway.

**Table 1** Solvent study data for the catalytic isomerization of 2M3BN to other isomers by $[Ni(dippe)H]_2$[a]

| | | Dielectric | | % Products | | | | | |
|---|---|---|---|---|---|---|---|---|---|
| | | constant | % | *Z*- | *E*- | *cis*- | *trans*- | *trans*- | Linear: |
| Entry | Solvent | (ε) | Conv | 2M2BN | 2M2BN | 2PN | 2PN | 3PN[b] | branched[c] |
| 1 | Decane | 2 | 96.9 | 2.6 | 4.9 | 4 | 6.1 | 82.5 | 12.4:1 |
| 2 | Benzene | 2.3 | 95.9 | 3.8 | 9.2 | 6.3 | 9.9 | 66.8 | 6.7:1 |
| 3 | 2,2,5,5-Tetra-methyl THF | 5 | 96.9 | 3.1 | 9 | 3.7 | 5.5 | 78.7 | 7.3:1 |
| 4 | THF | 7.5 | 87.2 | 5.7 | 14.7 | 5.5 | 8.6 | 65.5 | 3.9:1 |
| 5 | Trifluoro-toluene | 9.2 | 67.4 | 8.3 | 30.5 | 4.5 | 6 | 49[d] | 1.6:1 |
| 6 | Di-*tert*-butyl ketone | 10 | 99.6 | 9 | 50.6 | 5.7 | 9.3 | 22.5[d] | 1:1.5 |
| 7 | Acetone | 21 | 99.6 | 22.3 | 70 | 1.1 | 1.6 | 5 | 1:12 |
| 8 | Pivalonitrile | 21.1 | 98.6 | 8.9 | 31 | 4.8 | 6.8 | 47.1[d] | 1.5:1 |
| 9 | $CH_3CN$ | 36.6 | 100 | 24.3 | 73.9 | 0.6 | 0.8 | 2.2 | 1:27.3 |
| 10 | DMF | 38.3 | 95.8 | 16.4 | 73.4 | 0 | 0.7 | 5.4 | 1:14.9 |

[a]Reaction conditions: [2M3BN] = 1 mM; [Ni] = 0.105 mM; equiv. 2M3BN = 10; $T = 100°C$; time = 180 min

[b]1–2% *cis*-3PN seen in all samples

[c]Linear:branched ratio calculated with linear products *cis*-2PN, *trans*-2PN, *cis*-3PN, *trans*-3PN, and 4PN vs branched products *Z*-2M2BN and *E*-2M2BN

[d]~2% 4-PN also observed

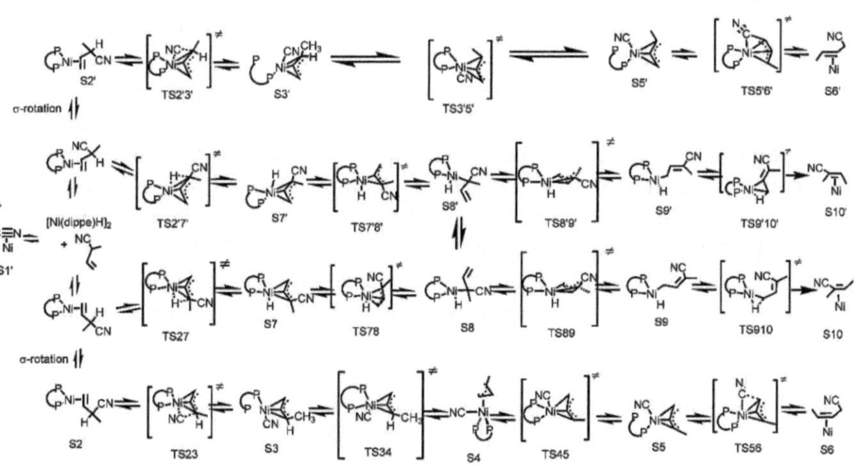


**Scheme 11** DFT calculated pathways for 2-methyl-3-butenenitrile isomerizations with [Ni(dmpe)]

These calculations agree well with experiment, and show clearly the preference of C–CN cleavage over C–H cleavage (Fig. 4) [53].

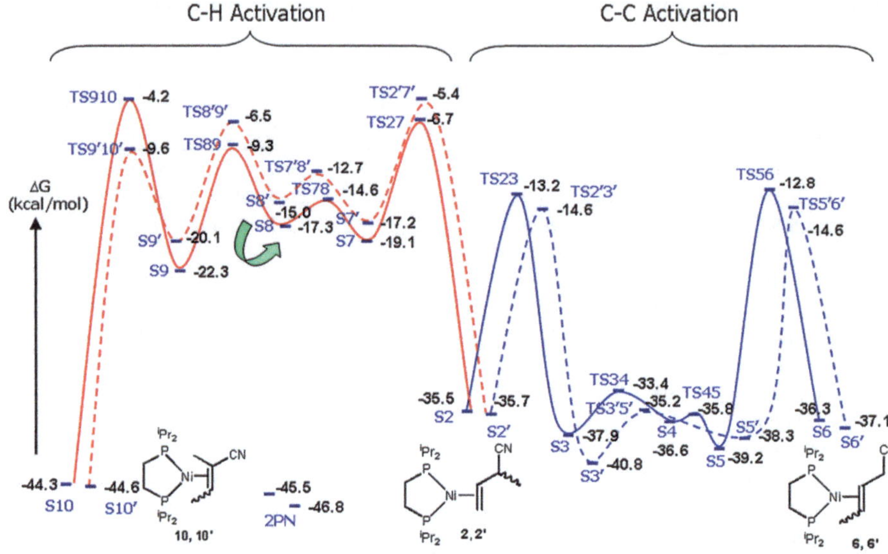


**Fig. 4** Energies of C–C and C–H activations of 2-methyl-3-butenenitrile by [Ni(dmpe)] (free energies in kcal/mol) relative to the total energies of fragments ([Ni(dmpe)] and 2-methyl-3-butenenitrile) (PCM corrected in THF) (*solid lines*: *Z*-isomer; *dashed lines*: *E*-isomer; *blue*: linear isomer, *red*: branched isomer). B3LYP/6-31G(d,p). Reprinted with permission from [53], Copyright (2011) American Chemical Society

Several other systems have been investigated to determine the nature of intermediates in the 2M3BN rearrangements. Garcia used bis-diphenylphosphinoferrocene complexes of nickel(0) to obtain 3PN in 83% yield. $ZnCl_2$ was actually found to inhibit the conversion, and an unreactive adduct (bis-diphenylphosphinoferrocene) Ni(π-butenyl)(CN–$ZnCl_2$) was obtained and structurally characterized [54]. Other diphosphinoferrocene nickel(0) derivatives were also examined, including mono- and bis-*tert*-butylphosphino derivatives and a P–N derivative. All of these showed lower selectivity for isomerization to 3PN than the parent bis-diphenylphosphinoferrocene complex [55]. Garcia also investigated triphos as a ligand for Ni(0) catalyzed isomerization of 2M3BN, but this catalyst gave mostly *cis*- and *trans*-2M2BN [56]. Several NHC complexes of nickel(0) were examined but showed similar disappointing results [57].

One final example of C–CN cleavage has appeared that involves rhodium instead of nickel [58]. Reduction of $[Rh(dippe)Cl]_2$ with potassium graphite led to a species assigned as $[Rh(dippe)]_2K_2(THF)_2$. Here the rhodium is formally in the $-1$ oxidation state, so the complex is isoelectronic with $d^{10}$ [Ni(dippe)]. The complex reacts with benzonitrile to give the C–CN cleavage product, just as seen with nickel. Use of labeled substrate $Ph^{13}CN$ gives a product that can be readily characterized as $K^+[Rh^I(dippe)(Ph)(^{13}CN)]^-$ using $^1H$, $^{13}C$, and $^{31}P$ NMR spectroscopy. Further study of this reaction was not possible, as sources of protons readily lead to the known complex $[(dippe)Rh]_2$(μ-H)(μ-N=CHPh) reported by Fryzuk [59].

# 4 C–C Cleavage of C–C≡C Bonds

In the previous section the cleavage of $sp^2$–sp C–C≡N bonds was extensively described. We began to wonder whether other $sp^2$–sp bonds such as those in C–C≡C could also be cleaved. To gain the maximum energy in terms of bond strengths in the products, platinum(0) complexes were investigated for this purpose. A series of alkyne derivatives containing bis(diisopropylphosphino)ethane (dippe), bis(dicyclohexylphosphino)ethane (dcpe), and diisopropylphosphinodimethylaminoethane (dippdmae) ligands were prepared. Heating the diphenylacetylene derivatives does not result in any observable reaction. However, irradiation with UV light (>300 nm) leads to the quantitative formation of Ph–C≡C bond cleavage products (26). Similar results were obtained with Pt(0) diphenylacetylene complexes with bis(di-*tert*-butylphosphino)methane, and bis(diisopropylphosphino) methane, ligands [60].

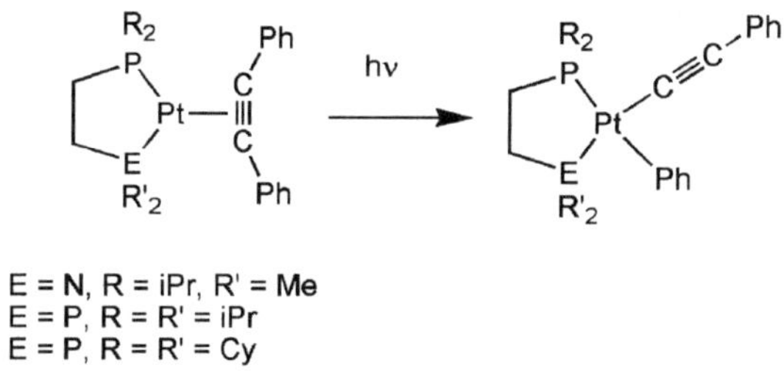


(26)

Furthermore, it was found that heating the oxidative addition products to 100–125°C resulted in reductive elimination to regenerate the $\eta^2$-alkyne complexes. The fact that the *reverse* reaction is spontaneous at this temperature means that the C–C cleavage in diphenylacetylene is thermodynamically uphill. This leads to the conclusion that the sum of the Pt–phenyl and Pt–acetylide bond strengths is less than the sum of the Ph–C≡C bond strength plus the Pt–$\eta^2$-acetylene bond strength. This also means that, even with the formation of strong platinum–carbon bonds, C–C cleavage will not occur spontaneously.

In an effort to make this reaction exothermic, the alkyne substituents were modified to try to strengthen the Pt–C bonds that are formed. Both electron rich and electron poor substituents were examined, with the latter being expected to form the stronger Pt–C bonds. Alkynes examined included di-(3,5-tolyl)acetylene, di-(*p*-fluorophenyl)acetylene, di(pentafluorophenyl)acetylene, and some mixed alkynes with these substituents [61]. In all cases, irradiation of the π-alkyne complex was required to effect $C_{aryl}$–C≡C activation. Thermolysis led to reversion back to the π-complex. The barrier for reversion was found to vary depending upon the substituent(s) present, from 47.3 kcal $mol^{-1}$ for bis(pentafluorophenyl)acetylene down to 31.3 for bis-3,5-dimethylphenyl)acetylene. Furthermore, irradiation of 4-fluorophenyl-*p*-tolylacetylene led to a ~1:1 mixture of the two possible C–C cleavage products. However, the rate of the back reaction was about five times

**Fig. 5** Free energy picture for C–C≡C bond activation

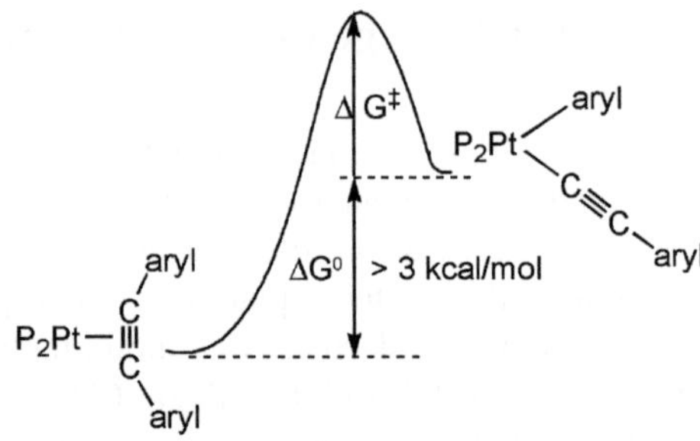


faster for the tolyl product vs the 4-fluorophenyl product (27). These studies all show that there is a thermodynamic preference for the more electron deficient aryl group being attached to platinum. Apparently, however, this preference is not sufficiently large to change the thermodynamics enough to render C–C cleavage favorable (Fig. 5). Several alkyl–phenyl alkynes were also examined (alkyl=Me, *t*-Bu, $CF_3$; dippe, dtbpe, dippdmae), but only trifluoromethylphenylacetylene underwent $C_{aryl}$–C≡C activation [62].

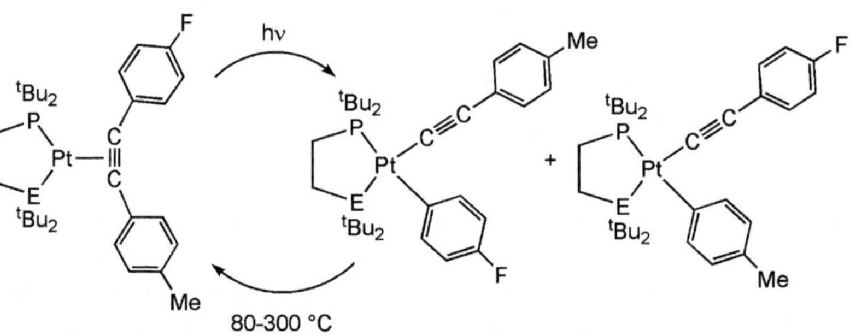


(27)

## 5 C–C Cleavage of Aryl–$CH_3$ Bonds

One other system in which substantial mechanistic work has been done on $sp^2$–$sp^3$ C–C activation is the rhodium pincer complexes that Milstein has investigated. The first discovery of this class of activation was made when he tried to attach a chelating phosphine to Rh(I). The chelate was found to undergo initial C–H activation, but continued heating under hydrogen led to cleavage of the methyl–aryl bond and loss of methane (28). The mechanism was suggested to be (1) substitution of the chelate for two $PPh_3$ ligands, (2) oxidative addition of the methyl C–H bond, (3) reductive elimination of $H_2$, (4) readdition of $H_2$, (5) reductive elimination of C–H, (6) oxidative addition of C–$CH_3$, and (7) reductive elimination of $CH_4$.

**Scheme 12** C–H and C–C activation in rhodium-pincer complexes

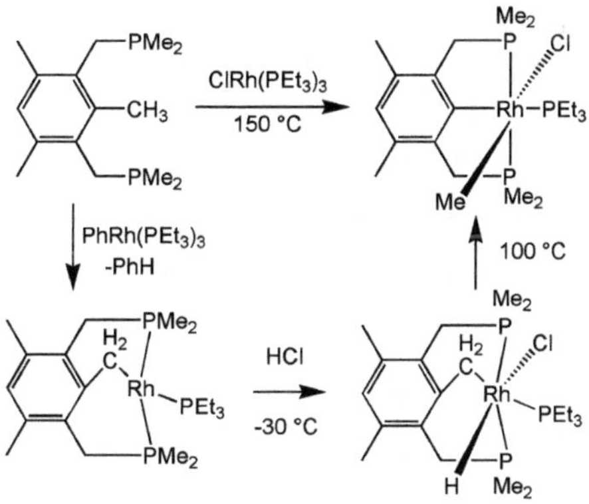


Therefore the C–H activation is kinetically favored, but reversible. The C–C cleavage occurs more slowly, but the elimination of methane is irreversible, leading to the thermodynamic PCP product [63].

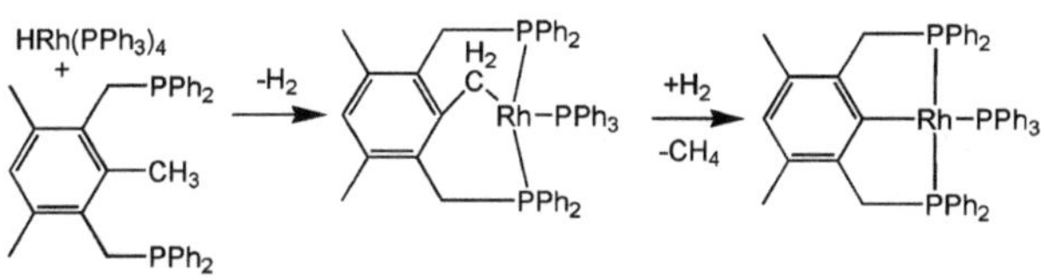


(28)

Many studies were carried out on derivatives of this system. For example, reaction of the $PMe_2$ variant of the ligand with $Rh(PEt_3)_3Cl$ led directly to the octahedral Rh(III) C–$CH_3$ addition product (Scheme 12). The C–H activation product could be synthesized by reaction of $Rh(PEt_3)_3Ph$ with the ligand to lose benzene, then reaction with HCl to give the Rh(III) octahedral C–H addition product. Heating this species induced rearrangement to the C–C cleavage product, as expected [64]. Here, if the $PMe_2$ groups are replaced with $PPh_2$ groups, C–H activation of the methyl group occurs, but there is no rearrangement to the C–C cleavage product as seen in (28). Use of the bulky, electron rich $P^tBu_2$ group led to both C–H and C–C activation using $[RhCl(C_2H_4)_2]_2$ at room temperature (1.25:1), with eventual formation of only the C–C insertion product [65]. With $[IrCl(coe)_2]_2$, both C–H and C–C activation are seen at room temperature (1.75:1), but the C–H insertion product must be heated to induce rearrangement to the C–C insertion product. Since the C–H insertion product does not convert into the C–C insertion product under the reaction conditions, it is not an intermediate in the C–C cleavage mechanism. Rather, the two activations occur independently of one another. The C–H/C–C selectivity is not very solvent-dependent. The ratio is 1.75 in benzene and 2.29 in THF. Para-substituents on the arene ring have little effect upon the ratio [66]. The original publication of this work reported that the free energy barrier is lower for C–C activation than for C–H activation, but this is incorrect [65]. The statistical correction applied by Milstein is actually part of the free energy (the entropic part) and should not be subtracted out [67]. The free energy barrier for

**Table 2** Influence of phosphine substituents on the activation of C–C vs C–H bonds in the PCP and PCN ligands. Adapted from Rybtchinski and Milstein [66]

| | Ligand | | C–C activation | | C–H activation | |
|---|---|---|---|---|---|---|
| Metal | Phosphine substituent | Additional phosphine | Thermo.[a] product | Kinetic product | Thermo.[a] product | Kinetic product |
| Rh | $-PMe_2$, $-PMe_2$ | $PEt_3$ | Yes | | | Yes |
| Rh | $-PPh_2$, $-PPh_2$ | $PPh_3$ | | | Yes | Yes |
| Rh | –P-*i*-$Pr_2$, –P-*i*-$Pr_2$ | – | Yes | | | |
| Rh | –P-*t*-$Bu_2$, –P-*t*-$Bu_2$ | – | Yes | | | |
| Ir | –P-*t*-$Bu_2$, –P-*t*-$Bu_2$ | – | Yes | | | |
| Rh | –P-*t*-$Bu_2$, $-NEt_2$ | – | Yes | | | |

[a]Thermodynamic

C–H activation is indeed slightly lower than for C–C activation in these systems, but the difference is very small ($RT \ln (1.25) = {\sim}0.1$ kcal mol$^{-1}$ for Rh, $RT \ln (1.75) = {\sim}0.3$ kcal mol$^{-1}$ for Ir). $[PtCl(coe)_2]_2$ was also found to undergo C–H and C–C cleavage of the analogous P-*i*-$Pr_2$ pincer complex [68].

Milstein was able to extend this chemistry to provide an interesting example of *catalytic* C–C cleavage. Using hydrogen with this pincer ligand, he demonstrated catalytic C–C cleavage of the methyl group using $[RhCl(coe)_2]_2$ catalyst (100 t.o.) (29) [69]. Triethoxysilane could also be used in a catalytic fashion to cleave this bond.

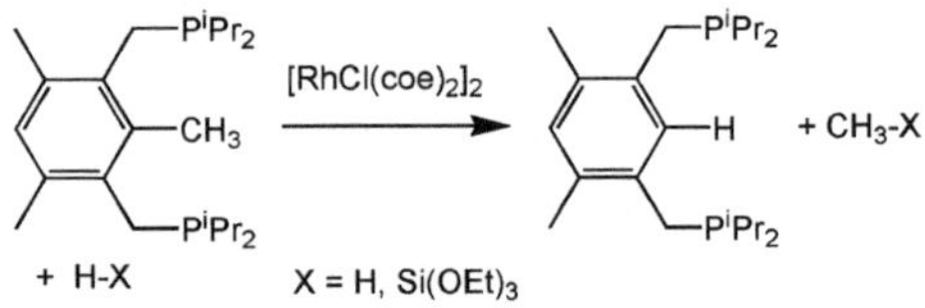


(29)

A slight variation in the ligand led to a surprising observation. Replacement of one P$^t$Bu$_2$ group by an $NEt_2$ group led to the observation of direct C–C cleavage (30). No evidence for competitive C–H activation could be seen, even at low temperatures [70]. The C–H/C–C chemistry of this class of compounds can be summarized as shown in Table 2.

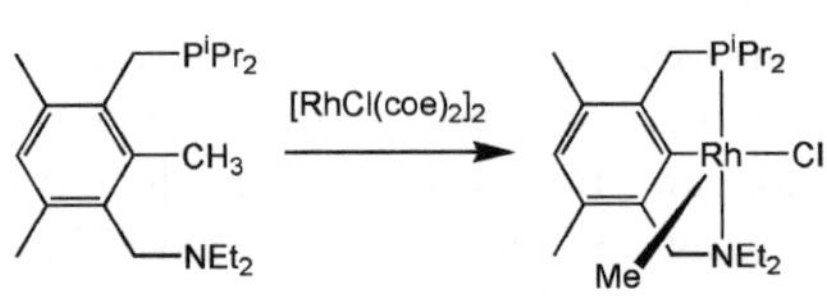


(30)

## 6 Conclusion

In conclusion, this contribution has provided some key mechanistic insight into the cleavage of carbon–carbon bonds. One common feature that can be seen is obvious – the C–C bond must get close to the metal for C–C cleavage to occur. However, DFT calculations go further to show that the metal–carbon bonds must be substantially formed before the C–C bond can be substantially broken. This requires that the metal be unhindered by ligands, and the examples cited here have only two or three atoms attached to the metal when the C–C bond is broken. This may be a general requirement for C–C cleavage, although additional examples will have to be studied in order to determine the veracity of this hypothesis. Further reactions of these C–C activated species have not been mentioned in this chapter, as other contributions to this volume will deal explicitly with some of the carbon–carbon bond cleavages mentioned above, as well as other C–C cleavage reactions.

## References

1. Jones WD (1993) Nature 364:676
2. Perthuisot C, Edelbach BL, Zubris DL, Simhai N, Iverson CN, Müller C, Satoh T, Jones WD (2002) J Mol Catal A Chem 189:157
3. Luo Y-R (2007) Comprehensive handbook of bond energies. CRC, Boca Raton
4. Simões JAM (1990) Beauchamp. Chem Rev 90:629
5. Atkinson ER, Levins PL, Dickelman TE (1964) Chem Ind (Lond) 934
6. Eisch JJ, Piotrowski AM, Han KI, Krüger C, Tsay YH (1985) Organometallics 4:224
7. Schwager H, Spyroudis S, Vollhardt KPC (1990) J Organomet Chem 382:191
8. Lu Z, Jun C-H, de Gala SR, Sigalas M, Eisenstein O, Crabtree RHJ (1993) Chem Soc Chem Commun 1877
9. Jones WD, Feher FJ (1989) Acc Chem Res 22:91
10. Perthuisot C, Jones WD (1994) J Am Chem Soc 116:3647
11. Chin RM, Dong L, Duckett SB, Jones WD (1992) Organometallics 11:871
12. Wick DD, Jones WD (2009) Inorg Chim Acta 362:4416
13. Perthuisot C, Edelbach BL, Zubris DL, Jones WD (1997) Organometallics 16:2016
14. Edelbach BL, Lachicotte RJ, Jones WD (1998) J Am Chem Soc 120:2843
15. Edelbach BE, Vicic DA, Lachicotte RJ, Jones WD (1998) Organometallics 17:4784
16. Simhai N, Iverson CN, Edelbach BE, Jones WD (2001) Organometallics 20:2759
17. Hirschler MM, Taylor RJ (1980) Chem Soc Chem Commun 967
18. Edelbach BL, Lachicotte RJ, Jones WD (1999) Organometallics 18:4040
19. Müller C, Lachicotte RJ, Jones WD (2002) Organometallics 21:1975
20. Edelbach BL, Lachicotte RJ, Jones WD (1999) Organometallics 18:4660
21. Satoh T, Jones WD (2001) Organometallics 20:2916
22. Iverson CN, Jones WD (2001) Organometallics 20:5745
23. Abba M, Yamamoto T (1997) J Organomet Chem 532:267
24. Favero G, Movillo A, Turco A (1983) J Organomet Chem 241:251
25. Morvillo A, Turco A (1981) J Organomet Chem 208:103

26. Parshall GW (1974) J Am Chem Soc 96:2360
27. Churchill D, Shin JH, Hascall T, Hahn JM, Bridgewater BM, Parkin G (1999) Organometallics 18:2403
28. Gerlach DH, Kane AR, Parshall GW, Jesson JP, Muetterties EL (1971) J Am Chem Soc 93:3543
29. Burmeister JL, Edwards LM (1971) J Chem Soc A 1663
30. Favero G, Gaddi M, Morvillo A, Turco A (1978) J Organomet Chem 149:395
31. Cassar L (1973) J Organomet Chem 54:C57
32. Tsuji Y, Kusui T, Kojima T, Sugiura Y, Yamada N, Tanaka S, Ebihara M, Kawamura T (1998) Organometallics 17:4835
33. Luo F-H, Chu C-I, Cheng C-H (1998) Organometallics 17:1025
34. Huang J, Haar CM, Nolan SP, Marcone JE, Moloy KG (1999) Organometallics 18:297
35. Marcone JE, Moloy KG (1998) J Am Chem Soc 120:8527
36. Garcia JJ, Jones WD (2000) Organometallics 19:5544
37. Garcia JJ, Brunkan NM, Jones WD (2002) J Am Chem Soc 124:9547
38. Garcia JJ, Arévalo A, Brunkan NM, Jones WD (2004) Organometallics 23:3997
39. Atesin TA, Li T, Lachaize S, Brennessel WW, Garcia JJ, Jones WD (2007) J Am Chem Soc 129:7562
40. Atesin TA, Li T, Lachaize S, Brennessel WW, Garcia JJ, Jones WD (2008) Organometallics 27:3811
41. Li T, Garcia JJ, Brennessel WW, Jones WD (2010) Organometallics 29:2430
42. Brauer DJ, Kruger C (1977) Inorg Chem 16:884
43. McKinney RJ, Roe DC (1985) J Am Chem Soc 107:261
44. Tolman CA, Kinney RJ, Seidel WC, Druliner JD, Stevens WR (1985) Adv Catal 33:1
45. Tolman CA, Seidel WC, Druliner JD, Domaille PJ (1984) Organometallics 3:33
46. Bini L, Müller CM, Wilting J, von Chrzanowski L, Spek AL, Vogt D (2007) J Am Chem Soc 129:12622
47. Brunkan NM, Jones WD (2003) J Organomet Chem 683:77
48. Brunkan NM, Brestensky DM, Jones WD (2004) J Am Chem Soc 126:3627
49. Chaumonnot A, Lamy F, Sabo-Etienne S, Donnadieu B, Chaudret B, Barthelat JC, Galland JC (2004) Organometallics 23:3363
50. Wilting J, Müller C, Hewat AC, Ellis DD, Tooke DM, Spek AL, Vogt D (2005) Organometallics 24:13
51. Swartz BD, Reinartz NM, Garcia JJ, Jones WD (2008) J Am Chem Soc 130:8548
52. Acosta-Ramirez A, Flores-Gaspar A, Munoz-Hernandez M, Arevalo A, Jones WD, Garcia J (2007) J Organometallics 26:1712
53. Li T, Jones WD (2011) Organometallics 30:547
54. Acosta-Ramirez A, Munoz-Hernandez M, Jones WD, Garcia JJ (2006) J Organomet Chem 691:3895
55. Acosta-Ramirez A, Munoz-Hernandez M, Jones WD, Garcia J (2007) J Organometallics 26:5766
56. Acosta-Ramirez A, Flores-Alamo M, Jones WD, Garcia J (2008) J Organometallics 27:1834
57. Acosta-Ramirez A, Morelas-Morelas D, Serrano-Becerra JM, Arevalo A, Jones WD, Garcia JJ (2008) J Mol Catal A Chem 288:14
58. Grochowski MR, Morris J, Brennessel WW, Jones WD (2011) Organometallics 30:5604
59. Fryzuk MD, Piers WE, Rettig S (1992) J Can J Chem 70:2381
60. Müller C, Lachicotte RJ, Jones WD (2001) J Am Chem Soc 123:9718
61. Gunay A, Jones WD (2007) J Am Chem Soc 129:8729
62. Gunay A, Müller C, Lachicotte RJ, Brennessel WW, Jones WD (2009) Organometallics 28:6524
63. Gozin M, Weisman A, Ben-David Y, Milstein D (1993) Nature 364:699

64. Liou S-Y, Gozin M, Milstein D (1995) J Am Chem Soc 117:9774
65. Rybtchinski B, Vigalok A, Ben-David Y, Milstein D (1996) J Am Chem Soc 118:12406
66. Rybtchinski B, Milstein D (1999) Angew Chem Int Ed 38:870
67. Subtraction of the statistical part of the free energy, RT ln ($k_1/k_2$), would only be appropriate in comparing the enthalpy of reaction
68. van der Boom ME, Kraatz H-B, Ben-David Y, Milstein S (1996) Chem Commun 2167
69. Liou S-Y, van der Boom ME, Milstein D (1998) Chem Commun 687
70. Gandelman M, Vigalok A, Shimon LJW, Milstein D (1997) Organometallics 16:3981

Top Curr Chem (2014) 346: 33–58
DOI: 10.1007/128_2013_494
© Springer-Verlag Berlin Heidelberg 2014
Published online: 20 February 2014

# Catalytic C–CN Bond Activation

**Yoshiaki Nakao**

**Abstract** Synthetic organic reactions through C–CN activation by transition metal catalysis are reviewed. C–CN bond activation by metal complexes proceeds mainly via two pathways; oxidative addition and C–CN cleavage accompanied by silylisonitrile formation. Both the elemental reactions have been successfully applied to the catalytic reactions, including hydrodecyanation of nitriles, cross-coupling using nitriles as electrophiles, cyanation of aryl halides and arenes using organic nitriles as cyanating agents, and carbocyanation of unsaturated compounds.

**Keywords** Carbocyanation · C–CN activation · Copper · Cross-coupling · Cyanation · Hydrodecyanation · Nickel · Rhodium

## Contents



Y. Nakao (✉)
Department of Material Chemistry, Graduate School of Engineering, Kyoto University,
Katsura, Nishikyo-ku, Kyoto 615-8510, Japan
e-mail: nakao.yoshiaki.8n@kyoto-u.ac.jp

## Abbreviations

| | |
|---|---|
| Ac | Acetyl |
| acac | Acetylacetonate |
| Alk | Alkyl |
| aq | Aqueous solution |
| Ar | Aryl |
| Bn | Benzyl |
| Bu | Butyl |
| Bz | Benzoyl |
| cat | Catalyst |
| cod | Cyclooctadiene |
| Cp | Cyclopentadienyl |
| Cp* | Pentamethylcyclopentadienyl |
| Cy | Cyclohexyl |
| DABCO | 1,4-Diazabicyclo[2.2.2]octane |
| DIBALH | Diisobutylaluminum hydride |
| DME | 1,2-Dimethoxyethane |
| DMF | Dimethylformamide |
| DMPU | 1,3-Dimethyl-3,4,5,6-tetrahydro-2(1H)-pyrimidinone |
| dppb | Bis(diphenylphosphino)butane |
| dppe | Bis(diphenylphosphino)ethane |
| ee | Enantiomer excess |
| equiv | Equivalent(s) |
| Et | Ethyl |
| h | Hour(s) |
| *i*-Pr | Isopropyl |
| Me | Methyl |
| Mes | Mesityl 2,4,6-trimethylphenyl (not methanesulfonyl) |
| Ph | Phenyl |
| phen | Phenanthroline |
| phth | Phthalate |
| pin | Pinacolato |
| Pr | Propyl |
| rt | Room temperature |
| SPhos | 2-Dicyclohexylphosphino-2′,6′-dimethoxybiphenyl |
| TBDMS | *tert*-Butyldimethylsilyl |
| *t*-Bu | *tert*-Butyl |
| THF | Tetrahydrofuran |
| TMEDA | *N,N,N′,N′*-Tetramethylethylenediamine |
| TMS | Trimethylsilyl |
| TON | Turnover number |
| Ts | Tosyl 4-toluenesulfonyl |
| Xantphos | 4,5-Bis(diphenylphosphino)-9,9-dimethylxanthene |

## 1 Introduction

Nitriles are common and ubiquitous organic compounds. They act as different functional molecules such as pharmaceuticals, pesticides, organic materials, and polymers; they are also important building blocks in organic synthesis. Though cyano group can be readily converted to carbonyl and aminoalkyl groups by a broad range of methods, it has rarely been regarded as a leaving group in organic syntheses, except for acyl and alkoxycarbonyl cyanides and some reactions involving electron transfer and/or addition/elimination pathways to result in the release of cyanide. C–CN bonds generally tolerate the various reaction conditions of many organic transformations, owing partly to their high bond dissociation energies (>100 kcal/mol). Nevertheless, low-valence transition metal complexes have shown that C–CN bonds of nitriles can be cleaved. Nitriles coordinate to a metal center either in a $\eta^1$- or in a $\eta^2$-manner. High-valence Lewis acidic metal complexes favor $\eta^1$-coordination of nitrogen atom in nitriles, whereas low-valence metal complexes often show $\eta^2$-coordination which can be strengthened through π-back donation (Scheme 1) [1]. In many cases, C–CN bond activation is initiated by $\eta^2$-coordination. Two main pathways have been revealed for the cleavage step: oxidative addition and formation of silylisonitrile complexes. Oxidative addition was already reported in 1971, when a group in DuPont described how benzonitrile adds to nickel(0) species at room temperature (Scheme 2) [2]. This elemental reaction of benzonitrile and other nitriles has been studied extensively [2–27] and is discussed in more detail in Chap. 1. The DuPont team has studied the catalytic isomerization of 2-methyl-3-butenenitrile (2M3BN) to 3- and/or 4-pentenenitrile (3PN, 4PN) via oxidative addition of the C–CN bond of 2M3BN to form a π-allylnickel intermediate (Scheme 3) [28]. This reaction is a part of DuPont's adiponitrile (ADN) process and represents a very early example of catalytic C–CN bond activation. The catalytic isomerization is still a topic of recent research by several groups [29–37] but will not be discussed further in this chapter. The DuPont team [38] and more recently Jones [14] have also revealed the effect of Lewis acid additives on the oxidative addition of C–CN bonds.

C–CN bond activation through the formation of silylisonitrile metal complexes has been disclosed for rhodium complexes by Brookhart and coworkers for the first time (Scheme 4) [39, 40]. Silyliron [41], silylene–iron [41], and silylene–ruthenium [42] complexes have also been demonstrated to activate C–CN bonds. Detailed mechanistic studies have been performed on the C–CN bond activation of this type to show the intermediacy of $\eta^2$-iminoacyls, for which C–CN cleavage takes place leading to the formation of silylisonitriles.

In spite of the rich chemistry of stoichiometric C–CN bond activation by various transition metal complexes via the different pathways described above, their application in catalytic transformations of nitriles directed to organic synthesis has rarely emerged until the last 10 years. This review features the progress of catalytic reactions via C–CN activation (for a previous review on this topic see [43]). Particular focus of this review is on C–CN activation by metal complexes to give

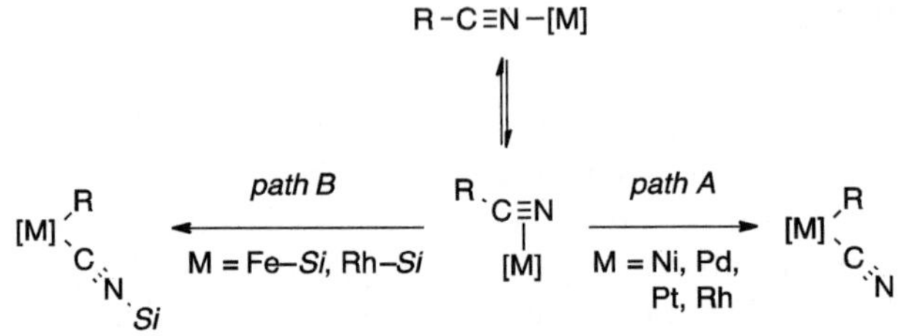


**Scheme 1** C–CN activation by transition metal complexes

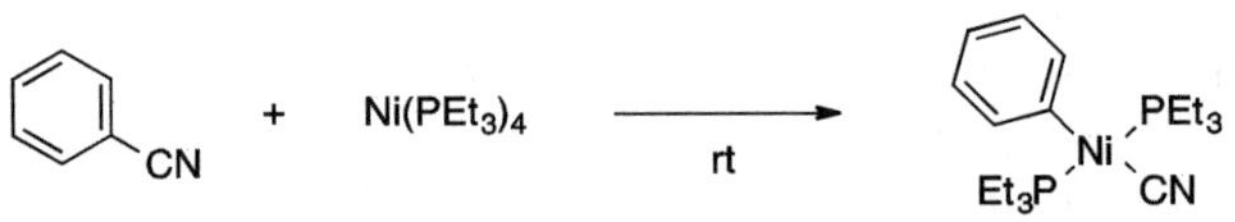


**Scheme 2** Oxidative addition of benzonitrile to $Ni(PEt_3)_4$

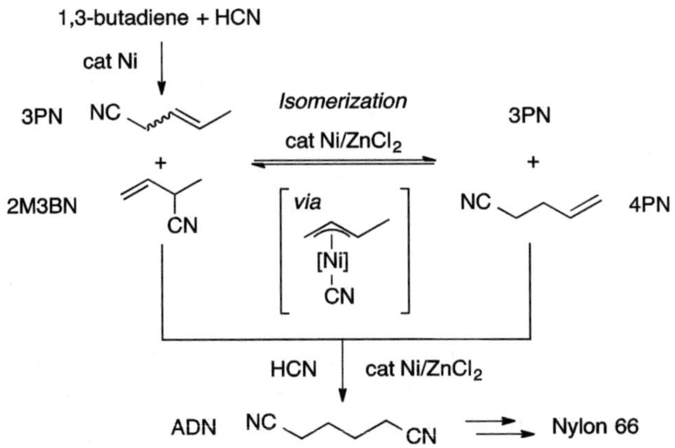


**Scheme 3** Isomerization of 2M3BN via C–CN activation catalyzed by $Ni/ZnCl_2$ in the DuPont's ADN process

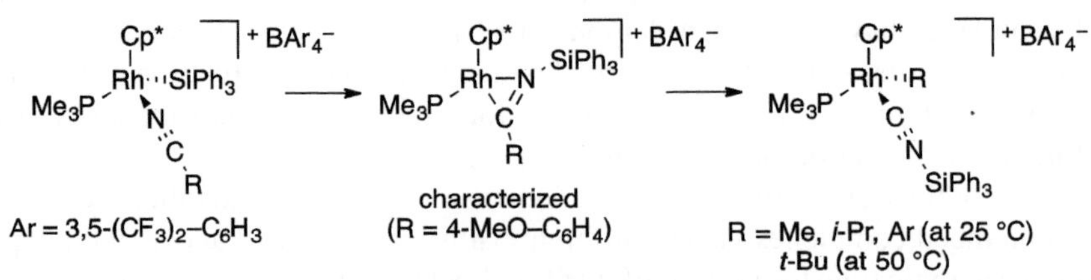


**Scheme 4** C–CN activation by silylrhodium(I) complexes

possibly organometallic intermediates bearing organic and/or cyano groups bound to a metal center, and thus, conventional synthetic transformations involving C–CN activation by electron transfer and/or addition/elimination pathways are not covered even when metal catalysts are involved.

## 2 Coupling Reactions via C–CN Bond Activation

Metal-catalyzed cross-coupling reaction is undoubtedly one of the most important methodologies in modern organic synthesis. This transformation typically employs aryl and heteroaryl halides and a wide range of nucleophiles to construct typically substituted arenes. Recent studies have also shown the use of aryl sulfonates and esters as aryl pseudohalides. These aryl electrophiles are known to undergo oxidative addition to palladium(0) and nickel(0) species to initiate a catalytic cycle for cross-coupling. Thus, the application of C–CN bond activation to cross-coupling catalysis, particularly through oxidative addition, can be envisaged, making aryl cyanides an alternative for aryl electrophiles in cross-coupling reactions.

### *2.1 Hydrodecyanation Reactions*

Oxidative addition of C–CN bonds to nickel(0) can be followed by transmetalation with various main-group organometallic reagents, and subsequent reductive elimination can result in the functionalization of C–CN bonds of nitriles (Scheme 5). As the simplest case, C–CN bonds can be transformed to C–H bonds via transmetalation with metal hydrides. Indeed, nickel-catalyzed hydrodecyanation of various aromatic and aliphatic nitriles proceeds with tetramethyldisiloxane as a hydride donor (Scheme 6) [44]. While a wide range of nitriles can be decyanated by this protocol, a relatively high amount of catalyst is required in this process, presumably because of the formation of catalytically inactive $(PCy_3)_2Ni(CN)_2$. The use of $AlMe_3$ as a Lewis acid is effective in some cases to promote the C–CN bond activation. Under these reaction conditions, the relative reactivity order of different aryl electrophiles is estimated: Ar–SMe>Ar–CN>Ar–OAr>Ar–OMe.

Alternatively, iron- or rhodium-catalyzed hydrodecyanation reaction is proposed to be initiated by the C–CN activation of aromatic and aliphatic nitriles with silylmetal species to give iminoacylmetal intermediates and then silylisonitrile metal complexes, which produce decyanated products and silyl cyanides upon the reaction with hydrosilanes to reproduce the catalytically active silylmetal intermediates (Scheme 7). Irradiation is required to generate coordinatively unsaturated iron complexes (Scheme 8) [45, 46]. A similar reaction mechanism also operates

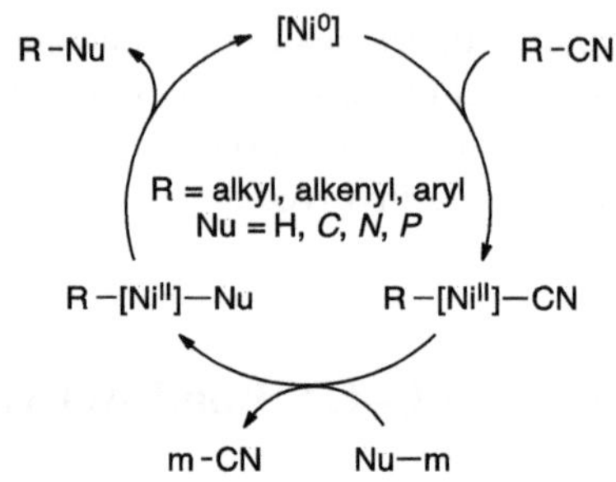


**Scheme 5** A general catalytic cycle for nickel-catalyzed coupling reactions via C–CN activation

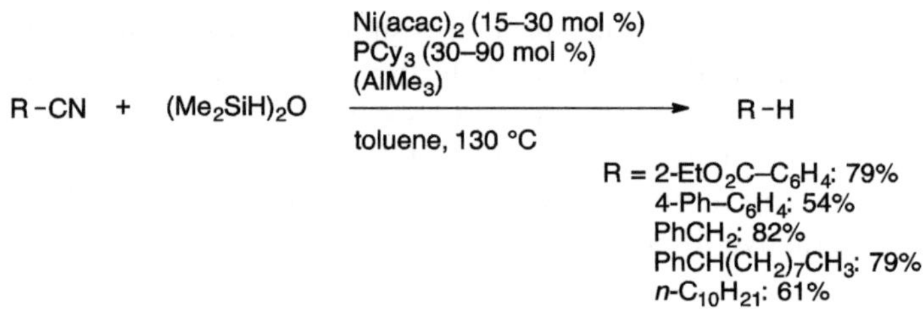


**Scheme 6** Hydrodecyanation of nitriles with hydrosilanes catalyzed by nickel

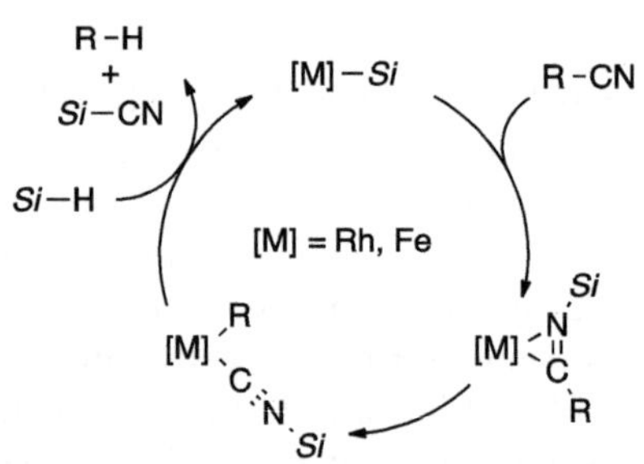


**Scheme 7** A plausible catalytic cycle for hydrodecyanation of nitriles with hydrosilanes

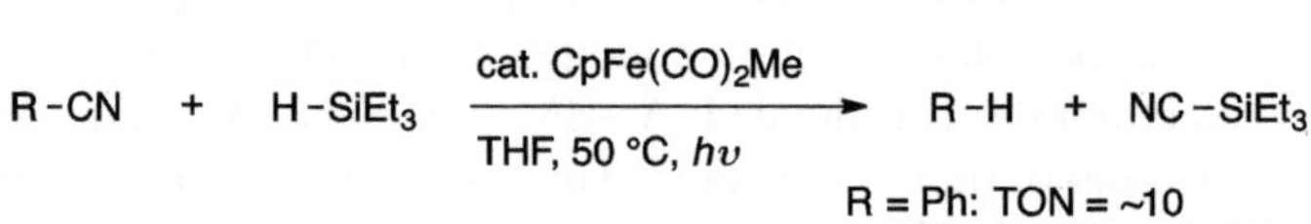


**Scheme 8** Hydrodecyanation of nitriles with hydrosilanes catalyzed by iron

with rhodium catalysis (Scheme 9) [47, 48]. Rhodium-catalyzed reaction tolerates a wide range of nitriles, including tertiary alkyl cyanides, using triisopropylsilane as a reducing agent. These catalytic hydrodecyanation reactions can be nicely combined with conventional transformations of nitriles, demonstrating that cyano group can serve as a "removable directing group" (Scheme 10).

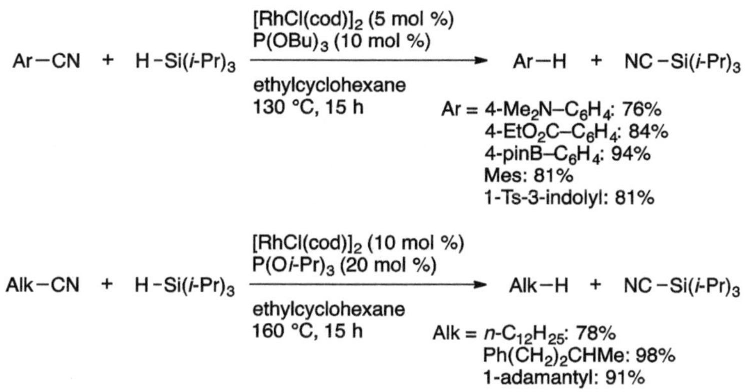


**Scheme 9** Hydrodecyanation of nitriles with hydrosilanes catalyzed by rhodium

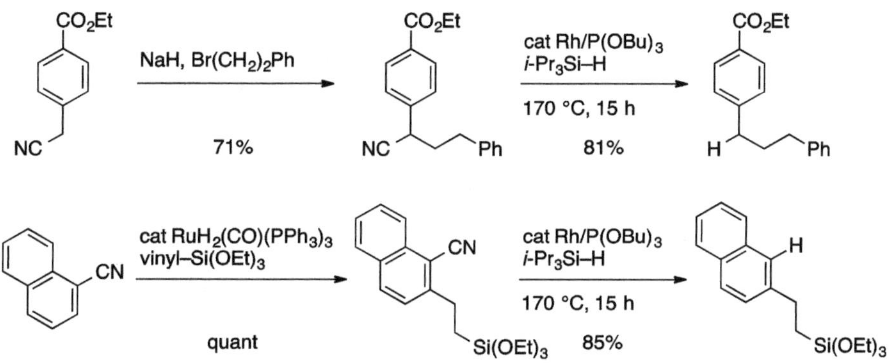


**Scheme 10** Synthetic applications utilizing rhodium-catalyzed hydrodecyanation

## *2.2 Cross-Coupling Reactions*

The transmetalation mentioned above for metal hydrides can naturally be extended to the use of main-group organometallic reagents to perform cross-coupling reactions using aryl cyanides instead of aryl halides. This reaction was first demonstrated with modified aryl, alkenyl, and alkyl Grignard reagents in the presence of Ni/$PMe_3$ catalyst (Scheme 11) [49, 50]. Alkynylzinc reagents also cross-couple with aryl cyanides to give various disubstituted acetylenes (Scheme 11) [51]. More recently, milder nucleophiles, such as arylboron reagents, have been introduced to undergo the cross-coupling reaction (Scheme 11) [52].

Arylrhodium species generated upon the cleavage of an Ar–CN bond by a silylrhodium intermediate, derived from the reaction of rhodium(I) with disilanes (see below), can be trapped by aryl halides in an intramolecular manner to give dibenzofurans and carbazoles (Scheme 12) [53]. The arylrhodium species can also

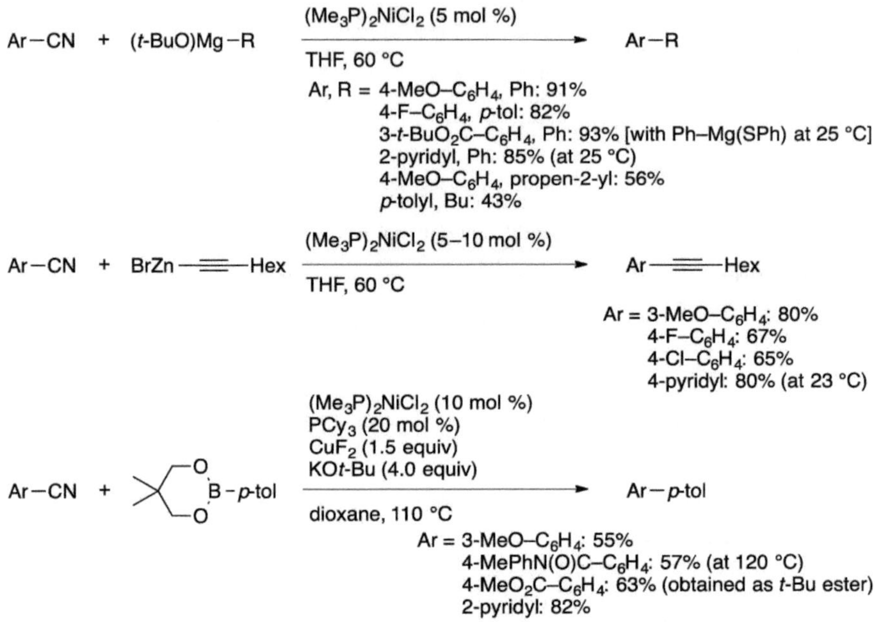

**Scheme 11** Cross-coupling reaction of aryl cyanides with carbon nucleophiles catalyzed by nickel

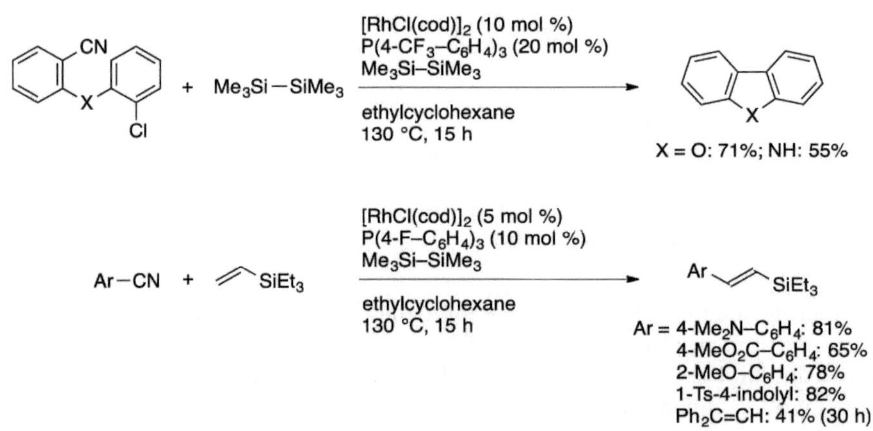

**Scheme 12** C–C bond-forming coupling reactions of aryl cyanides catalyzed by rhodium

react intermolecularly with vinylsilanes to give 2-arylethenylsilanes as a Heck-type product (Scheme 12) [54].

In addition to carbon nucleophiles, nitrogen-based nucleophiles can be used for the nickel-catalyzed cross-coupling with aryl cyanides in the presence of CsF, the role of which is yet to be clarified (Scheme 13) [55]. Silylphosphines are reported to

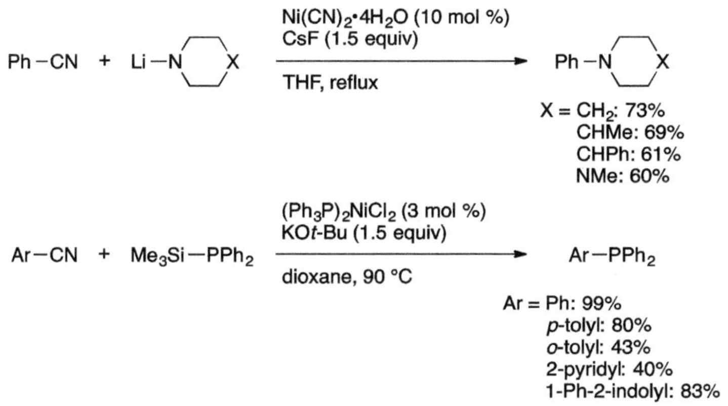

**Scheme 13** Cross-coupling reaction of aryl cyanides with nitrogen- and phosphorus-based nucleophiles catalyzed by nickel

serve as a phosphorus-based nucleophile to give a variety of aryl(diphenyl)phosphines by nickel-catalyzed cross-coupling reaction with aryl cyanides (Scheme 13) [56]. The use of a stoichiometric amount of strong bases such as KO*t*-Bu and NaOMe is crucial, and nucleophilic $MPPh_3$ ($M = K, Na$) may be generated in situ through the reaction of the bases with the silylphosphine reagents.

## 2.3 Silylation and Borylation Reactions

C–CN bond activation by silylrhodium species can be applied to a catalytic cycle for silylative decyanation of nitriles, when rhodium intermediates active in C–CN activation are generated from disilanes instead of hydrosilanes (Scheme 14). The reaction proceeds with a range of nitriles, including aryl, alkenyl, and alkyl cyanides, to give the corresponding tetraorganosilanes (Scheme 15) [53, 57], although the yields of tetra-alkylsilanes are modest.

More recently, decyanative borylation of nitriles is found to proceed with diboranes to give organoboron compounds through C–CN activation (Scheme 16) [58, 59]. A borylrhodium(I) species is expected to be responsible for the C–CN activation in a manner similar to that by a silylrhodium(I) species based on theoretical calculations (Scheme 17) [60]. The intermediates subsequently react with diborane reagents via oxidative addition to give rhodium(III) intermediates, which undergo reductive elimination to give cyanoboranes and borylation products and to regenerate the catalytically active borylrhodium(I) species.

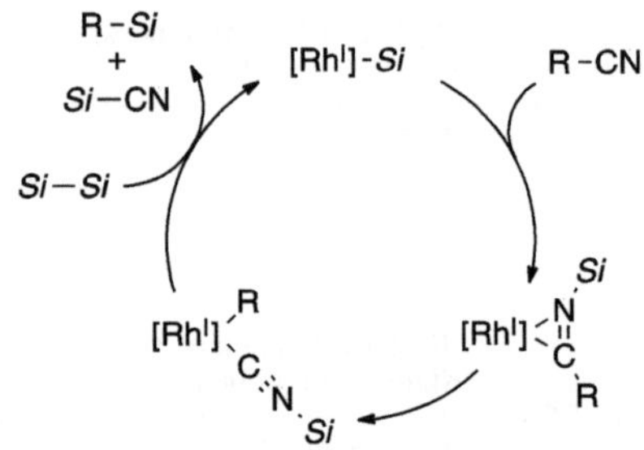


**Scheme 14** A plausible mechanism for rhodium-catalyzed decyanative silylation of nitriles

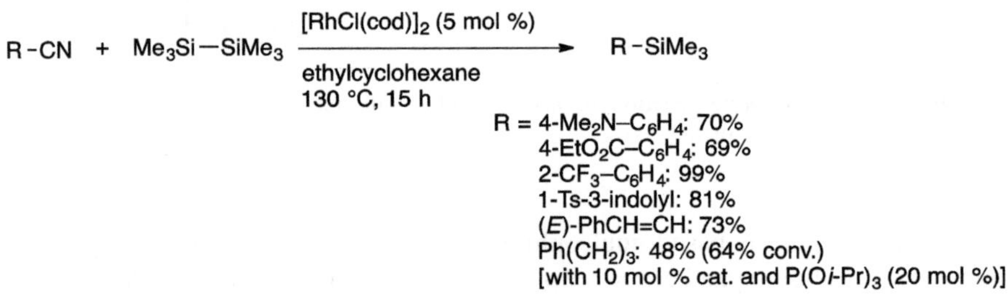


**Scheme 15** Decyanative silylation of nitriles catalyzed by rhodium

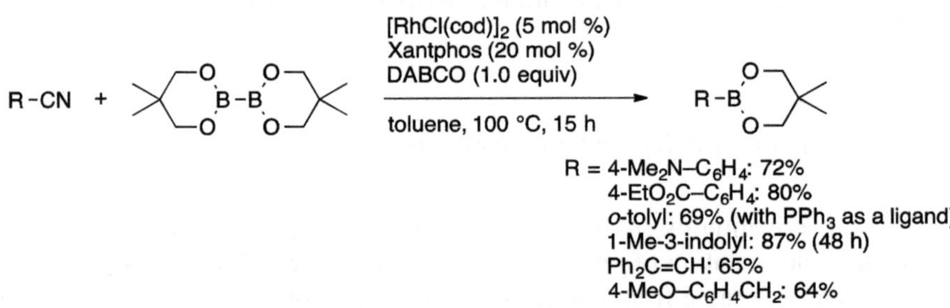


**Scheme 16** Decyanative borylation of nitriles catalyzed by rhodium

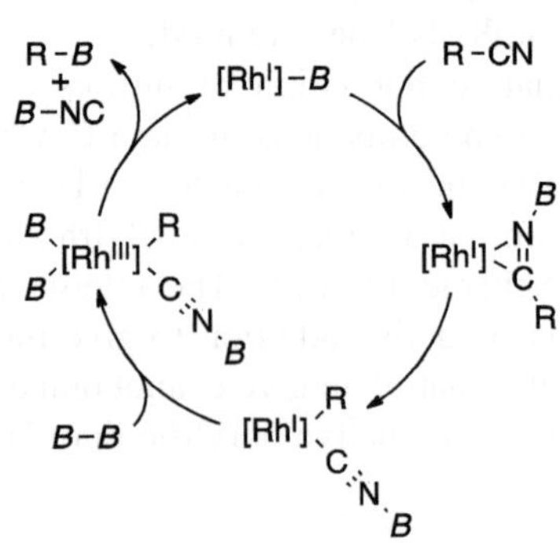


**Scheme 17** A plausible mechanism for rhodium-catalyzed decyanative borylation of nitriles

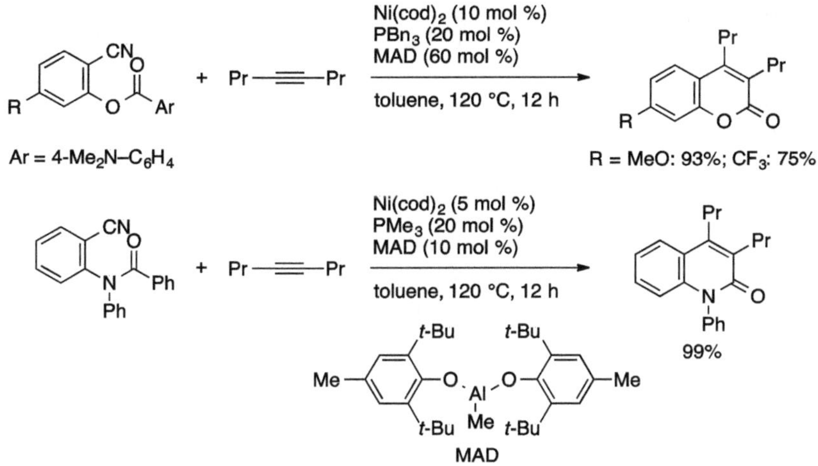


**Scheme 18** Cycloaddition reactions of aryl cyanides with alkynes via C–CN and acyl–Ar activation

## 2.4 *Cycloaddition Reactions*

C–CN activation via oxidative addition can be followed by the activation of another C–C bond to develop cycloaddition reactions. The reaction of *o*-arylcarboxybenzonitrile with alkynes proceeds in this manner to give coumarins, aryl cyanides, and an alkyne-arylcyanation product in the presence of catalytic amounts of nickel and aluminum-based Lewis acid (Scheme 18) [61]. Likewise, *o*-cyanophenylbenzamides undergo the transformation to give quinolones (Scheme 18 [62]. A catalytic cycle involving a five-membered nickelacycle intermediate, generated possibly by the oxidative addition of Ar–CN bonds, and the subsequent C–C bond activation [63] is proposed (Scheme 19).

# 3 Cyanation Reactions via C–CN Bond Activation

C–CN activation by metal complexes often generates metal cyanides, which can serve as cyanating agents to give nitriles as a product. Because many of the cyanation reactions have conventionally been performed by using a stoichiometric amount of generally highly toxic metal cyanides and/or hydrogen cyanides, cyanation reactions via metal cyanides generated catalytically in situ through C–CN activation can be less toxic. Thus, they can be practically useful alternative protocols to introduce a cyano functionality into organic molecules using commonly available less toxic nitriles. Moreover, if both organic and cyano groups of nitriles can be introduced at the same time through C–CN activation, nitriles having

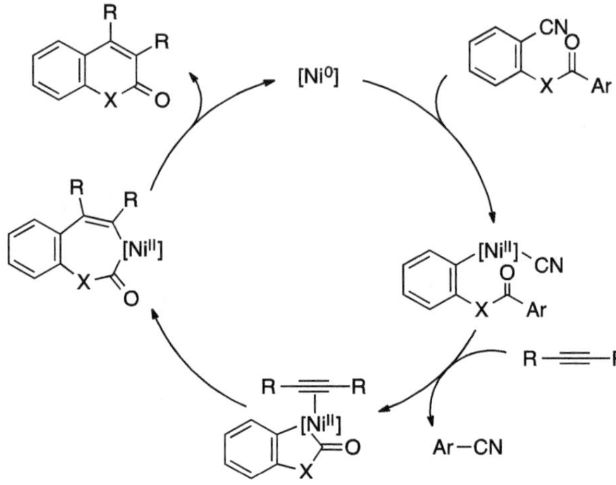


**Scheme 19** A plausible mechanism for nickel-catalyzed cycloaddition via C–CN and acyl–Ar activation

higher complexity can be readily accessed in a single operation. These cyanation reactions via C–CN activation are described in this section.

## 3.1 *Cyanation of Aryl Halides and Arenes*

Copper has been demonstrated to mediate cyanation of aryl bromides and iodides through the activation of C–CN bonds (Scheme 20). Phenylacetonitrile [64], malononitrile [65], and even acetonitrile as a reaction solvent [66] have been reported to serve as cyanating agents. 2-Phenylpyridines [67] and indoles [68] are directly cyanated by copper-mediated cyanation reaction using phenylacetonitrile, which is supposedly oxidized first at its benzylic position to give benzoyl cyanide, which further reacts with copper complexes to generate a cyanocopper species responsible for the cyanation event. Nevertheless, detailed mechanisms of these cyanation reactions remain elusive.

Palladium complexes have also been reported to catalyze the cyanation of aryl halides via C–CN activation of nitriles, such as phenylacetonitrile [69] and ethyl cyanoacetate [70] (Scheme 21). An excess amount of acetonitrile can also be activated to serve as a cyanating agent in the presence of palladium catalyst [71]. The mechanism of C–CN activation in these palladium-catalyzed cyanation reactions is also yet to be understood.

## 3.2 *Carbocyanation of Unsaturated Bonds*

The oxidative addition of C–CN bonds of nitriles can be followed by migratory insertion of unsaturated compounds into C–metal bonds and subsequent reductive

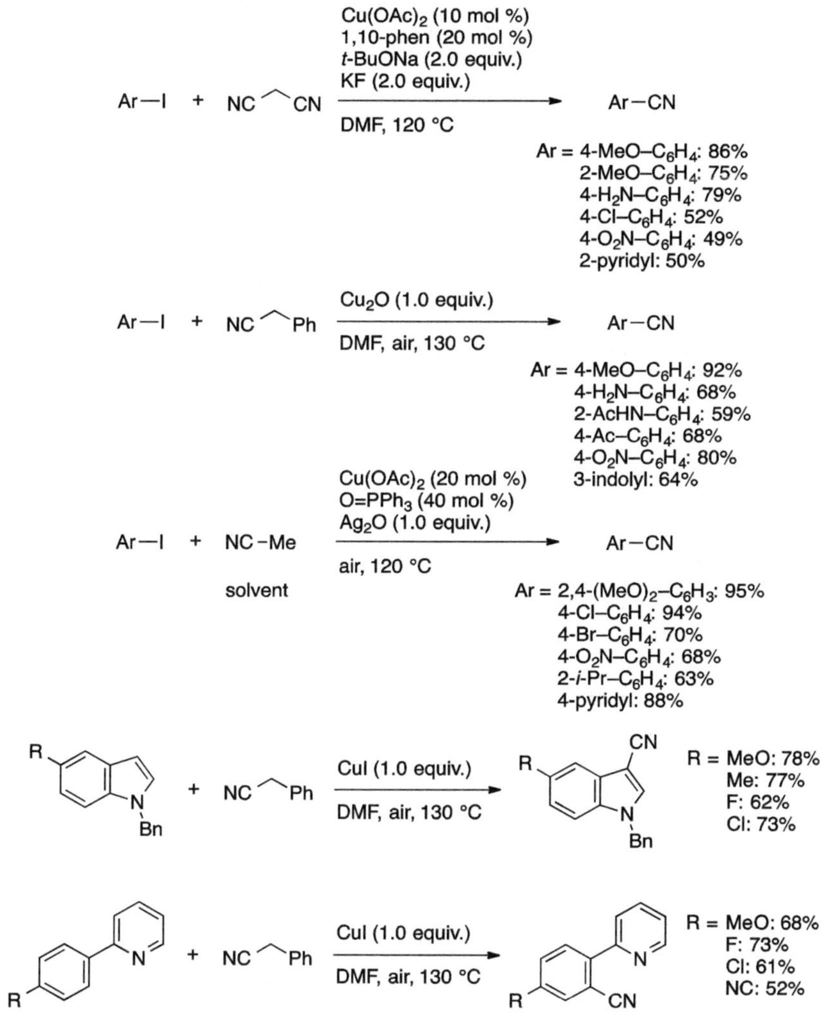


**Scheme 20** Cyanation of aryl halides and arenes with nitriles catalyzed or mediated by copper

elimination to develop addition reactions of organic and cyano groups of nitriles across unsaturated compounds through C–CN cleavage, namely by carbocyanation reaction (Scheme 22) [72, 73]. Initially, these synthetically novel transformations were attempted by using benzoyl cyanide and alkynes in the presence of a palladium catalyst (Scheme 23) [74, 75]. The reaction proceeds to give *cis*-adducts but with a mechanistic scenario different from that shown in Scheme 22. Thus, arylacetylenes are first acylated by the nitrile substrate, and the thus generated HCN adds across the aroyl(aryl)acetylenes to give formal *trans*-aroylcyanation products, which finally isomerize under the reaction conditions to give *cis*-adducts.

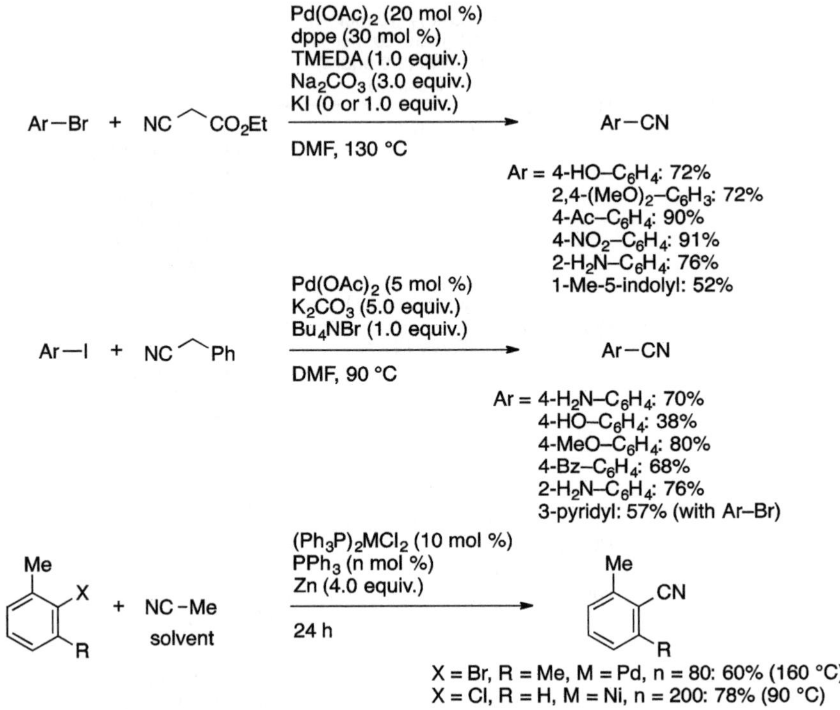


**Scheme 21** Cyanation of aryl halides with nitriles catalyzed by palladium

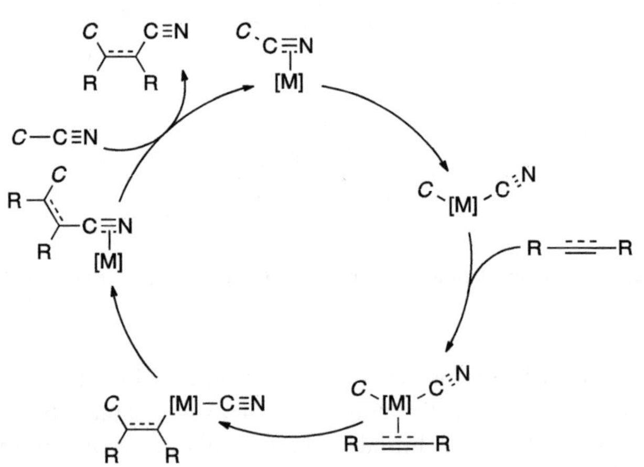


**Scheme 22** A possible catalytic cycle for carbocyanation of unsaturated bonds

Given a number of examples of stoichiometric studies on the oxidative addition of C–CN bonds to nickel(0), nickel catalysts were envisaged to catalyze the carbocyanation reaction. Indeed, aryl cyanides add across alkynes in the presence

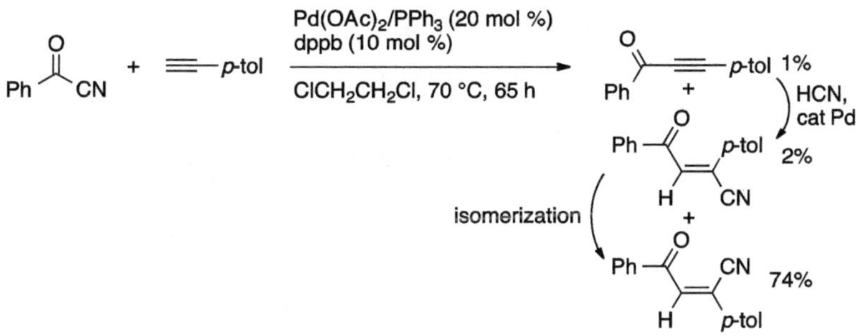


**Scheme 23** Benzoylcyanation of arylacetylenes catalyzed by palladium

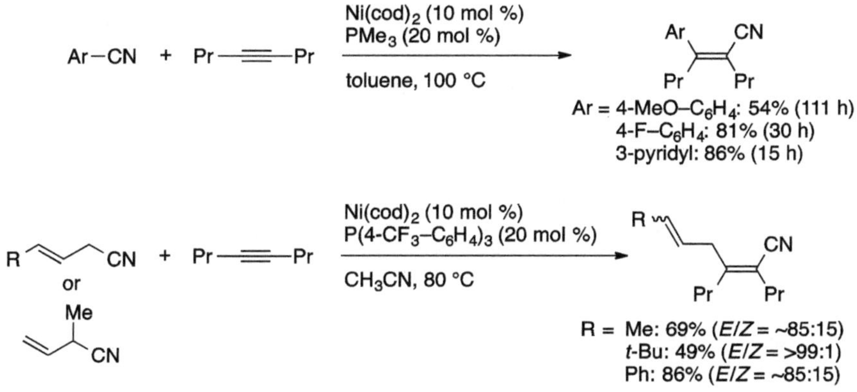


**Scheme 24** Aryl- and allylcyanation of alkynes catalyzed by nickel

of Ni/$PMe_3$ catalyst to give various tetra-substituted olefins (Scheme 24) [76, 77]. Allyl cyanides are also viable nitrile substrates to undergo alkyne-carbocyanation reaction by using less electron-donating $P(4\text{-}CF_3\text{–}C_6H_4)_3$ as a ligand [78]. Nevertheless, the scope of nitriles is limited and the catalyst loadings of the original protocols are quite high for nickel-catalyzed carbocyanation. DFT calculations of the arylcyanation of alkynes have revealed that the oxidative addition of Ar–CN bonds to nickel(0) via $\eta^2$-arene nickel intermediates is the rate-determining step [79]. Therefore, it has been envisioned that the promotion of this step could overcome the limitations.

As mentioned above, the presence of Lewis acidic additives, such as triorganoaluminums and -aluminums, are known to facilitate the oxidative addition of C–CN bonds to nickel(0) species through the coordination of the nitrogen atom of cyano group to the Lewis acids [14, 38]. The arylcyanation of alkynes is indeed significantly accelerated by using aluminum-based Lewis acid cocatalysts (Scheme 25) [80, 81]. By cooperative nickel/Lewis acid catalysis, the scope of aryl cyanides has been improved to include those having labile bromo and chloro groups as well as sterically demanding substrates. The reaction conditions of allylcyanation can also

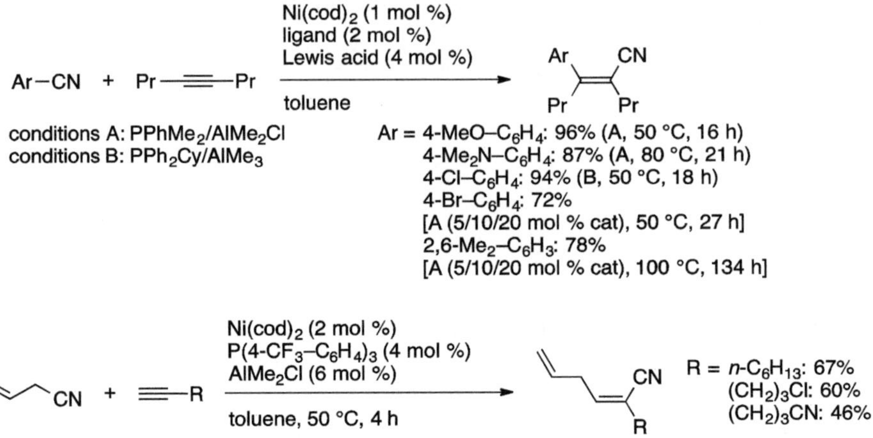


**Scheme 25** Aryl- and allylcyanation of alkynes catalyzed cooperatively by nickel/aluminum

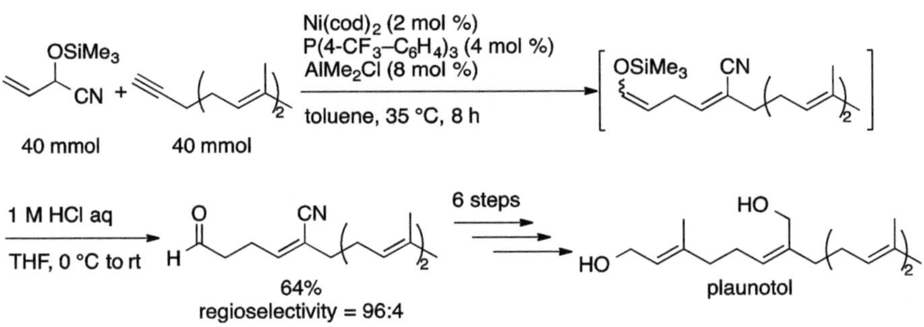


**Scheme 26** Total synthesis of plaunotol through allylcyanation of alkynes

be made milder by using $AlMe_2Cl$ as a Lewis acid cocatalyst, allowing highly stereoselective preparation of tri-substituted alkenes bearing a bulkier substituent at the cyano-substituted carbon [82]. Similar regioselectivity is also observed in the arylcyanation of alkynes and can be ascribed to aryl- or allylnickelation proceeding preferentially at sterically less hindered carbons of coordinated alkynes [79]. The protocol employing α-siloxyallyl cyanides affords tri-substituted ethenes bearing a formyl functionality upon the hydrolysis of silyl enol ether products, and can thus be used for the synthesis of functionalized multi-substituted olefins such as plaunotol, a diterpene known for antibacterial activity against *Helicobacter pylori* (Scheme 26).

The scope of nitriles for the carbocyanation reaction of alkynes can be expanded by cooperative nickel/Lewis acid catalysis. Alkenyl [80, 81] and alkynyl cyanides [83, 84] also participate in the addition reaction to give highly conjugated nitrile products by nickel/$BPh_3$ catalysis (Scheme 27). The use of aluminum-based Lewis acids causes isomerization of the double bond of alkenylcyanation products, whereas alkynylcyanation is sluggish with the aluminum reagents.

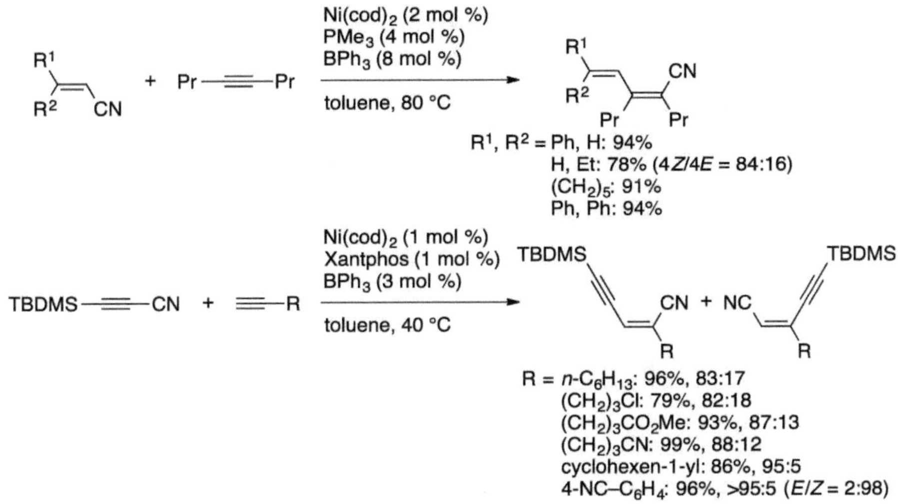


**Scheme 27** Alkenyl- and alkynylcyanation of alkynes catalyzed by nickel/$BPh_3$

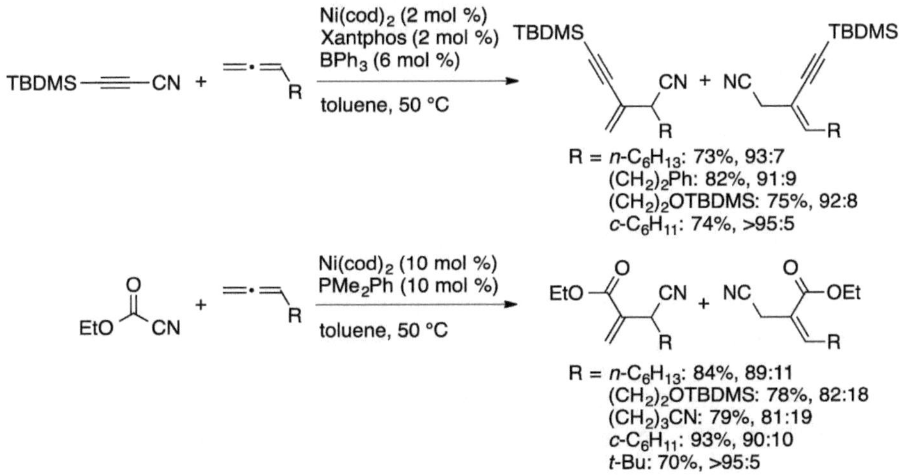


**Scheme 28** Carbocyanation of 1,2-dienes using alkynyl cyanides and cyanoformates

Some nitriles also add across 1,2-dienes (Scheme 28). Alkynylcyanation takes place predominantly across the internal double bond of 1,2-dienes to give selectively cyanoalkyl-substituted enynes [83, 84]. Cyanoformates also add across 1,2-dienes in a similar manner in the presence of nickel catalyst alone to give cyanoalkyl-substituted acrylates [85, 86], whereas carbocyanation of 1,2-dienes with other nitriles remains unexplored. The 1,2-diene–carbocyanation can be initiated by oxidative addition followed by the coordination of 1,2-dienes at the terminal double bond, and subsequent migratory insertion into the C–Ni bond

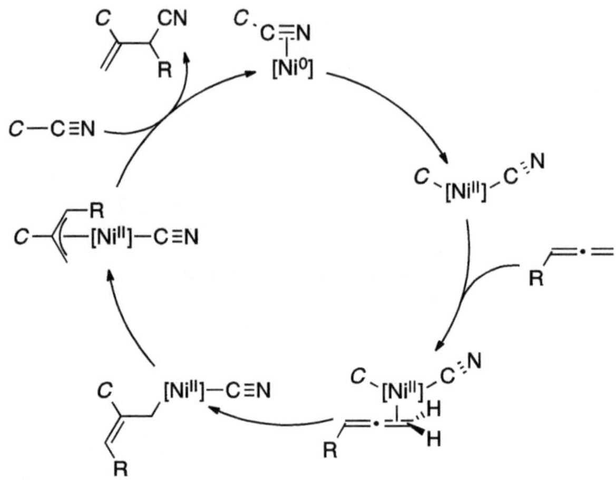


**Scheme 29** A plausible catalytic cycle for carbocyanation of 1,2-dienes

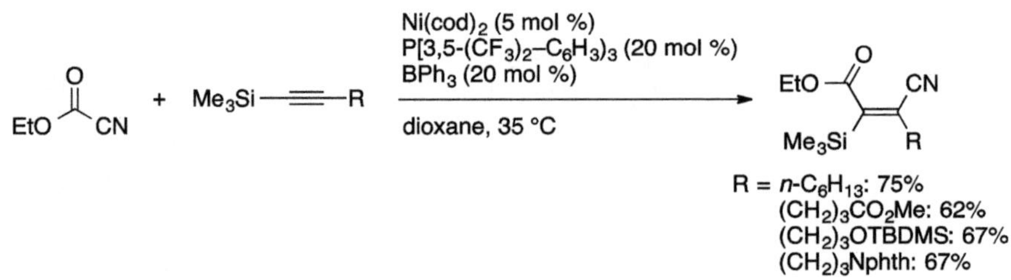


**Scheme 30** Cyanoesterification of silyl-substituted alkynes catalyzed by nickel/$BPh_3$

(Scheme 29). The subsequently formed allylnickel intermediates can be isomerized to give π-allylnickel species, which is likely responsible for the carbocyanation of the internal olefins.

Alkynes can also be functionalized both stereoselectively and regioselectively by cooperative catalysis (Scheme 30) [87]. Cyanoesterification of silyl-substituted alkynes proceeds to give β-cyanoesters with silyl group at α-position as a single product. The regiochemistry in contrast to other carbocyanation reactions may be derived from interaction of carbonyl with the silyl group prior to the carbonickelation event. A catalytic cycle through the oxidative addition of C–CN bonds is supported by the stoichiometric reaction of cyanoformamide with nickel (0) complex and $BPh_3$ to give an oxidative adduct and its further reaction with an alkyne, as well as by its use as a catalyst, both giving the corresponding cyanocarbamoylation product (Scheme 31). Palladium complexes, on the other hand, have been reported to catalyze intramolecular cyanoesterification of alkynes to give lactone products (Scheme 32) [88].

Alkanenitriles including acetonitriles and propionitriles undergo a reaction across alkynes to give *cis*-methylcyanation and ethylcyanation products,

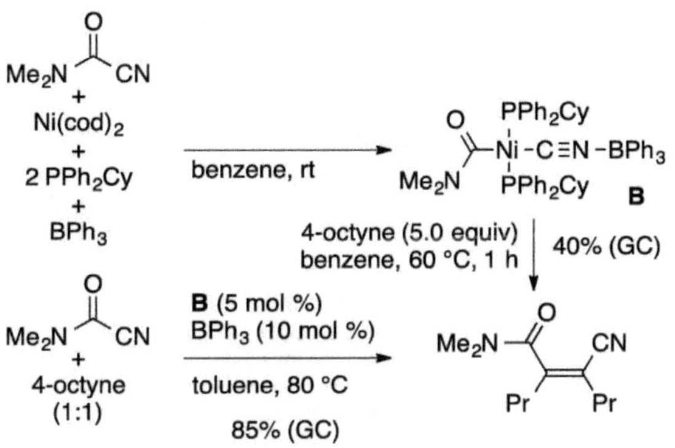


**Scheme 31** Oxidative addition of cyanoformamides to nickel(0) assisted by $BPh_3$ and its intermediacy in cyanocarbamoylation of alkynes

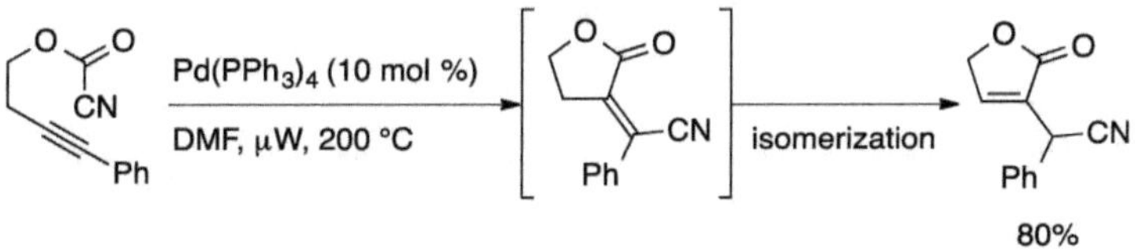


**Scheme 32** Intramolecular cyanoesterification of alkynes catalyzed by palladium

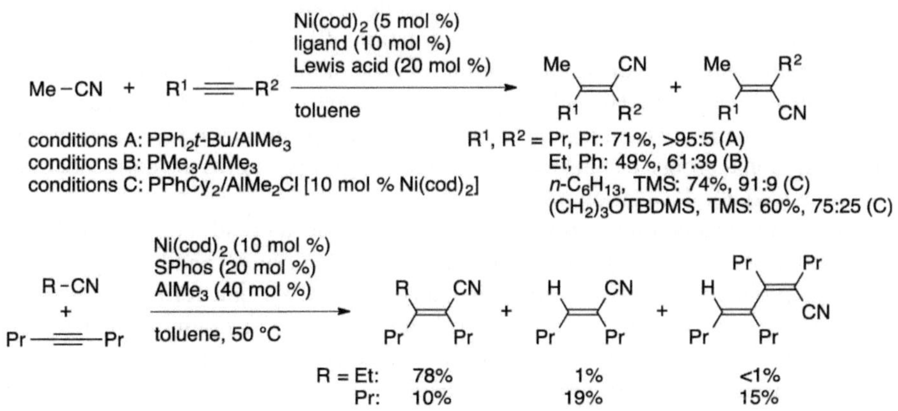


**Scheme 33** Alkylcyanation of alkynes catalyzed by nickel/aluminum

respectively (Scheme 33) [80, 89, 90]. Although the methylcyanation proceeds with excellent regioselectivity, formal *trans*-adducts derived from the isomerization of double bonds are contaminated. A trace amount of hydrocyanation product is observed in the ethylcyanation reaction, possibly through the β-hydride elimination from ethylnickel species, which can be generated upon the oxidative addition of Et–CN bond to nickel(0). Byproducts derived from this unwanted pathway can be observed in much higher amounts with alkyl cyanides having higher alkyl chains.

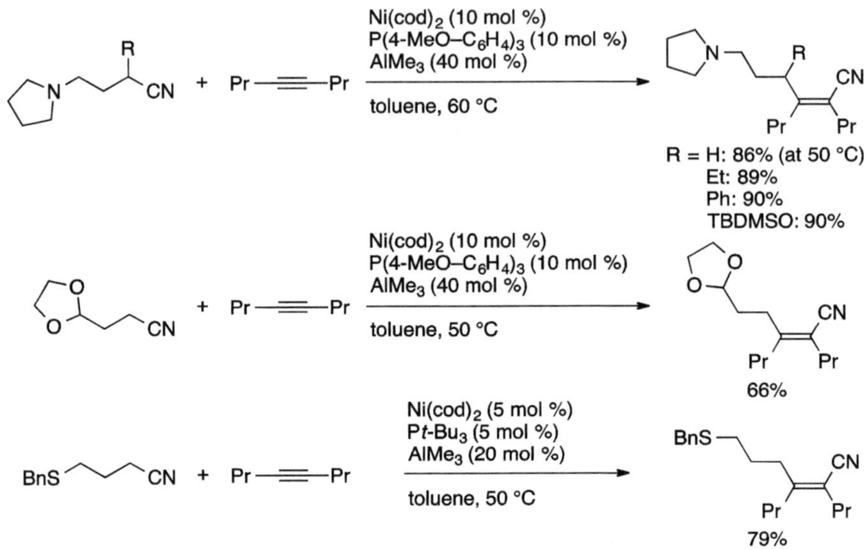


**Scheme 34** Heteroatom-directed alkylcyanation of alkynes catalyzed by nickel/$AlMe_3$

When alkanenitriles having a heteroatom functionality at the γ-position are used for the alkylcyanation, hydrocyanation byproducts can be suppressed (Scheme 34) [91]. With these particular nitriles, oxidative adducts can possibly form metallacycle intermediates through intramolecular coordination of heteroatom functionalities to retard the unwanted β-hydride elimination because of the absence of vacant coordination sites. Even secondary alkyl cyanides participate in the addition reaction in good yields when they have an amino group at their γ-position. Oxygen- and sulfur-based functional groups can also serve as directing groups to give the corresponding functionalized alkylcyanation products.

While carbocyanation reactions across alkynes show a broad scope of nitriles as described above, intermolecular reactions across simple alkenes are generally sluggish, probably because of the reluctant C($sp^3$)–CN bond-forming reductive elimination, which competes with β-hydride elimination. For example, the reaction of aryl cyanides with vinylsilanes gives a Heck-type product in modest yield, possibly through migratory insertion of the alkenes into the Ar–Ni bond followed by β-hydride elimination (Scheme 35). Bicyclic alkenes, typically norbornene and norbornadiene, on the other hand, successfully undergo the nickel-catalyzed carbocyanation to give *exo-cis* adducts exclusively (Scheme 36) [81, 84, 92]. Palladium can catalyze the cyanoesterification of norbornene and norbornadiene [93–95]. The observed *exo*-selectivity can be ascribed to the higher electron density of the *exo*-face [96]. These functionalized norbornenes can be monomers for ring-opening metathesis polymerization to give functional polymer materials [97–101]. Successful C($sp^3$)–CN bond-forming reductive elimination can be

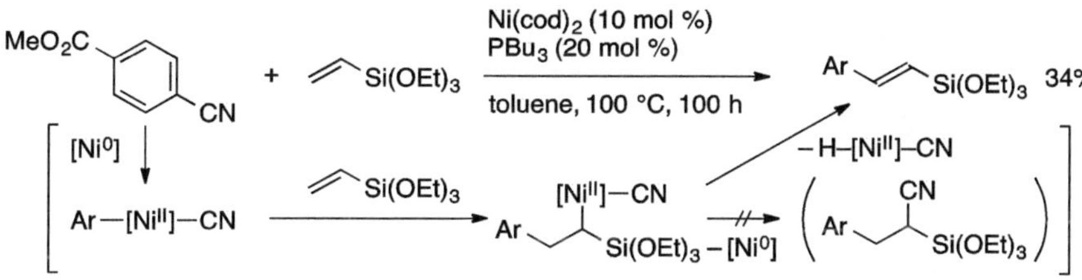


**Scheme 35** Attempted arylcyanation of vinylsilanes to result in a Heck-type coupling

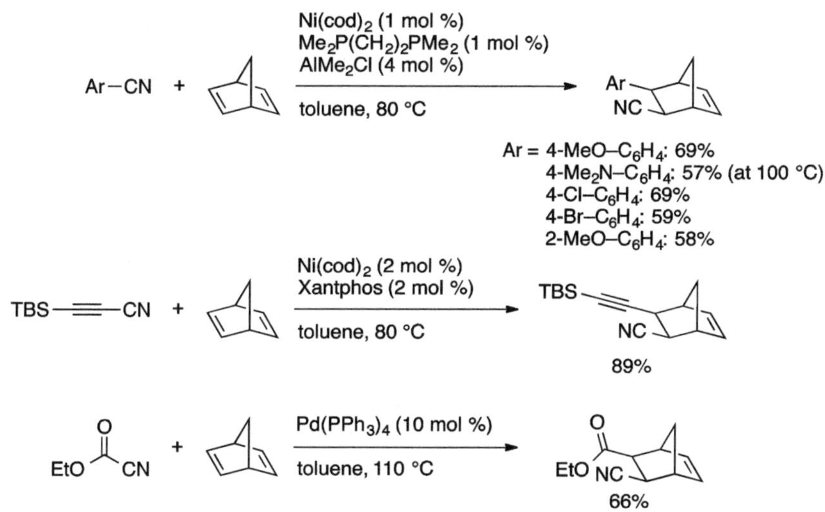


**Scheme 36** Carbocyanation of norbornadiene

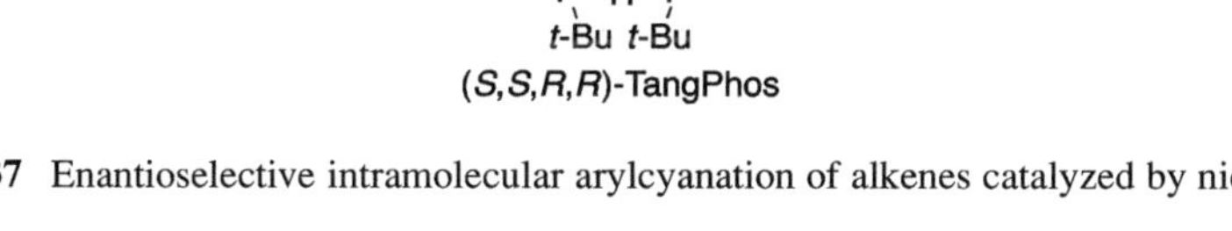


**Scheme 37** Enantioselective intramolecular arylcyanation of alkenes catalyzed by nickel/$BPh_3$

possible because of unfavorable β-hydride elimination to give rise to *anti*-Bredt olefin products.

Intramolecular arylcyanation of simple alkenes proceeds smoothly by cooperative nickel/Lewis acid catalysis (Schemes 37, 38, and 39) [102–104], whereas palladium catalysts have been shown to be useful for intramolecular

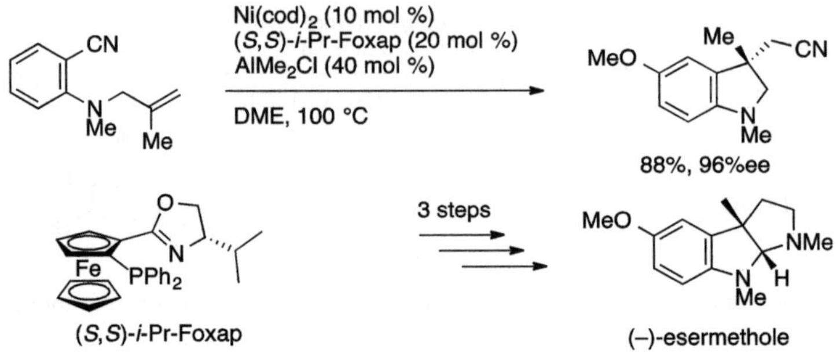


**Scheme 38** Total synthesis of (–)-esermethole through enantioselecive intramolecular arylcyanation of alkenes

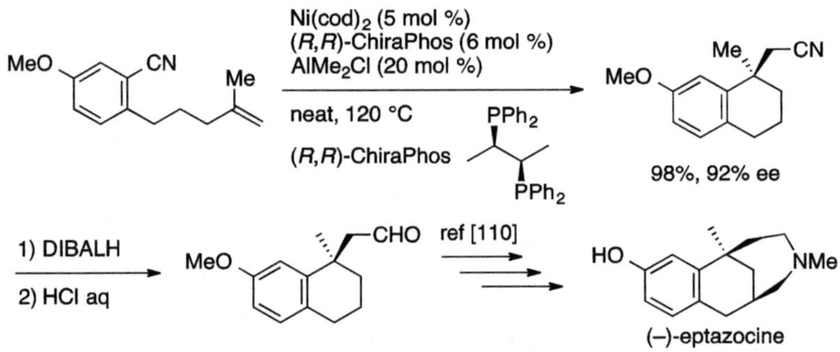


**Scheme 39** Formal total synthesis of (–)-eptazocine through enantioselecive intramolecular arylcyanation of alkenes

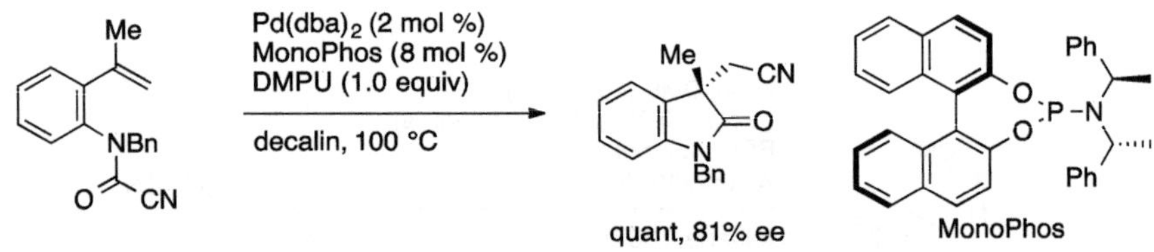


**Scheme 40** Enantioselective intramolecular cyanocarbamoylation of alkenes catalyzed by palladium

cyanocarbamoylation (Scheme 40) [105–109]. Proper chiral phosphorus ligands have been identified for each substrate structure to allow for the access to nitriles having a quaternary stereocenter with high enantiomeric excess. Some of these optically active nitrile products serve as synthetic precursors for biologically active natural products (Schemes 38 and 39) [110]. The transformation can be an alternative to asymmetric intramolecular Heck reactions, which have been applied extensively for natural product syntheses [111]. The intramolecular carbocyanation of alkenes can be advantageous in terms of the retention of cyano functionality in a

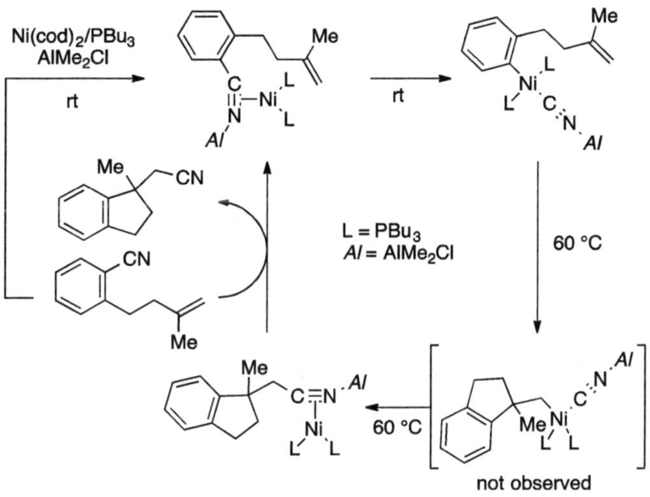


**Scheme 41** $\eta^2$-Coordination and oxidative addition of aryl cyanides to nickel(0) assisted by $AlMe_2Cl$ and its intermediacy in intramolecular arylcyanation of alkenes

product, while (pseudo)halogen functionalities in starting materials are lost in the Heck cyclization.

The reaction intermediates of the intramolecular arylcyanation are fully characterized by NMR and X-ray analyses (Scheme 41). The oxidative addition of an Ar–CN bond to nickel(0) assisted by $AlMe_2Cl$ takes place directly from the Lewis acid adduct of $\eta^2$-nitrile complexes at room temperature [103]. The oxidative adduct undergoes cyclization and reductive elimination to give another $\eta^2$-nitrile complex derived from the intramolecular C–C bond forming event upon heating at 60°C. Treatment of the $\eta^2$-nitrile complex with the starting aryl cyanide gives the initially observed $\eta^2$-nitrile complex through the exchange of nitrile ligand. The overall scheme shows that the rate-determining step of the intramolecular arylcyanation is either the exchange of bound phosphorus by the tethered alkene or the migratory insertion step rather than oxidative addition of Ar–CN bond.

# 4 Summary, Conclusions, Outlook

Synthetic reactions involving C–CN activation by metal catalysis are reviewed in this chapter. All the transformations presented herein provide the synthetic community with novel reaction modes of nitriles, and thus with new ideas and strategies for the syntheses of target molecules. In addition, the developments in novel metal catalysis for the activation of unreactive C–CN bonds should be of interest to the organometallic community. These catalyst designs will further stimulate the development of more active catalysts as well as novel catalysts for the activation of other unreactive bonds. At the moment, each reaction has a different maturity. Hydrodecyanation reaction covers a wide range of nitrile substrates including the most challenging alkanenitriles using nickel and rhodium catalysis, whereas cross-coupling reactions using nitriles as electrophiles are limited to aryl cyanides. Given the recent extensive studies on the cross-coupling of alkyl electrophiles, the use of alkanenitriles as alkyl electrophiles for cross-coupling should be an interesting and important direction for this particular transformation. Carbocyanation reaction also covers a broad range of nitriles for the addition across alkynes, whereas alkene-carbocyanation is totally underexplored except for intramolecular reactions. Because $C(sp^3)$–CN bond-forming reductive elimination seems unfavorable compared with competitive β-hydride elimination of alkylnickel species, different metal systems and/or different mechanistic scenarios have to be envisioned to develop the potentially highly valuable C–C bond forming reaction.

**Acknowledgement** Financial support through Grant-in-Aids for Young Scientists (Nos. 19750076 and 21685023), Priority Areas "Chemistry of Concerto Catalysis" (Nos. 19028030 and 20037035) and "Molecular Theory for Real Systems" (Nos. 19029024 and 20038027), and Scientific Research on Innovative Areas "Molecular Activation Directed toward Straightforward Synthesis" (No. 22105003) by MEXT and JSPS are gratefully acknowledged.

## References

1. Storhoff BN, Lewis HC Jr (1997) Coord Chem Rev 23:1
2. Gerlach DH, Kane AR, Parshall GW, Jesson JP, Muetterties EL (1971) J Am Chem Soc 93:3543
3. Burmeister JL, Edwards LM (1971) J Chem Soc A 1663
4. Parshall GW (1974) J Am Chem Soc 96:2360
5. Clarke DA, Hunt MM, Kemmitt DW (1979) J Organomet Chem 175:303
6. Morvillo A, Turco A (1981) J Organomet Chem 208:103
7. Favero G, Morvillo A, Turco A (1983) J Organomet Chem 241:251
8. Bianchini C, Masi D, Meli A, Sabat M (1986) Organometallics 5:1670
9. Abla M, Yamamoto T (1997) J Organomet Chem 532:267
10. Churchill D, Shin JH, Hascall T, Hahn JM, Bridgewater BM, Parkin G (1999) Organometallics 18:2403
11. García JJ, Jones WD (2000) Organometallics 19:5544
12. García JJ, Brunkan NM, Jones WD (2002) J Am Chem Soc 124:9547
13. Liu Q-X, Xu F-B, Li Q-S, Song H-B, Zhang Z-Z (2004) Organometallics 23:610
14. Brunkan NM, Brestensky DM, Jones WD (2004) J Am Chem Soc 126:3627
15. García JJ, Arévalo A, Brunkan NM, Jones WD (2004) Organometallics 23:3997
16. Li X, Sun H, Yu F, Flörke U, Klein H-F (2006) Organometallics 25:4695
17. Ateşin TA, Li T, Lachaize S, Brennessel WW, Carcía JJ, Jones WD (2007) J Am Chem Soc 129:7562
18. Schaub T, Döring C, Radius U (2007) Dalton Trans 1993
19. Acosta-Ramírez A, Flores-Álamo M, Jones WD, García JJ (2008) Organometallics 27:1834
20. Atesin TA, Li T, Lachaize S, Carcía JJ, Jones WD (2008) Organometallics 27:3811
21. Swartz BD, Reinartz NM, Brennessel WW, García JJ, Jones WD (2008) J Am Chem Soc 130:8548
22. Li T, García JJ, Brennessel WW, Jones WD (2010) Organometallics 29:2430

23. Evans ME, Li T, Jones WD (2010) J Am Chem Soc 132:16278
24. Tanabe T, Evans ME, Brennessel WW, Jones WD (2011) Organometallics 30:834
25. Swartz BD, Brennessel WW, Jones WD (2011) Organometallics 30:1523
26. Evans ME, Jones WD (2011) Organometallics 30:3371
27. Grochowski MR, Brennessel WW, Jones WD (2011) Organometallics 30:5604
28. McKinney RJ (1992) In: Parshall GW (ed) Homogeneous catalyst. Wiley, New York, pp 42–50
29. Chaumonnot A, Lamy F, Sabo-Etienne S, Donnadieu B, Chaudret B, Barthelat J-C, Galland J-C (2004) Organometallics 23:3363
30. van der Vlugt JI, Hewat AC, Neto S, Sablong R, Mills AM, Lutz M, Spek AL, Müller C, Vogt D (2004) Adv Synth Catal 346:993
31. Wilting J, Müller C, Hewat AC, Ellis DD, Tooke DM, Spek AL, Vogt D (2005) Organometallics 24:13
32. Acosta-Ramírez A, Muñoz-Hernández M, Jones WD, García JJ (2006) J Organomet Chem 619:3895
33. Acosta-Ramírez A, Flores-Gaspar A, Muñoz-Hernández M, Arévalo A, Jones WD, García JJ (2007) Organometallics 26:1712
34. Acosta-Ramírez A, Muñoz-Hernández M, Jones WD, García JJ (2007) Organometallics 26:5766
35. Acosta-Ramírez A, Morales-Morales D, Serrano-Becerra JM, Arévalo A, Jones WD, García JJ (2008) J Mol Catal A Chem 288:14
36. Tauchert ME, Kaiser TR, Göthlich APV, Rominger F, Warth DCM, Hofmann P (2010) ChemCatChem 2:674
37. Li T, Jones WD (2011) Organometallics 30:547
38. Tolman CA, Seidel WC, Druliner JD, Domaille PJ (1984) Organometallics 3:33
39. Taw FL, White PS, Bergman RG, Brookhart M (2002) J Am Chem Soc 124:4192
40. Taw FL, Mueller AH, Bergman RG, Brookhart M (2003) J Am Chem Soc 125:9808
41. Nakazawa H, Kawasaki T, Miyoshi K, Suresh CH, Koga N (2004) Organometallics 23:117
42. Hashimoto H, Matsuda A, Tobita H (2006) Organometallics 25:472
43. Tobisu M, Chatani N (2008) Chem Soc Rev 37:300
44. Patra T, Agasti S, Maiti A, Maiti D (2013) Chem Sci 49:69
45. Nakazawa H, Kamata K, Itazaki M (2005) Chem Commun 4004
46. Nakazawa H, Itazaki M, Kamata K, Ueda K (2007) Chem Asian J 2:882
47. Tobisu M, Nakamura R, Kita Y, Chatani N (2009) J Am Chem Soc 131:3174
48. Tobisu M, Nakamura R, Kita Y, Chatani N (2010) Bull Kor Chem Soc 31:582
49. Miller JA (2001) Tetrahedron Lett 42:6991
50. Miller JA, Dankwardt JW (2003) Tetrahedron Lett 44:1907
51. Penney JM, Miller JA (2004) Tetrahedron Lett 45:4989
52. Yu D-G, Yu M, Guan B-T, Li B-J, Zheng Y, Wu Z-H, Shi Z-J (2009) Org Lett 11:3374
53. Tobisu M, Kita Y, Ano Y, Chatani N (2008) J Am Chem Soc 130:15982
54. Kita Y, Tobisu M, Chatani N (2010) Org Lett 12:1864
55. Miller JA, Dankwardt JW, Penney JM (2003) Synthesis 1643
56. Sun M, Zhang H-Y, Han Q, Yang K, Yang S-D (2011) Chem Eur J 35:9566
57. Tobisu M, Kita Y, Chatani N (2006) J Am Chem Soc 128:8152
58. Tobisu M, Kinuta H, Kita Y, Rémond E, Chatani N (2012) J Am Chem Soc 134:115
59. Kinuta H, Kita Y, Rémond E, Tobisu M, Chatani N (2012) Synthesis 44:2999
60. Jian Y-Y, Yu H-Z, Fu Y (2013) Organometallics 32:926
61. Nakai K, Kurahashi T, Matsubara S (2011) J Am Chem Soc 133:11066
62. Nakai K, Kurahashi T, Matsubara S (2013) Org Lett 15:856
63. Shukla P, Cheng CH (2006) Org Lett 8:2867
64. Wen Q, Jin J, Mei Y, Lu P, Wang Y (2013) Eur J Org Chem 4032
65. Jiang Z, Huang Q, Chen S, Long L, Zhou X (2012) Adv Synth Catal 354:589
66. Li J-H, Song R-J, Wu J-C, Liu Y, Deng G-B, Wu C-Y, Wei W-T (2012) Synlett 23:2491

67. Jin J, Wen Q, Lu P, Wang Y (2012) Chem Commun 48:9933
68. Yuen OY, Choy PY, Chow WK, Wong WT, Kwong FY (2013) J Org Chem 78:3374
69. Wen Q, Jin J, Hu B, Lu P, Wang Y (2012) RSC Adv 2:6167
70. Zheng S, Yu C, Shen Z (2012) Org Lett 14:3644
71. Luo FH, Chu CI, Cheng CH (1998) Organometallics 17:1025
72. Nakao Y, Hiyama T (2008) Pure Appl Chem 80:1097
73. Nakao Y (2012) Bull Chem Soc Jpn 85:731
74. Nozaki K, Sato N, Takaya H (1994) J Org Chem 59:2679
75. Nozaki K, Sato N, Takaya H (1996) Bull Chem Soc Jpn 69:1629
76. Nakao Y, Oda S, Hiyama T (2004) J Am Chem Soc 126:13904
77. Nakao Y, Oda S, Yada A, Hiyama T (2006) Tetrahedron 62:7567
78. Nakao Y, Yukawa T, Hirata Y, Oda S, Satoh J, Hiyama T (2006) J Am Chem Soc 128:7116
79. Ohnishi Y, Nakao Y, Sato H, Nakao Y, Hiyama T, Sakaki S (2009) Organometallics 28:2583
80. Nakao Y, Yada A, Ebata S, Hiyama T (2007) J Am Chem Soc 129:2428
81. Yada A, Ebata S, Zhang D, Nakao Y, Hiyama T (2010) Bull Chem Soc Jpn 83:1170
82. Hirata Y, Yukawa T, Kashihara N, Nakao Y, Hiyama T (2009) J Am Chem Soc 131:10694
83. Nakao Y, Hirata Y, Tanaka M, Hiyama T (2008) Angew Chem Int Ed 47:385
84. Hirata Y, Tanaka M, Yada A, Nakao Y, Hiyama T (2009) Tetrahedron 65:5037
85. Nakao Y, Hirata Y, Hiyama T (2006) J Am Chem Soc 128:7420
86. Hirata Y, Inui T, Nakao Y, Hiyama T (2009) J Am Chem Soc 131:6624
87. Hirata Y, Yada A, Morita E, Nakao Y, Hiyama T, Ohashi M, Ogoshi S (2010) J Am Chem Soc 132:10070
88. Rondla NR, Levi SM, Ryss JM, Berg RAV, Douglas CJ (2011) Org Lett 13:1940
89. Yada A, Yukawa T, Nakao Y, Hiyama T (2009) Chem Commun 3931
90. Yada A, Yukawa T, Idei H, Nakao Y, Hiyama T (2010) Bull Chem Soc Jpn 83:619
91. Nakao Y, Yada A, Hiyama T (2010) J Am Chem Soc 132:10024
92. Nakao Y, Yada A, Satoh J, Ebata S, Oda S, Hiyama T (2006) Chem Lett 35:790
93. Nishihara Y, Inoue Y, Itazaki M, Takagi K (2005) Org Lett 7:2639
94. Nishihara Y, Inoue Y, Izawa S, Miyasaka M, Tanemura K, Nakajima K, Takagi K (2006) Tetrahedron 62:9872
95. Nishihara Y, Miyasaka M, Inoue Y, Yamaguchi T, Kojima M, Takagi K (2007) Organometallics 26:4054
96. Inagaki S, Fujimoto H, Fukui K (1976) J Am Chem Soc 98:4054
97. Yoshida Y, Goto K, Komiya Z (1997) J Appl Polym Sci 66:367
98. Nishihara Y, Inoue Y, Nakayama Y, Shiono T, Takagi K (2006) Macromolecules 39:7458
99. Nishihara Y, Inoue Y, Saito AT, Nakayama Y, Shiono T, Takagi K (2007) Polym J 39:318
100. Nishihara Y, Izawa S, Inoue Y, Nakayama Y, Shiono T, Takagi K (2008) J Polym Sci Part A Polym Chem 46:3314
101. Nishihara Y, Doi Y, Izawa S, Li H-Y, Inoue Y, Kojima M, Chen J-T, Takagi K (2010) J Polym Sci Part A Polym Chem 48:485
102. Watson MP, Jacobsen EN (2008) J Am Chem Soc 130:12594
103. Nakao Y, Ebata S, Yada A, Hiyama T, Ikawa M, Ogoshi S (2008) J Am Chem Soc 130:12874
104. Hsieh J-C, Nakao Y, Hiyama T (2010) Synlett 1709
105. Kobayashi Y, Kamisaki H, Yanada R, Takemoto Y (2006) Org Lett 8:2711
106. Kobayashi Y, Kamisaki H, Takeda H, Yasui Y, Yanada R, Takemoto Y (2007) Tetrahedron 63:2978
107. Yasui Y, Kamisaki H, Takemoto Y (2008) Org Lett 10:3303
108. Yasui Y, Kinugawa T, Takemoto Y (2009) Chem Commun 4275
109. Yasui Y, Kamisaki H, Ishida T, Takemoto Y (2010) Tetrahedron 66:1980
110. Takemoto T, Sodeoka M, Sasai H, Shibasaki M (1993) J Am Chem Soc 115:8477
111. Dounay AB, Overman LE (2003) Chem Rev 103:2945

Top Curr Chem (2014) 346: 59–84
DOI: 10.1007/128_2013_493
© Springer-Verlag Berlin Heidelberg 2013
Published online: 31 October 2013

# Metal–Organic Cooperative Catalysis in C–C Bond Activation

**Chul-Ho Jun and Jung-Woo Park**

**Abstract** This review describes recent advances that have been made in studies of transition metal-promoted metal–organic cooperative C–C single bond activation reactions of unstrained organic substances, which use 2-aminopicoline as a temporal chelating ligand. In addition, metal–organic cooperative C–C double bond and C–C triple bond cleavage processes are discussed in association with transition metal-catalyzed C–H bond activation. Recent progress made in these areas has opened up the new paradigms in synthetic organic chemistry for the construction of organic frameworks by structural reorganization of organic backbones. Among the many strategies devised, chelation-assisted C–C bond cleavage reactions, which operate through cooperation between metal complexes and organic substances, have attracted perhaps the greatest attention. Utilizing this approach, efficient C–C single bond activation reactions have been developed for a variety of substrates, including linear alkyl ketones, secondary alcohols, primary amines, and cycloalkanones. In addition, reactions that lead to cleavage of C–C double and triple bonds can be facilitated by using the metal–organic cooperation strategy. C–C double bonds in α,β-enones can be cleaved by addition of cyclohexylamine, which facilitates Michael addition and *retro*-Mannich type fragmentation cascades proceeding via β-aminoketimine intermediates. Aldehydes, which serve as one of the fragmentation products of these processes, can be trapped by chelation-assisted hydroacylation reactions to produce ketones. Finally, C–C triple bond cleavage of alkynes can be achieved through hydroacylation reactions with aldehydes and subsequent C–C double bond cleavage of the resulting α,β-enones.

**Keywords** C–C bond activation · Chelation-assistant strategy · Cooperative catalysis · Ketone · Organic catalyst · *Retro*-Mannich fragmentation · Transition metal

---

C.-H. Jun (✉) and J.-W. Park
Department of Chemistry, Yonsei University, 120-749 Seoul, South Korea
e-mail: junch@yonsei.ac.kr

**Contents**



# Abbreviations

| | |
|---|---|
| COE | Cyclooctene |
| Cp* | Pentamethylcyclopentadienyl |
| Cy | Cyclohexyl |
| equiv | Equivalent(s) |
| Et | Ethyl |
| h | Hour(s) |
| $^{i}$Pr | Isopropyl |
| Me | Methyl |
| mol | Mole(s) |
| Ph | Phenyl |
| pyr | Pyridine |
| THF | Tetrahydrofuran |

# 1 Introduction

In spite of the intrinsic inertness of carbon–carbon bonds, the development of methods for their activation has gradually attracted the attention of synthetic chemists whose efforts focus on routes for the preparation of complex organic substances [1–10]. Over the past several decades, a variety of C–C bond cleavage reactions have been developed by using a variety of transition metal-based strategies.

## *1.1 Strategies for C–C Bond Cleavage*

The first strategy devised for promoting carbon–carbon bond cleavage involves the use of ring strain as a driving force. Thus, transition metal complexes, derived from highly strained ring systems such as those present in cyclopropane and cyclobutane derivatives, typically undergo C–C bond cleavage to form more stable four- and five-membered ring metallacyclic complexes [11, 12].

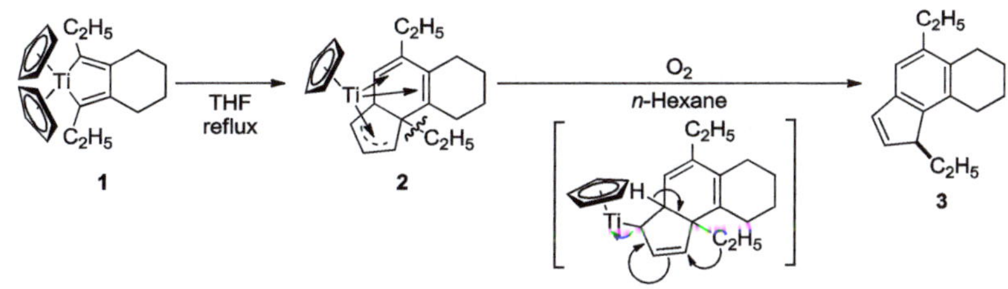


**Scheme 1** An example of C–C bond cleavage promoted by aromatization

The energy provided by transforming pre-aromatic to aromatic substances is another element for the design of effective transition metal-mediated C–C bond cleavage reactions. The most widely investigated processes that are based on this strategy involve transition metal complexes with pre-aromatic compounds, such as cyclopentadiene [13] and dihydrobenzene derivatives [14]. A recent example of C–C bond cleavage reaction driven by the release of aromatization energy is the transformation of cyclopentadienyl titanacyclopentadiene complexes **2**, prepared by heating bis(cyclopentadienyl)titanacyclopentadiene **1** in THF (Scheme 1) [15]. It was observed that **2** reacts with oxygen to produce the tricyclic indene derivative **3** by a pathway involving C–C bond cleavage, inducing migration of the ethyl group to the β-carbon of the allylic titanium moiety, which results in concurrent conversion of the cyclohexadiene ring in **2** to the benzene ring in **3**.

The third type of C–C bond cleavage strategy takes advantage of a driving force provided by generation of strong metal–carbon bonds, which promotes skeletal reorganization. A process that is representative of this strategy is β-carbon elimination to form a thermodynamically more stable metal–carbon bond from a metal–heteroatom bond [1, 10]. An example of a process guided by this phenomenon is found in studies by Hartwig et al. which demonstrate that C–C bond cleavage occurs in the β-carbon elimination/phenyl migration reaction of triphenylmethoxyrhodium(I) complex **4** that produces the phenyl-rhodium complex **5** and benzophenone (Scheme 2a) [16]. Shi et al. also described a novel rhodium promoted β-carbon elimination driven C–C bond cleavage reaction of a 2-phenylpyridine derivative that possesses an *ortho*-benzylic alcohol group on the phenyl ring (Scheme 2b) [17]. Specifically, reaction of 1-phenyl-(4-methyl-2-(pyridin-2-yl))benzyl alcohol (**6**) with styrene (**7**) in the presence of a mixture of $[Cp^*RhCl_2]_2$ and $Ag_2CO_3$ generates the corresponding stilbene **8**. In the mechanistic pathway followed in this process, C–C bond cleavage takes place in the seven-membered rhodacyclic complex **10** via β-carbon elimination to form thermodynamically favored five-membered rhodacyclic complex **11** and benzaldehyde (**9**).

Among those that have been developed, the potentially most powerful C–C bond activation strategy involves participation by chelating ligands located close to the targeted C–C bond. In processes of this type, properly positioned chelating ligands participate in energy releasing formation of stable metallacyclic complexes through

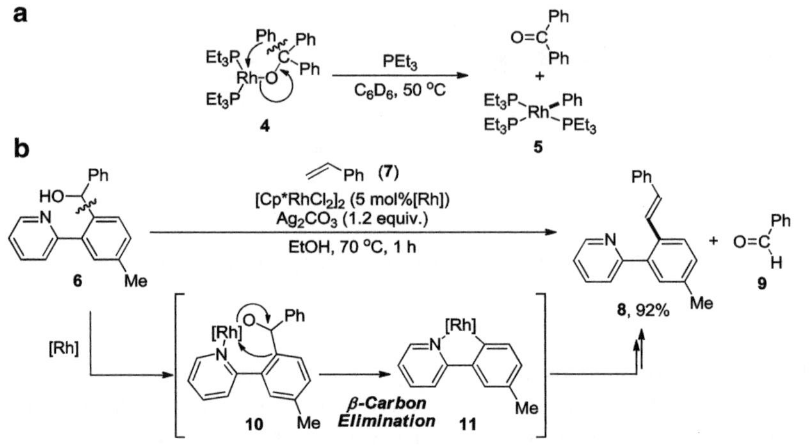

**Scheme 2** Examples of C–C bond activation through β-carbon elimination

the coordination-directed C–C bond cleavage. This chelation-assistance strategy is particularly attractive for activating C–C bond present in unstrained organic molecules, where relief of ring strain is not a driving force.

An example of a C–H bond activation process that employs this strategy is found in the pioneering work by Suggs et al. Specifically, Rh(I)-promoted activation of an α-C–C bond to ketone in 8-quinolinyl ketone **12** takes place via formation of a stable five-membered ring acylrhodium(III) complex (Scheme 3a) [18, 19]. Examples of chelation-assisted C–C bond activation processes are found in reactions of the bidentate pincer-type compound **13** [20] (Scheme 3b) and *ortho*-acyl 2-phenyloxazole derivative **14** [21] (Scheme 3c).

## *1.2 Metal–Organic Cooperative Strategy for C–H Bond Cleavage*

Although taking place with high efficiencies, most chelation-assisted transition metal promoted C–C bond cleavage reactions require substrates that contain preinstalled coordinating ligands and, therefore, they often result in the formation of products that possess these undesired groups, which need to be removed in later steps. A possible solution to this problem would involve the use of readily convertible or removable chelating ligands in these C–C bond activation processes. This strategy has led to an expansion in the scope of these reactions to include common organic substances that do not contain any coordinating ligands. For example, the removable chelating ligand approach has been applied to the intermolecular hydroacylation reactions of aldimine **15** with alkenes (Scheme 4) by Suggs and his coworkers [22, 23]. This process takes advantage of the fact that aldimine **15**, preformed from and serving as a surrogate for benzaldehyde, contains a readily

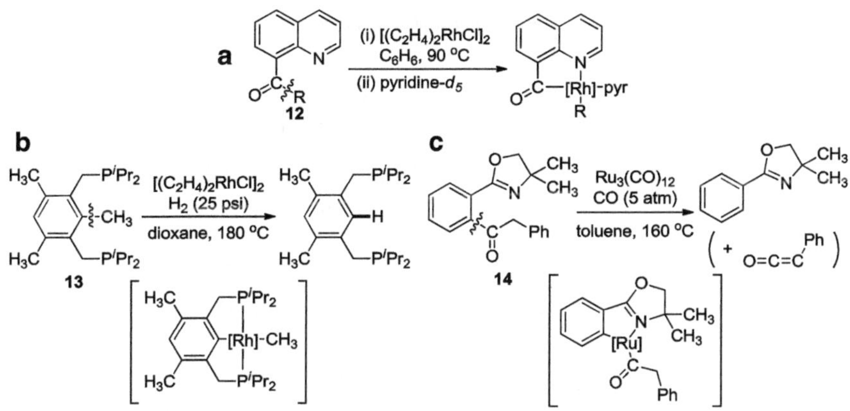


**Scheme 3** Several modes of C–C bond activation promoted by chelation-assistance

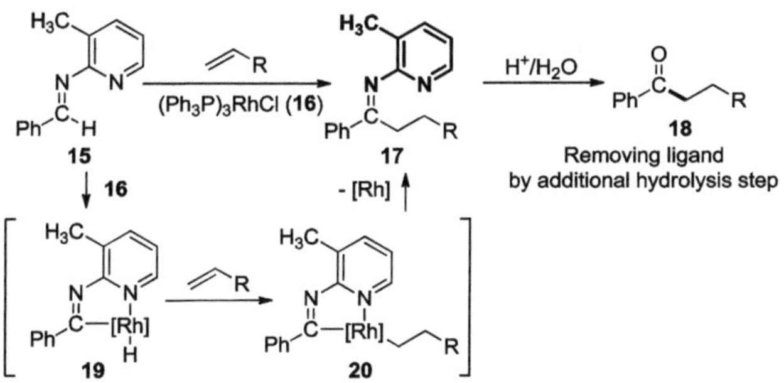


**Scheme 4** C–H bond activation using a substrate with preinstalled chelating ligand

removed pyridine coordinating group that assists in the bond forming and cleavage steps involved in formation of the rhodacycle intermediate **20** and product imine **17**. The reaction follows a pathway in which initial C–H bond activation of **15** by Rh(I) **16** generates iminoacyl-rhodium(III) hydride **19**, which reacts with the alkene to form ketimine **17** through reductive elimination of the resulting alkyl Rh(III) intermediate **20**. The pyridine chelating ligand in the resulting ketimine **17** is then readily removed by acid-promoted hydrolysis to produce corresponding ketone **18**.

This strategy has been expanded and made more general by demonstrating that only catalytic amounts of chelating ligand are required and that the active substrate can be formed in situ [24] (Scheme 5). In the new protocol it is unnecessary to preform the reactive aldimine **15**. Thus, treatment of a mixture containing benzaldehyde (**9**) and 2-amino-3-picoline (**21**) with Rh(I)-catalyst **16** and an alkene leads to direct formation of the desired ketone product **18** along with regenerated **21**. In this process, **15** generated in situ by condensation of benzaldehyde (**9**) with **21** reacts to produce ketimine **17** which then undergoes hydrolysis to form **18** and

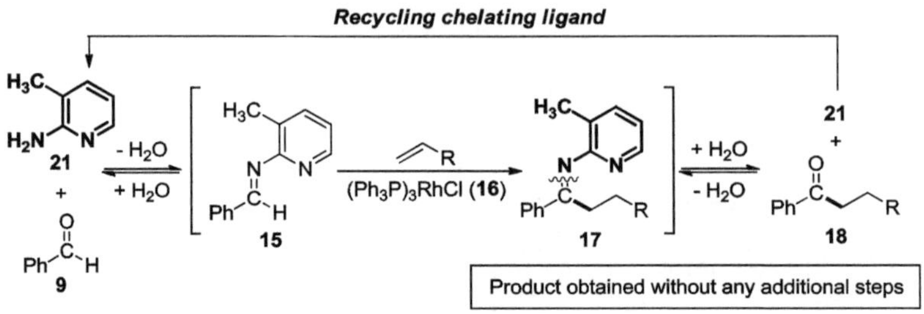


**Scheme 5** An example of metal–organic cooperation strategy for C–H bond activation

regenerate the chelating ligand **21**. Because Rh(I) serves as an organometallic catalyst and 2-amino-3-picoline acts as an organic catalyst in this reaction, the combination of these two catalysts can be viewed as a "metal–organic cooperative catalyst."

In the remaining sections of this chapter, advances made in the study of metal–organic cooperative C–C single bond activation reactions of unstrained organic molecule, which utilize 2-amino-3-picoline as a temporal chelating ligand, will be discussed. In addition, examples of the use of this methodology to promote C–C double and triple bond cleavage processes will be presented in association with C–H bond activation reactions.

# 2 Main Text

## *2.1 Metal–Organic Cooperative C–C Single Bond Cleavage*

The first example of a metal–organic cooperative C–C bond activation reaction is found in the transition metal-catalyzed transformations of alkyl ketones, such as benzylacetone (**22**), that possess β-hydrogens (Scheme 6) [25]. Reaction of **22** with 1-hexene (**23**) in the presence of a catalytic amount of $(Ph_3P)_3RhCl$ (**16**) and 2-amino-3-picoline (**21**) at 150°C was found to produce 2-octanone (**24**) in a high yield along with styrene (**25**). The first step in the mechanism for this process involves condensation of **22** with **21** to form the corresponding ketimine **26**. Rh(I)-mediated chelation-assisted C–C bond cleavage in **26**, followed by subsequent β-hydride elimination in the resulting complex **27**, takes place to give rhodium hydride intermediate **28**. Insertion of **28** into olefin **23** to form alkylrhodium complex **29** and ensuing reductive elimination then affords ketimine **30**, which undergoes hydrolysis to produce ketone **24** along with liberation of **21**. In this metal–organic cooperative C–C bond activation process, a condensation reaction occurs to install the chelation-auxiliary and the auxiliary is regenerated through a simple hydrolysis reaction.

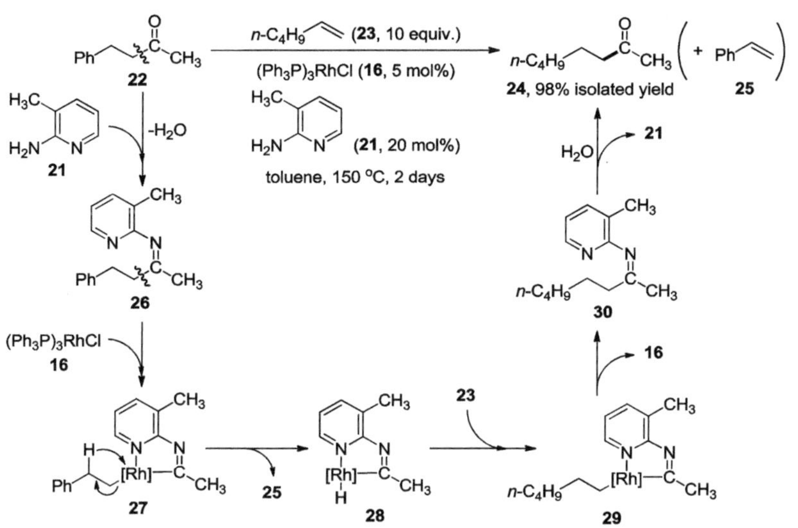


**Scheme 6** Metal–organic cooperative C–C bond activation of benzylacetone (**22**)

An interesting application of this strategy is the crossover reaction between two dialkyl ketones in which the alkyl groups are exchanged (Scheme 7). Treatment of 1,5-diphenyl-3-pentanone (**31**) and 5-nonanone (**32**) with Rh(I) **16** and aminopicoline **21** at 150°C leads to formation of 1-phenyl-3-heptanone (**33**) in 21% isolated yield. In this alkyl group exchange process, styrene (**25**) and 1-butene (**36**) are produced by C–C bond activation of complexes **34** and **35** and then consumed in subsequent hydroimination reactions of the resulting complexes **37** and **38** that yield metal–alkyl complexes **39** and **40**. The final steps of these parallel processes involve reductive elimination of **39** and **40** and hydrolysis to afford the alkyl group exchange product **33**.

This type of C–C bond activation reaction was observed to be facilitated by using microwave irradiation [26]. Under these conditions, the reaction proceeds with significantly enhanced efficiency as compared to that promoted by using conventional heating methods [27]. In addition, the reaction efficiency was found to increase when cyclohexylamine (**41**) is used as a co-catalyst. For example, reaction of benzylacetone (**22**) with norbornene (**42**) in the presence of 2-amino-3-picoline (**21**), Rh(I) catalyst **16**, and cyclohexylamine (**41**) under microwave irradiation conditions results in the formation of ketone **43** in an 85% isolated yield in a 5-min reaction time (Scheme 8a). In contrast, **43** is formed in a 16% yield when conventional thermal reaction conditions are employed. In addition, the microwave promoted reaction, carried out in the absence of cyclohexylamine (**41**), produces **43** in only a 44% isolated yield. This yield enhancement effect caused by microwave irradiation appears to be associated with the fact that the rate-determining step of the reaction involves formation of imine intermediates

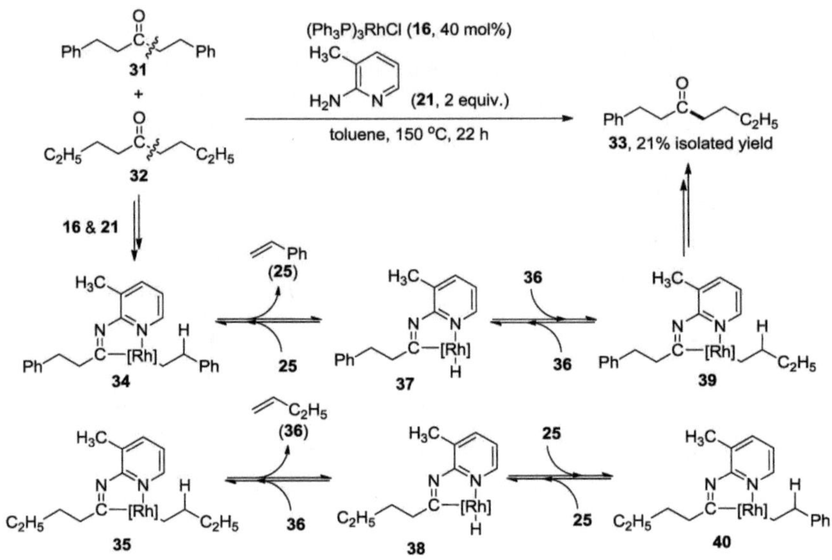


**Scheme 7** Application of metal–organic cooperative C–C bond activation to an alkyl group exchange reaction between ketones **31** and **32**

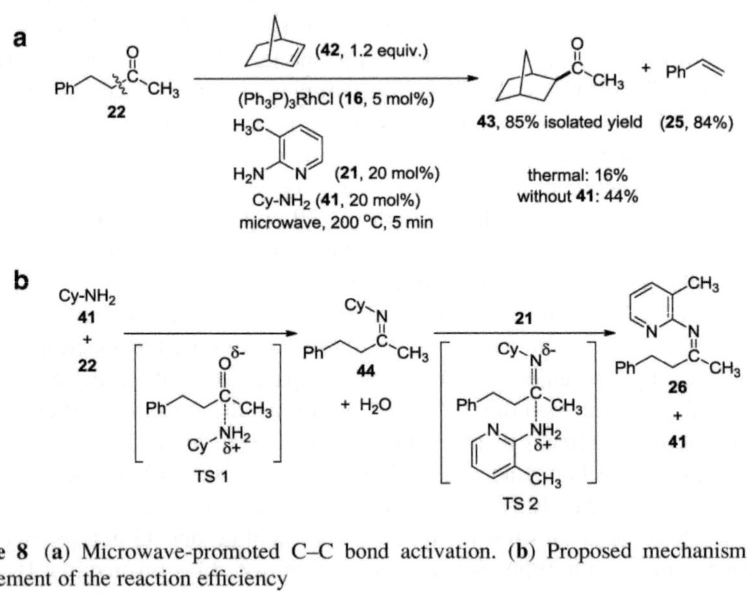


**Scheme 8** (**a**) Microwave-promoted C–C bond activation. (**b**) Proposed mechanism for the enhancement of the reaction efficiency

**44** and **26**. The polarity of the systems increases when the corresponding transition states TS1 and TS2 for imine formation are generated. Thus, the energies of these transitions states along with the activation energies of each step are lowered owing to enhanced dipole–dipole interactions with the electric field (Scheme 8b).

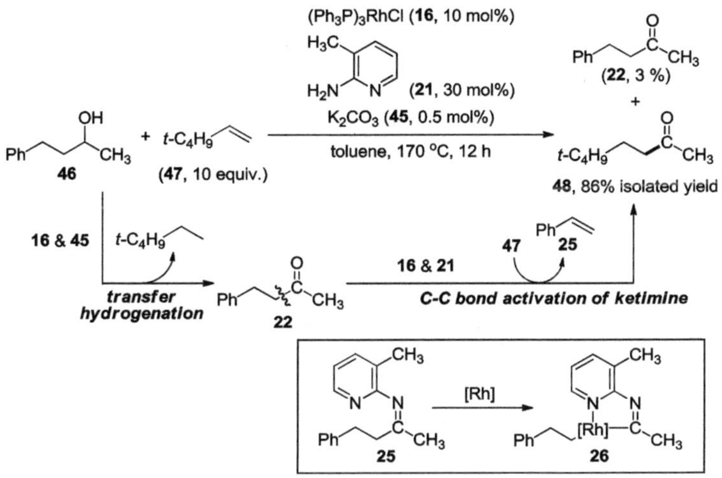

**Scheme 9** C–C bond activation of a secondary alcohol through sequential transfer hydrogenation and C–C bond activation

By employing this strategy, C–C bond activation of a secondary alcohol can be achieved by using a catalytic amount of a base such as $K_2CO_3$ (**45**) [28]. For example, reaction of 4-phenylbutan-2-ol (**46**) with 3,3-dimethyl-1-butene (**47**) in the presence of a catalyst mixture comprised of **16**, **21**, and **45** at 170°C produces 5,5-dimethylhexan-2-one (**48**) as the major product along with a trace amount of benzylacetone (**22**) (Scheme 9). In this process, C–C bond activation occurs on imine **25**, generated by Rh(I) and $K_2CO_3$ promoted transfer hydrogenation of **46** to form **22** and subsequent condensation of the resulting ketone with **21**.

Primary amines can be used as substrates for C–C bond activation reactions that consist of four independent transformations [29]. This process is exemplified by reaction of 3-phenylpropan-1-amine (**49**) with 3,3-dimethylbut-1-ene (**47**) in the presence of **16** and **21**, which produces both the symmetric dialkyl ketone **51** and unsymmetric ketone **50** (Scheme 10a). The route followed in this reaction (Scheme 10b) begins with rhodium mediated transfer hydrogenation between amine **49** and alkene **47** to generate phenethylimine **52**, which then undergoes transimination with **21** to yield the aminopicoline derived imine **53**. Chelation-assisted hydroimination of **53** with the olefin then forms ketimine **54**, which upon acid promoted hydrolysis produces ketone **50**. In a competing pathway, Rh(I)-catalyzed C–C bond activation of ketimine **54**, followed by subsequent addition of **47**, affords the symmetric dialkyl ketimine **55**, which is converted to symmetric dialkyl ketone **51** upon hydrolysis.

Interesting examples of chelation-assisted C–C bond activation are found in the Rh(I)-promoted skeletal rearrangements of cycloalkanones [30]. Reaction of cycloheptanone (**56**) with 2-amino-3-picoline (**21**) in the presence of $(Ph_3P)_3RhCl$ (**16**) at 150°C was observed to generate a mixture of α-methylcyclohexanone (**57**) and α-ethylcyclopentanone (**58**) (Scheme 11). In this process, cycloheptanone

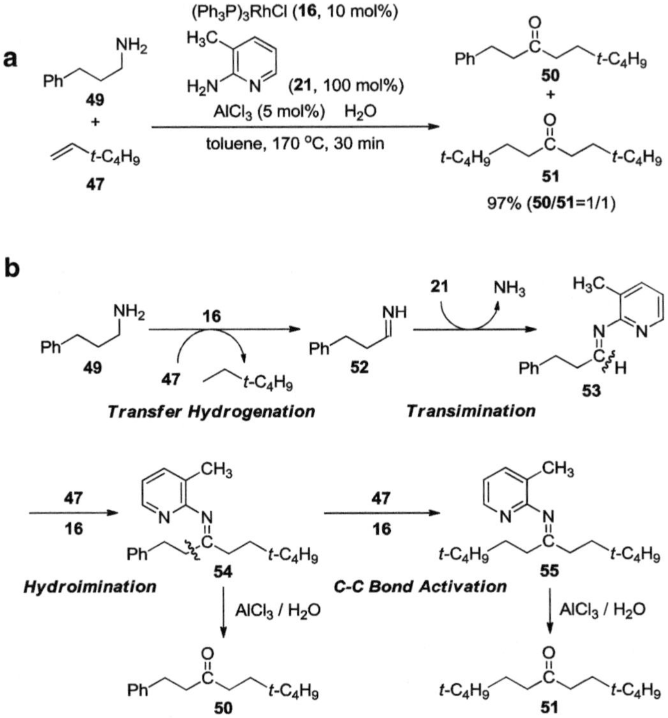


**Scheme 10** (**a**) C–C bond activation of a primary amine and (**b**) the mechanistic pathway followed

imine **59**, first generated in situ by condensation of **56** with **21**, undergoes Rh(I)-promoted C–C bond cleavage to give the five-membered metallacyclic complex **60**. β-Hydride elimination in **60** generates the metal hydride intermediate **61**, which participates in intramolecular metal hydride insertion to produce the ring contracted metallacyclic complex **62** that reacts to form α-methylcyclohexanone (**57**) through reductive elimination and hydrolysis. Ensuing reaction of **62**, involving β-hydride elimination and subsequent intramolecular metal hydride insertion forms **63**, which is converted to the ring contracted metallacyclic complex **64** that serves as the precursor of α-ethylcyclopentanone (**58**). It is interesting to note that the ring contraction process forms five- and six-membered cycloalkanones even when sterically congested complexes, **62** and **64**, with α-branched alkyl groups are formed as intermediates. The reverse reactions leading to ring expansion do not take place, implying that the process favors formation of thermodynamically more stable five- and six-membered cycloalkanones rather than seven-membered analogs.

When the polycyclic substrate, norcamphor ketimine **65**, is treated with $[(COE)_2RhCl]_2$ (**66**)/$PCy_3$ (**67**), a hexahydropentalenone derivative **68** is generated (Scheme 12a). Norcamphor ketimine **65** has two different C–C bonds that are α to the imine group. However, in the reaction C–C bond activation takes place

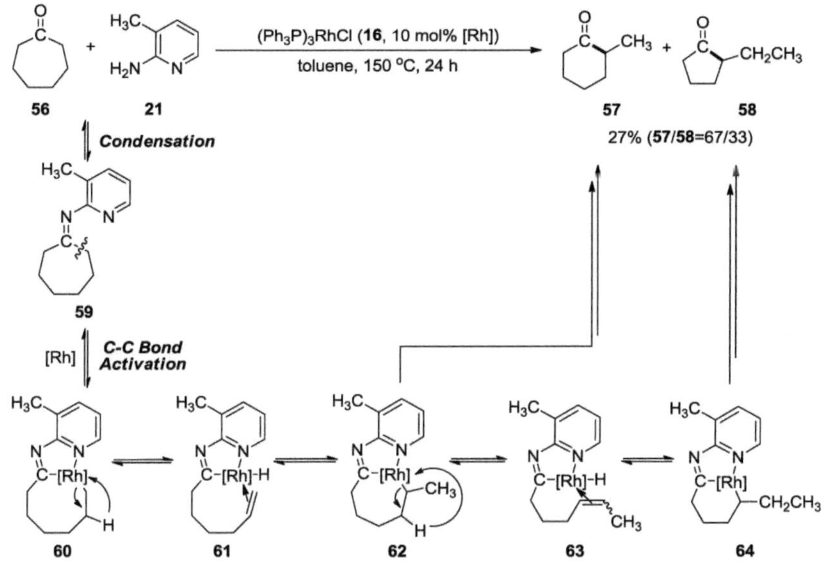


Scheme 11 Metal–organic cooperative skeletal rearrangement of cycloheptanone by Rh(I)

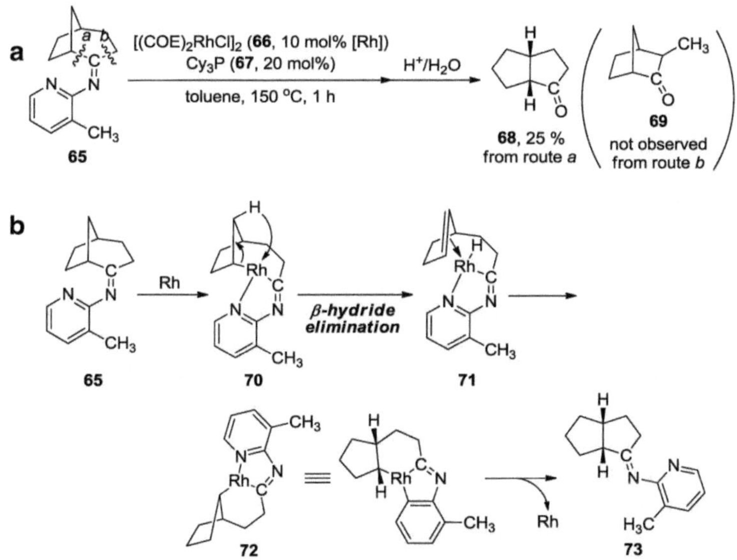


Scheme 12 Skeletal rearrangement of norcamphor imine **65** by Rh(I)

exclusively at the more substituted *a* site forming **68** rather than at the less substituted *b* site. The mechanism for this skeletal rearrangement reaction (Scheme 12b) likely involves Rh(I)-promoted C–C bond cleavage of the *a* C–C bond of **65** to generate the five-membered rhodacyclic complex **70**. Sequential

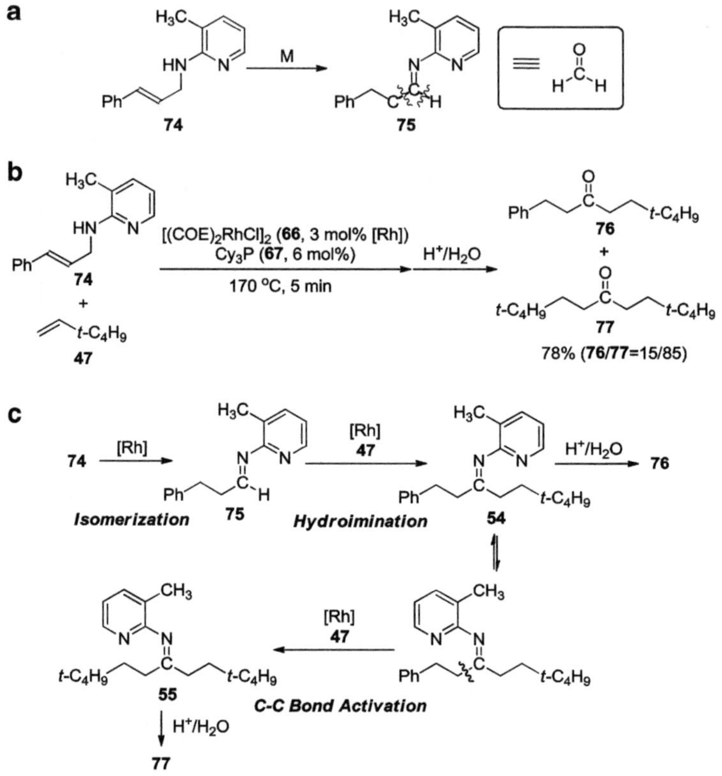


**Scheme 13** (**a**) Allylamine **74** as a synthetic equivalent of formaldehyde. (**b**) C–C bond activation of allylamine **74** through isomerization and sequential C–C bond activation. (**c**) Mechanistic pathway followed

β-hydride elimination takes place to afford complex **71**, which then undergoes intramolecular hydrometalation to form bicyclic rhodacyclic complex **72**. Reductive elimination in **72** affords hexahydropentalenone ketimine **73**, which is hydrolyzed to form **68**. It is interesting that C–C bond activation in **65** takes place to cleave the cyclohexanone moiety, a likely consequence of the inherent ring strain in the norcamphor skeleton.

C–C bond activation also occurs in reactions of allylamines. For example, allylamine **74** undergoes double bond isomerization to form the corresponding aldimine **75** under transition metal catalyzed conditions. In **75**, C–H and C–C bonds exist α to the imine moiety and, as a result, activation of both of these bonds enables this substance to serve as a synthetic surrogate of formaldehyde (Scheme 13a). Accordingly, treatment of allylamine **74** and olefin **47** with a catalytic amount of rhodium catalyst (**66**/**67**) leads to formation of the symmetric dialkyl ketone **77** along with a small amount of ketone **76** (Scheme 13b) [31]. In this process, **74** is first transformed to the corresponding imine **75**, which then

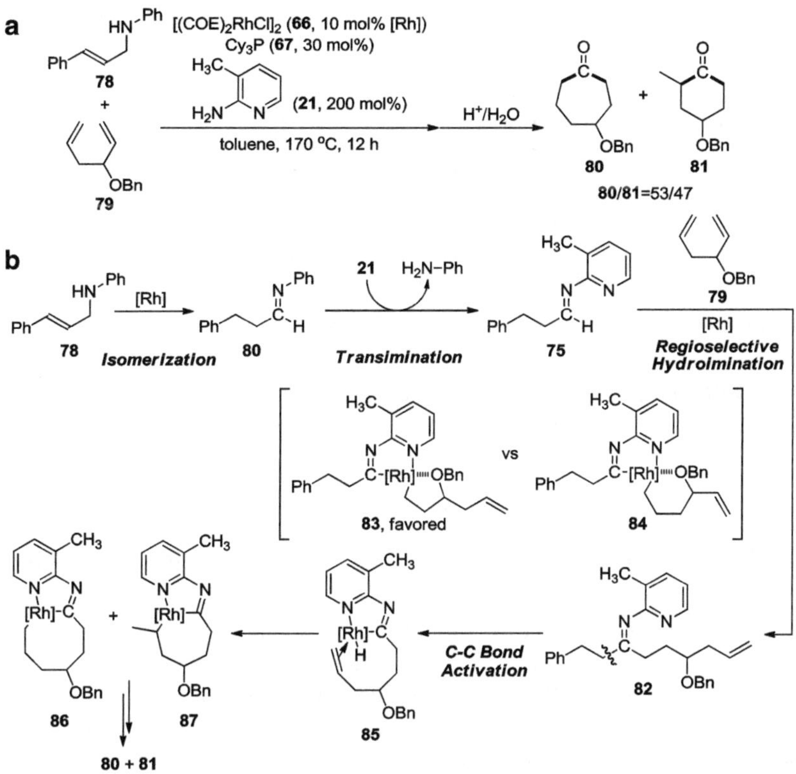


**Scheme 14** (**a**) C–C bond activation of allylamine in the preparation of cycloalkanones. (**b**) Proposed mechanism

undergoes chelation-assisted hydroimination in the presence of Rh(I) catalyst to produce ketimine **54**. Hydrolysis of **54** forms **76**. C–C bond activation of **54** and subsequent olefin insertion also take place to generate symmetric dialkyl ketimine **55**, which is hydrolyzed to give symmetric dialkyl ketone **77** (Scheme 13c).

The strategy for metal–organic cooperative C–H and C–C bond activation of allylamines has been applied to the preparation of cycloalkanones. For example, reaction of allylamine **78** with diene **79** in the presence of rhodium catalyst and 2-amino-3-picoline (**21**) produces, after acidic hydrolysis, a mixture of cycloheptanone **80** and α-methylcyclohexanone **81** (Scheme 14a) [32]. C–C bond activation in this reaction occurs on ketimine **82**, which is formed by a sequence (Scheme 14b) involving Rh(I)-catalyzed isomerization of allylamine **78** to generate corresponding imine **80** followed by transimination of **80** with **21** to produce imine **75**. Regioselective hydroimination of **75** with diene **79** then affords the more stable five-membered metallacyclic intermediate **83** rather than the six-membered counterpart, which then in a dual-chelation manner generates ketimine **82**. C–C bond activation of imine **82** and intramolecular hydrometalation of the resulting intermediate **85** leads to formation of seven- and eight-membered metallacyclic

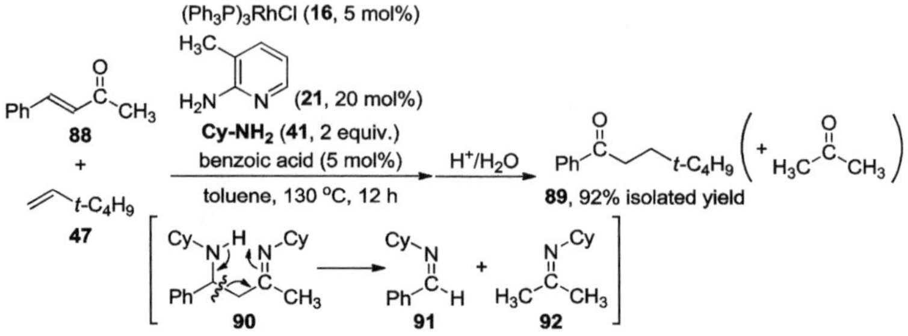

**Scheme 15** C–C double bond cleavage of benzalacetone (**88**)

complexes **86** and **87**, which undergo independent reductive elimination followed by hydrolysis to produce **80** and **81**, respectively.

## 2.2 *Metal–Organic Cooperative C–C Double Bond Cleavage*

Metal–organic cooperative C–C double bond cleavage processes have been observed to take place in reactions of α,β-enones, such as benzalacetone (**88**), with mono-substituted alkene **47** in the presence of a chelation-assisted hydroacylation catalyst system and cyclohexylamine (**41**). This reaction produces ketone **89** after hydrolysis [33]. In the mechanism for this process shown in Scheme 15, benzalacetone (**88**) reacts with **41** to produce β-aminoketimine **90** through a 1,4-conjugate addition reaction. *Retro*-Mannich type fragmentation of **90** produces aldimine **91** and ketimine **92**, the former of which is trapped by chelation-assisted hydroacylation with alkene **47** to give **89** after hydrolysis.

In the same manner, C–C double bond cleavage reactions of cycloalken-2-ones such as cyclohepten-2-one (**93**) also take place. For example, cyclohepten-2-one (**93**) was observed to undergo an interesting ring opening reaction with cyclohexylamine (**41**) through a route which involves 1,4-addition and subsequent *retro*-Mannich type fragmentation of the resulting β-aminoketimine **94** to give intermediate **95** (Scheme 16) [34]. When an olefin and other catalysts are not present, ring contraction occurs to form the more stable 1-acetylcyclopent-1-ene (**98**) through reaction of **96** and subsequent hydrolysis of the resulting β-aminoketimine **97**. However, when the reaction of **93** and **41** is carried out in the presence of alkene **47**, Rh catalyst **16** and 2-amino-3-picoline (**21**), diketone **101** is generated. In this process, intermediate **95** is trapped by **21** to form **99**, which then undergoes chelation-assisted hydroimination with olefin **47** to produce ketimine **100**, which is hydrolyzed to give **101**.

As described above, C–C double bond cleavage reactions of α,β-enones are promoted by cyclohexylamine [33, 34]. In these processes, more than 1 equiv of cyclohexylamine based on the starting α,β-enone is required and a separate step is needed for hydrolysis of the resulting ketimine in order to generate the ketone

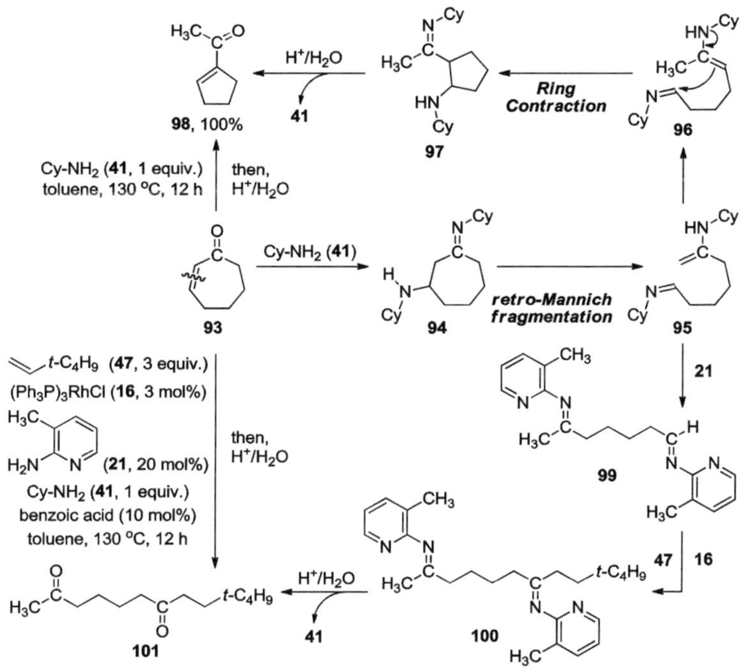


**Scheme 16** C–C double bond cleavage of cyclohepten-2-one (**93**) promoted by cyclohexylamine (**41**)

product because the presence of water in the reaction mixture suppresses the rate of the transition metal-catalyzed hydroacylation reaction. To avoid this problem, octadecyl group-immobilized silica spheres, prepared by catalytic reaction of silica with octadecyltrimethallylsilane in the presence of $Sc(OTf)_3$, have been developed as phase separators of the organic media and water [35]. The use of alkyl group-functionalized silica in the C–C double bond cleavage reaction has three major advantageous features, including sustaining the activity of the transition metal catalyst even in water, negating the need for a separate hydrolysis step, and requiring that only a catalytic amount of cyclohexylamine is used because it is regenerated in the hydrolysis step (Scheme 17a). An example of this strategy is found in the reaction of benzalacetone (**88**) with olefin **47** in the presence of octadecyl group-immobilized silica, catalytic amounts of cyclohexylamine (**41**), $(Ph_3P)_3RhCl$ (**16**), 2-amino-3-picoline (**21**), benzoic acid and water. This process results in the production of 4,4-dimethyl-1-phenylpentan-1-one (**89**) in a high yield, which contrast with the same reaction performed without using functionalized silica that gives ketone **89** in a much lower yield (Scheme 17b).

Chelation-assisted hydroacylation of olefins with aliphatic aldehydes possesses an intrinsic problem associated with competitive formation of aldol condensation products. A cooperative C–C double bond cleavage process has been developed as a solution to this problem [36]. For example, hydroacylation reactions of aliphatic

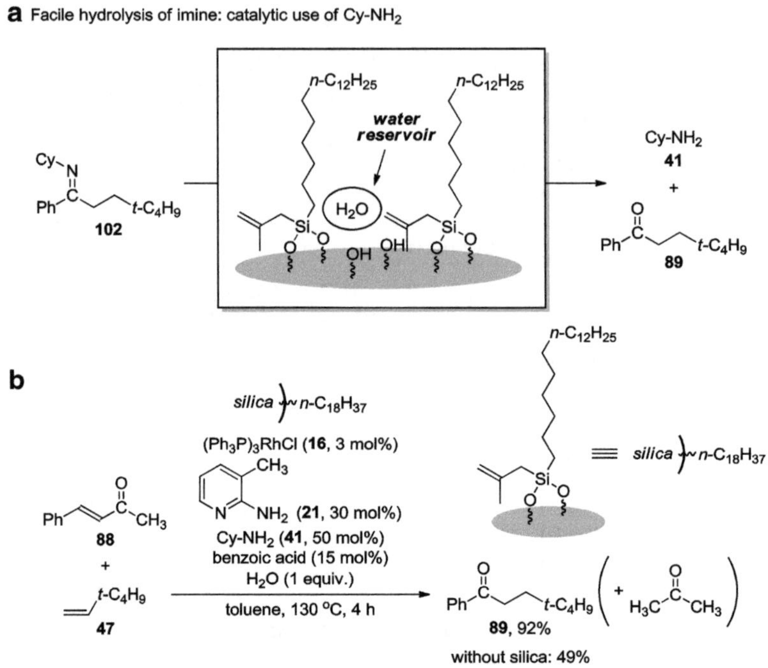


**Scheme 17** (**a**) Octadecyl group-immobilized silica as a phase separator of the organic media and water. (**b**) The reaction of benzalacetone (**88**) with olefin in the presence of octadecyl group-immobilized silica under catalytic amount of cyclohexylamine (**41**)

aldehydes such as hydrocinnamaldehyde (**103**) typically take place with low efficiencies as a consequence of the formation of homoaldol side products (e.g., **104**) (Scheme 18a), which occurs much more rapidly than the desired process. However, the efficiency of this reaction is enhanced by conditions which enable reconversion of the homoaldol side products **104** to **105**, which is the imine of the starting aldehyde **103**, through a route involving imine formation, conjugate addition of cyclohexylamine (**41**) and *retro*-Mannich type fragmentation.

This approach has been applied to the synthesis of two ketones from an α,β-unsaturated aldehyde and alkene by using a double hydroacylation process [37]. A prototypical example is seen in the reaction of α-methyl cinnamaldehyde (**106**) with 3,3-dimethylbut-1-ene (**47**) in the presence of a mixture of $(Ph_3P)_3RhCl$ (**16**), 2-amino-3-picoline (**21**) and *n*-hexylamine (**107**), which produces ketones **89** and **108** (Scheme 19a). Tracing the progress of reaction of **106** shows that formation of ketimine **109** through hydroacylation of the aldehyde moiety in **106** with alkene **47** (*path a*) takes place prior to C–C double bond cleavage through intermediate **110** (*path b*) (Scheme 19b). Transimination and 1,4-addition of **107** to the resulting ketimine **109** generate β-aminoketimine **111**, which undergoes *retro*-Mannich type fragmentation to give imines **112** and **113**. Aldimine **112** reacts with

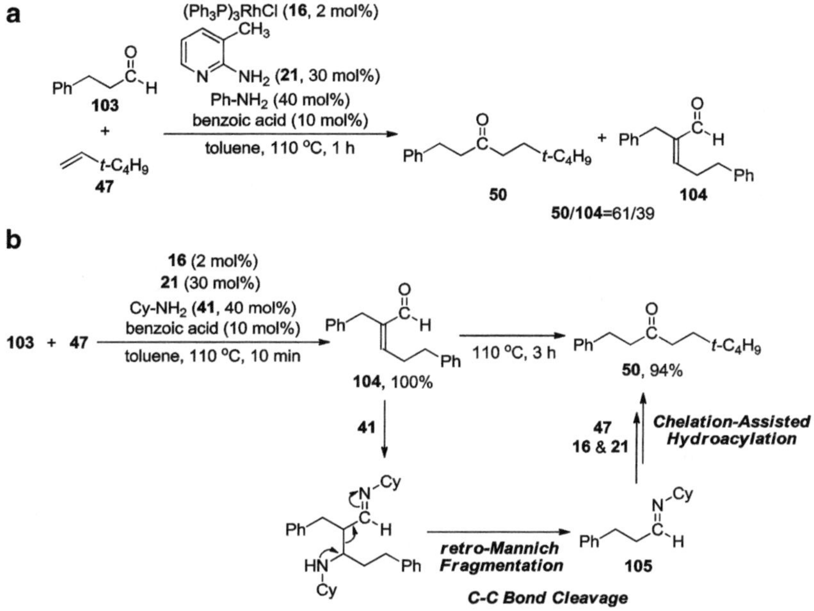


**Scheme 18** (**a**) Hydroacylation using an aliphatic aldehyde **103**. (**b**) Role played by cyclohexylamine (**41**) in chelation-assisted hydroacylation in aliphatic aldehyde

alkene **47** under chelation-assisted hydroacylation catalysis to produce **114**, which is hydrolyzed to give ketone **89**.

Because they can be regarded as synthetic equivalents of α,ω-dialdehydes (**116**), 1-cycloalkenecarboxaldehydes such as 1-cyclododecenecarboxaldehyde (**115**) are interesting substrates for double hydroacylation reactions that form diketones (Scheme 20). The reaction of **115** with **47** in the presence of a catalyst mixture consisting of **16**, **21**, **107** and *p*-$CF_3$-$C_6H_4CO_2H$ was found to give linear diketone **117**. In this process, cyclic α,β-enal **115** undergoes hydroacylation with **47** to produce imine **118**, which is then followed by addition of **107** that promotes transimination and 1,4-addition of **118** and subsequent *retro*-Mannich type fragmentation of β-aminoketimine **119** to give an intermediate **120**. The aldimine moiety in intermediate **120** participates in chelation-assisted hydroacylation with **47** to form diketone **117** after hydrolysis.

An interesting example is found in the double hydroacylation reactions of enantiopure myrtenal (**121**) with alkenes **47** and **123**, which afford interesting chiral products (Scheme 21). The first hydroacylation reaction occurring in this process transforms **121** and olefin **47** to the α,β-enone **122** ([α] = −20.68°), which is converted to chiral ketone (−)**124** by amine-assisted C–C double bond cleavage and subsequent chelation-assisted hydroacylation with olefin **123**. The specific rotation of chiral product **124** formed in this process was determined to be (−)3.5°.

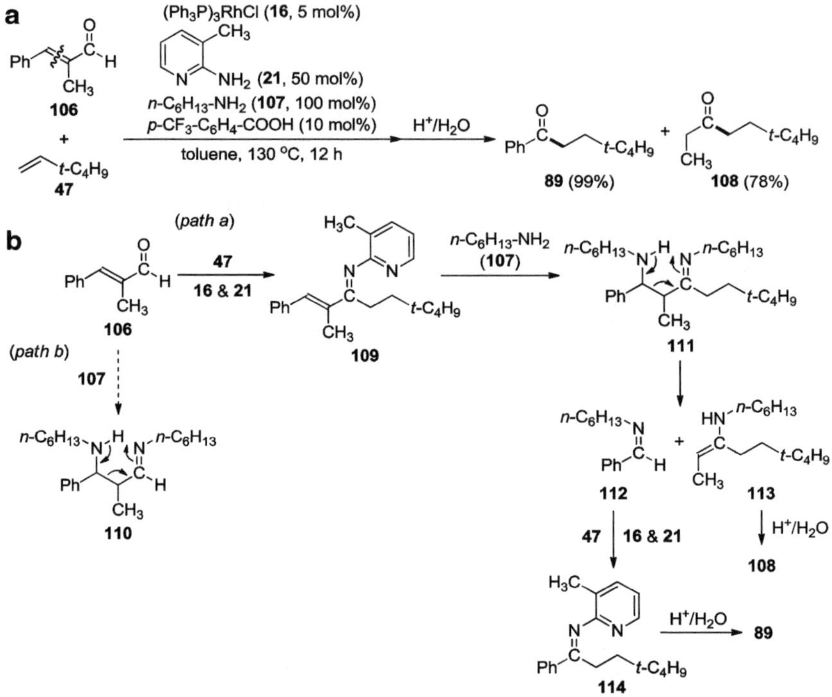

**Scheme 19** (**a**) Double hydroacylation of α,β-unsaturated aldehyde with olefin to produce two ketones. (**b**) Mechanism of the process

In contrast, when **121** is treated with the alkenes in the reverse order (i.e., first **123** then **47**) under identical conditions, the specific rotation of the product ((+)**124**) is (+)3.5°. This result shows that the sequential hydroacylation protocol can be employed to generate enantiomers whose absolute configurations are controlled by order in which hydroacylation reactions of two alkenes take place.

## 2.3 Metal–Organic Cooperative C–C Triple Bond Cleavage

The development of C–C triple bond cleavage reactions is an interesting issue in organometallic chemistry. As described above, double bonds of α,β-enones (or corresponding ketimines) are readily cleaved by sequential 1,4-additions of amine and *retro*-Mannich type fragmentations of the resulting β-aminoketimine. C–C triple bond cleavage can be achieved in a similar way. In this protocol, hydroacylation of alkynes is a key step because it results in formation of α,β-enones (or corresponding ketimines). Therefore, the basic approach to triple bond cleavage involves the combination of transition metal-catalyzed hydroacylation of alkynes with aldehydes and subsequent C–C double bond cleavage of the resulting

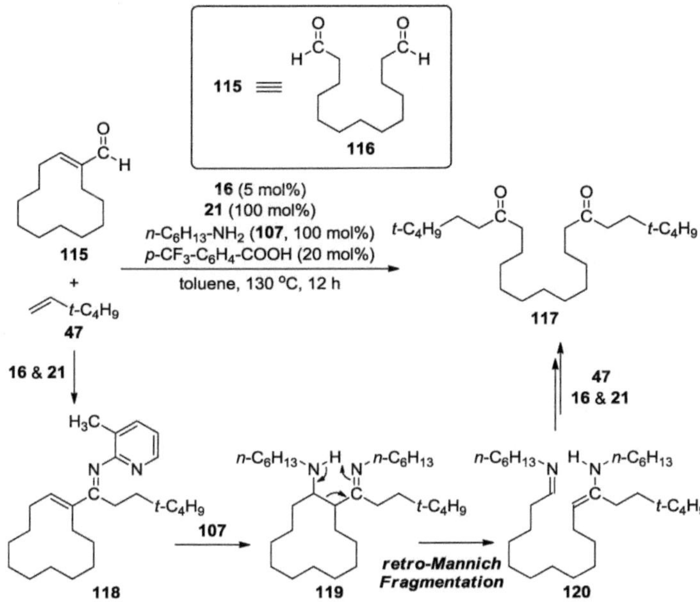


**Scheme 20** Double hydroacylation of cyclic α,β-unsaturated aldehyde **115** with olefin **47** to produce diketone **117**

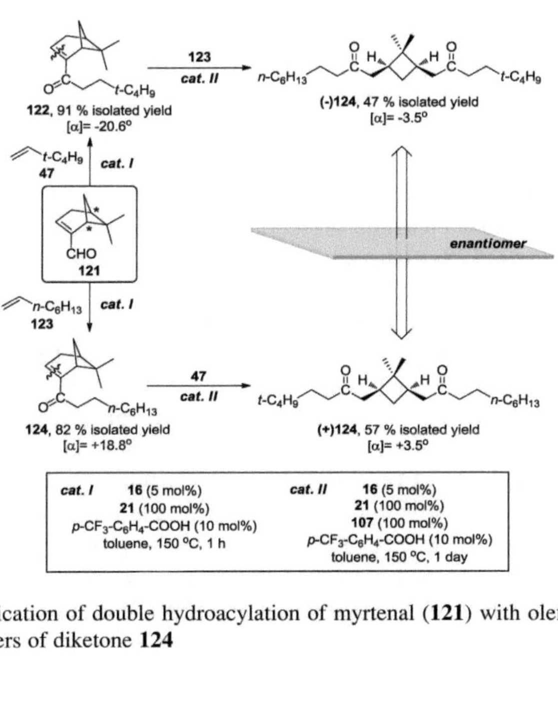


**Scheme 21** Application of double hydroacylation of myrtenal (**121**) with olefins **47** and **123** to produce enantiomers of diketone **124**

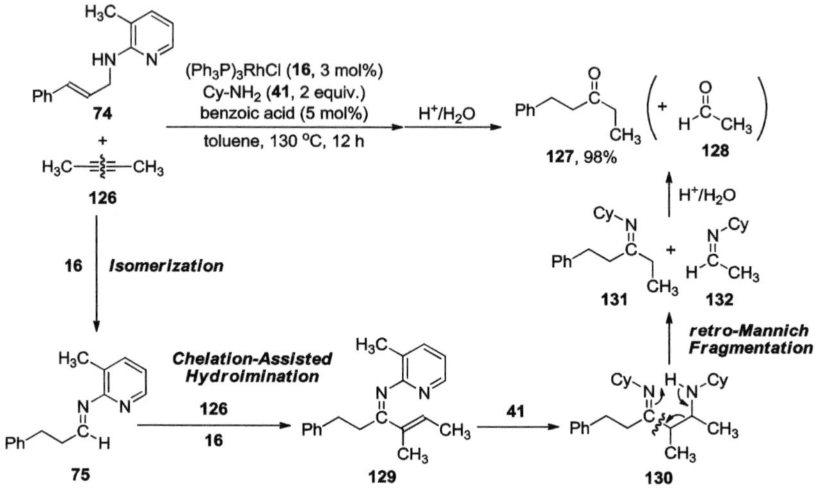


**Scheme 22** C–C triple bond cleavage in a reaction of allylamine **74** with internal alkyne **126**

α,β-enone derivatives [38]. This protocol is exemplified by the hydroacylation reaction of alkyne **126** with allylamine **74**, carried out in the presence of Rh(I) catalyst **16** and cyclohexylamine (**41**), which generates a mixture of ketone **127** and acetaldehyde (**128**) (Scheme 22). In this process, the allylamine **74** is transformed to an α,β-unsaturated ketimine through rhodium mediated isomerization to form aldimine **75** and subsequent chelation-assisted hydroimination with **47** occurs to give α,β-unsaturated ketimine **129**. C–C bond cleavage of **129** takes place through *retro*-Mannich type fragmentation of the β-aminoketimine **130**, generated by 1,4-addition of **41** to **129** to give ketimine **131** and aldimine **132**. Subsequent hydrolysis of the respective ketimines yield ketone **127** and aldehyde **128**.

When the C–C bond cleavage protocol is applied to alkyne **133** and a small amount of aldehyde **128** (Scheme 23), ketone **134** is produced in high yield [39]. This reaction is triggered by hydroacylation of **133** with **128** and follows a route in which internal alkyne **133** is initially converted to α,β-unsaturated ketimine **136** through chelation-assisted hydroimination with aldimine **135**, generated in-situ by condensation reaction with 2-amino-3-picoline (**21**). Addition of **41** to **136** then forms **137**, which subsequently undergoes *retro*-Mannich type fragmentation to yield aldimine **138** and ketimine **139**, the latter of which undergoes hydrolysis to form ketone **140**. Aldimine **138** participates in an ensuing catalytic cycle involving chelation-assisted hydroimination with **133** to form ketimine **141**, which then adds **41** to generate intermediate **142**. This catalytic cycle produces aldimine **138** and ketimine **143**. Then **143** is hydrolyzed to produce ketone **134** while aldimine **138** reenters the catalytic cycle until all alkyne **133** is consumed.

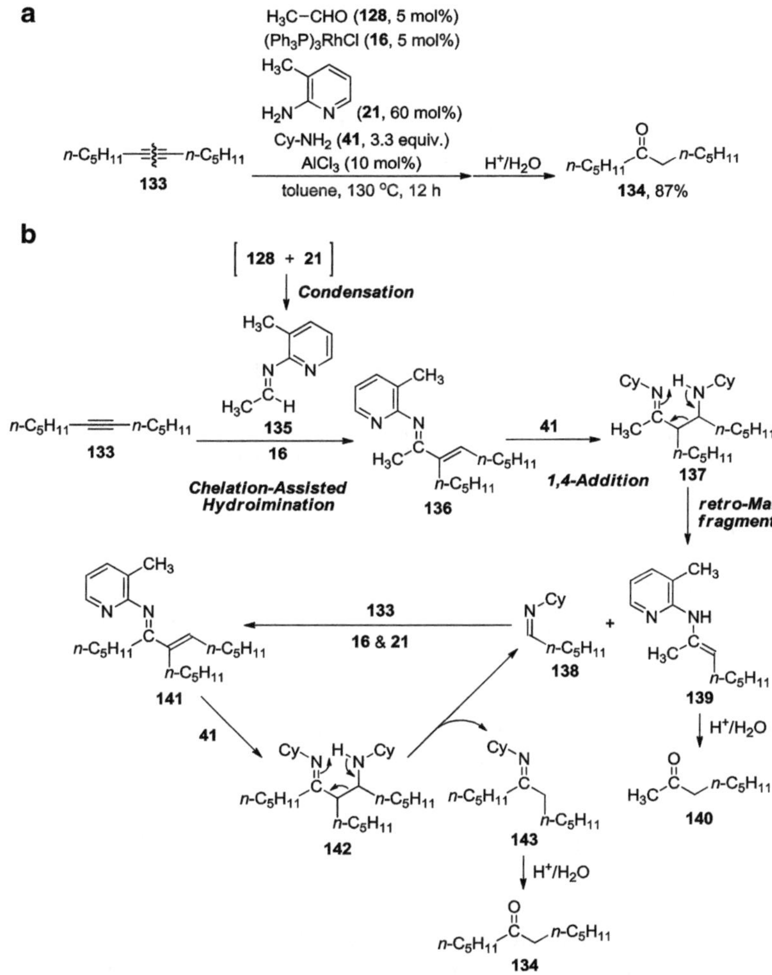


**Scheme 23** (**a**) Hydroacylation-triggered C–C triple bond cleavage reactions. (**b**) Proposed mechanism

The hydroacylation triggered C–C triple bond activation process has been applied to reactions of strain-free cycloalkynes such as cyclododecyne (**145**) that give rise to polyketone oligomers [39]. For example, reaction of cyclododecyne (**145**) with a small amount of phenylacetaldehyde (**144**), performed using a catalyst mixture containing **16**, **21**, **41** and $AlCl_3$, produces polyketone **146** (Scheme 24a). The catalytic cycle involved in this process, including chelation-assisted hydroacylation of **145** with **144**, transimination and 1,4-addition of ketimine **147** with **41**, and *retro*-Mannich type fragmentation of the resulting intermediate **148**, leads to formation of the ring-opened aldimine **149**. The aldimine moiety

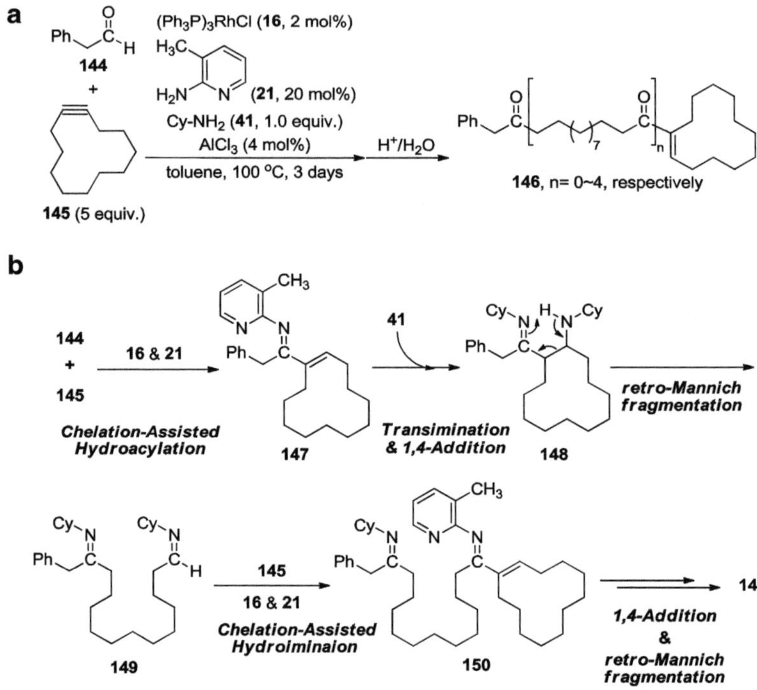


**Scheme 24** (**a**) Application of C–C triple bond cleavage of alkyne **145** to the ring-opening polymerization of **145** to produce polyketone **146**. (**b**) Mechanism for this process

in **149** reacts with **145** in the presence of **16** and **21** to generate ketimine **150**. Repetition of these reactions results in eventual formation of polyketone **146** (Scheme 24b).

The alkene and aldehyde functional groups present in 3-vinylbenzaldehyde (**151**) can be transformed to the respective ketones by utilizing this C–C triple bond cleavage strategy [40]. Accordingly, treatment of **151** with 3-hexyne (**152**) in the presence of **16**, **21**, *p*-trifluoromethylbenzoic acid and **107** leads to formation of diketone **153** (Scheme 25). This reaction takes place via a mechanism in which chelation-assisted hydroacylation of **152** with **151** gives ketimine **154**. Transimination and 1,4-addition of **107** to **154** generates **155**, which by *retro*-Mannich type fragmentation gives enamine **156** and aldimine **157**. These respective substances undergo transimination with **21** to afford ketimine **158** and aldimine **159**. Aldimine **159** reacts with the vinyl group of **158** in a chelation-assisted hydroacylation process to give diketimine **160**, which is hydrolyzed to yield **153**.

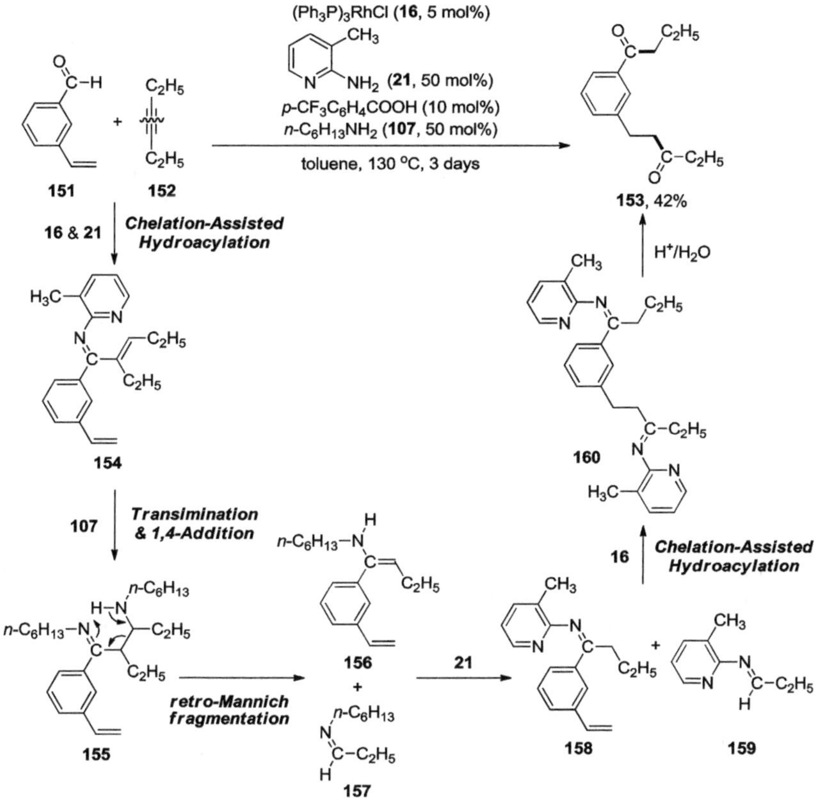


**Scheme 25** Application of C–C triple bond cleavage of alkyne **152** to the functionalization of 3-vinylbenzaldehyde (**151**)

# 3 Summary, Conclusions, Outlook

Recent progress made in the discovery of new C–C bond cleavage reactions opens up new paradigms in synthetic organic chemistry for the construction of organic frameworks through the structural reorganization of organic backbones. Gaining special attention in this regard are chelation-assisted C–C bond cleavage reactions that operate through cooperation between metal complexes and organic molecules. These processes have attracted great attention owing to their extraordinary efficiency and synthetic usefulness. The metal–organic cooperative strategy has been utilized for C–C single bond activation of a variety of substrates, including linear alkyl ketones, secondary alcohols, primary amines and cycloalkanones. Cleavage of C–C double and triple bonds can also be facilitated by metal–organic cooperation. For example, C–C double bonds of α,β-enones can be cleaved by addition of cyclohexylamine to the reaction mixtures, owing to the intervention of *retro*-Mannich fragmentation reactions of β-aminoketimine intermediates generated by

1,4-addition of cyclohexylamine to the α,β-enones. Aldehydes, which are one of the fragmentation products of these reactions, can be trapped by chelation-assisted hydroacylation reactions to give ketone products. C–C triple bond cleavage reactions of alkynes can be achieved through hydroacylation of alkynes with aldehydes and subsequent C–C double bond cleavage of the resulting α,β-enones.

Although the reactions described above have not yet been widely applied in synthetic organic chemistry, their use should significantly increase when their versatility is fully recognized.

## References


1. Crabtree RH (1985) The organometallic chemistry of alkanes. Chem Rev 85:245
2. Jones WD (1993) The fall of the C–C bond. Nature 364:676
3. Rybtchiski B, Milstein D (1999) Metal insertion into C–C bonds in solution. Angew Chem Int Ed 38:870
4. Murakami M, Ito Y (1999) Cleavage of carbon–carbon single bonds by transition metals. In: Murai S (ed) Activation of unreactive bonds and organic synthesis, topics in organometallic chemistry, vol 3. Springer, Berlin Heidelberg New York, p 97
5. van der Boom ME, Milstein D (2003) Cyclometalated phosphine-based pincer complexes: mechanistic insight in catalysis, coordination, and bond activation. Chem Rev 103:1759
6. Jun CH (2004) Transition metal-catalyzed carbon–carbon bond activation. Chem Soc Rev 33:610
7. Kondo T, Mitsudo TA (2005) Ruthenium-catalyzed reconstructive synthesis of functional organic molecules via cleavage of carbon–carbon bonds. Chem Lett 34:1462
8. Jun CH, Park JW (2007) Directed C–C bond activation by transition metal complexes. In: Chatani N (ed) Directed metallation, topics in organometallic chemistry. Springer, Berlin Heidelberg New York, p 117
9. Park YJ, Park JW, Jun CH (2008) Metal-organic cooperative catalysis in C–H and C–C bond activation and its concurrent recovery. Acc Chem Res 41:222
10. Ruhland K (2012) Transition-metal-mediated cleavage and activation of C–C single bonds. Eur J Org Chem 2683
11. Chaplin AB, Weller AS (2010) C–C bond activation of a cyclopropyl phosphine: isolation and reactivity of a tetrameric rhodacyclobutane. Organometallics 29:2332
12. Seiser T, Saget T, Tran DN, Cramer N (2011) Cyclobutanes in catalysis. Angew Chem Int Ed 50:7740
13. Crabtree RH, Dion RP (1984) Selective alkane C–C bond cleavage via prior dehydrogenation by a transition metal complex. Chem Commun 108:7222
14. Halcrow MA, Urbanos F, Chaudret B (1993) Organometallics 12:955
15. Takahashi T, Kuzuba Y, Kong F, Nakajima K, Xi Z (2005) Formation of indene derivatives from bis(cyclopentadienyl)titanacyclopentadienes with alkyl group migration via carbon–carbon bond cleavage. J Am Chem Soc 127:17188
16. Zhao P, Incarvito CD, Hartwig JF (2006) Direct observation of β-aryl eliminations from Rh (I) alkoxides. J Am Chem Soc 128:3124
17. Li H, Li Y, Zhang XS, Chen K, Wang X, Shi ZJ (2011) Pyridinyl directed alkenylation with olefins via Rh(III)-catalyzed C–C bond cleavage of secondary arylmethanols. J Am Chem Soc 133:15244
18. Suggs JW, Sharman DC (1982) Directed cleavage of $sp^2$–sp carbon–carbon bonds. J Organomet Chem 221:199

19. Suggs JW, Jun CH (1984) Directed cleavage of carbon–carbon bonds by transition metals: the α-bonds of ketones. J Am Chem Soc 106:3054
20. Liou SY, van der Boom ME, Milstein D (1998) Catalytic selective cleavage of a strong C–C single bond by rhodium in solution. Chem Commun 687
21. Chatani N, Ie Y, Kakiuchi F, Murai S (1999) $Ru_3(CO)_{12}$-catalyzed decarbonylative cleavage of a C–C bond of alkyl phenyl ketones. J Am Chem Soc 121:8645
22. Suggs JW (1978) Isolation of a stable acylrhodium(III) hydride intermediate formed during aldehyde decarbonylation. hydroacylation. J Am Chem Soc 100:640
23. Suggs JW (1979) Activation of aldehyde C–H bonds to oxidative addition via formation of 3-methyl-2-aminopyridyl aldimines and related compounds: rhodium based catalytic hydroacylation. J Am Chem Soc 101:489
24. Jun CH, Lee H, Hong JB (1997) Chelation-assisted intermolecular hydroacylation: direct synthesis of ketone from aldehyde and 1-alkene. J Org Chem 62:1200
25. Jun CH, Lee H (1999) Catalytic carbon–carbon bond activation of unstrained ketone by soluble transition-metal complex. J Am Chem Soc 121:880
26. Ahn JA, Chang DH, Park YJ, Yon YR, Loupy A, Jun CH (2006) Solvent-free chelation-assisted catalytic C–C bond cleavage of unstrained ketone by rhodium(I) complexes under microwave irradiation. Adv Synth Catal 348:55
27. Kappe CO (2004) Controlled microwave heating in modern organic synthesis. Angew Chem Int Ed 43:6250
28. Jun CH, Lee DY, Kim YH, Lee H (2001) Catalytic carbon–carbon bond activation of sec-alcohols by a rhodium(I) catalyst. Organometallics 20:2928
29. Jun CH, Chung KY, Hong JB (2001) C–H and C–C bond activation of primary amines through dehydrogenation and transimination. Org Lett 3:785
30. Jun CH, Lee H, Lim SG (2001) The C–C bond activation and skeletal rearrangement of cycloalkanone imine by Rh(I) catalysts. J Am Chem Soc 123:751
31. Jun CH, Lee H, Park JB, Lee DY (1999) Catalytic activation of C–H and C–C bonds of allylamines via olefin isomerization by transition metal complexes. Org Lett 1:2161
32. Lee DY, Kim IJ, Jun CH (2002) Synthesis of cycloalkanones from dienes and allylamines through C–H and C–C bond activation catalyzed by a rhodium(I) complex. Angew Chem Int Ed 41:3031
33. Lim SG, Jun CH (2004) C–C double bond cleavage of linear α,β-unsaturated ketones. Bull Kor Chem Soc 25:1623
34. Jun CH, Moon CW, Lim SG, Lee H (2002) Application of Rh(I)-catalyzed C–H bond activation to the ring opening of 2-cycloalkenones in the presence of amines. Org Lett 4:1595
35. Lee DH, Jo EA, Park JW, Jun CH (2010) One-pot catalytic C–C double bond cleavage of α, β-enones aided by alkyl group-immobilized silica spheres. Tetrahedron Lett 51:160
36. Jo EA, Jun CH (2009) The effects of amine and acid catalysts on efficient chelation-assisted hydroacylation of alkene with aliphatic aldehyde. Tetrahedron Lett 50:3338
37. Cha KM, Lee H, Park JW, Lee Y, Jo EA, Jun CH (2011) Double hydroacylation reactions of acyclic and cyclic α,β-unsaturated aldehydes. Chem Asian J 6:1926
38. Jun CH, Lee H, Moon CW, Hong HS (2001) Cleavage of carbon-carbon triple bond of alkyne via hydroiminoacylation by Rh(I) catalyst. J Am Chem Soc 123:8600
39. Lee DY, Hong BS, Cho EG, Lee H, Jun CH (2003) A hydroacylation-triggered carbon-carbon triple bond cleavage in alkynes via retro-Mannich type fragmentation. J Am Chem Soc 125:6372
40. Cha KM, Jo EA, Jun CH (2009) Tandem catalytic triple-bond cleavage of alkyne in association with aldehyde, alkene, and water. Synlett 2939

Top Curr Chem (2014) 346: 85–110
DOI: 10.1007/128_2013_523
© Springer-Verlag Berlin Heidelberg 2014
Published online: 20 March 2013

# Carbon–Carbon Bond Activation with 8-Acylquinolines

**Ashley M. Dreis and Christopher J. Douglas**

**Abstract** Synthetically relevant advances in the area of carbon–carbon sigma bond activation have been made possible by 8-acylquinoline directing groups. Stable rhodium metallacycle intermediates have been shown to undergo a variety of transformations, including carboacylation reactions, to produce value-added products containing all-carbon quaternary centers. The kinetic profile of such reactions has been shown to be substrate dependent.

**Keywords** 8-Acylquinoline · 8-Quinolinyl ketone · All-carbon quaternary centers · Bond insertion · Carboacylation · Carbon–carbon bond activation · Chelation · Cyclometalation · Directed metalation · Directing group · Metallacycle · Oxidative addition · Rhodium catalyst

## Contents



A.M. Dreis and C.J. Douglas (✉)
Department of Chemistry, University of Minnesota, 207 Pleasant St. SE, Minneapolis, MN 55455, USA
e-mail: cdouglas@umn.edu

## Abbreviations

| | |
|---|---|
| BDE | Bond dissociation energy |
| brsm | Based on recovered starting material |
| C–C | Carbon–carbon sigma bond |
| C–CN | Carbon–nitrile bond |
| C–H | Carbon–hydrogen bond |
| cod | Cyclooctadiene |
| MeCN | Acetonitrile |
| OAc | Acetate |
| PhMe | Toluene |
| py | Pyridine |
| THF | Tetrahydrofuran |

## 1 Introduction

Following the advent of carbon–hydrogen bond activation, the activation of carbon–carbon sigma bonds has gained increasing attention from the organometallic community over the past 2 decades. The ability to cleave carbon–carbon bonds selectively with subsequent functionalization would not only unveil unusual retrosynthetic disconnections, but would also impact industrial hydrocarbon processes. In essence, viewing the carbon–carbon sigma bond (C–C) as a functional group has the power to change the way chemists think about organic chemistry and how they approach synthetic challenges.

Both thermodynamic and kinetic barriers contribute to the overall lack in reactivity of C–C bonds. Several strategies have been employed to overcome such barriers. Coupling the activation with an exothermic process such as ring-strain relief, hydrogenolysis, or aromatization helps drive the energetically unfavorable oxidative addition of metal complexes into C–C bonds [1–7]. With C–H bonds typically outnumbering the sterically "hidden" C–C bonds, the preferential activation of a C–C bond can be made kinetically feasible through directed metalation with strategically appended directing groups – a process also known as cyclometalation.

Of the early reports on carbon–carbon bond activation reactions reported in the literature, many require a stoichiometric amount of transition metal, degrade molecular complexity, and/or fragment the starting material [1–7]. In order for a reaction to be viable in complex molecule synthesis, it must be capable of delivering value-added products while ideally being economically and environmentally viable. While it remains challenging to develop such transformations, the catalytic activation of unstrained carbon–carbon bonds utilizing an 8-acylquinoline directing group has thus far been a promising entry into such methodologies.

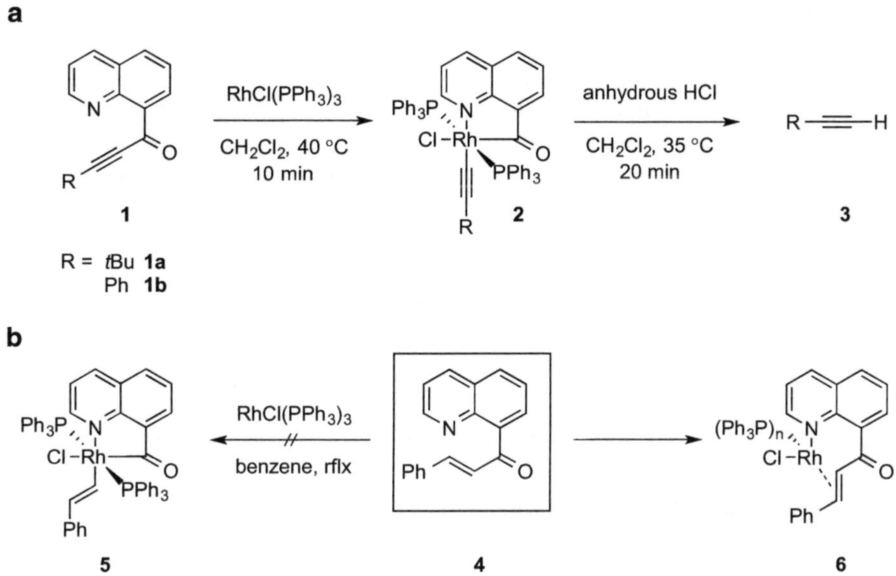


**Scheme 1** (**a**) $sp^2$–sp C–C bond activation and protonation of rhodium acetylide. (**b**) Inability to activate $sp^2$–$sp^2$ C–C bond due to alkene coordination

# 2 Stoichiometric Carbon–Carbon Bond Activation Reactions

## 2.1 *Directed Cleavage of* $sp^2$*–sp Carbon–Carbon Bonds*

In 1969, di-alkynyl ketones were shown to decarbonylate by treatment with RhCl $(PPh_3)_3$ in refluxing xylenes to yield 1,3-diynes and $RhCl(CO)(PPh_3)_2$ [8] (for a catalytic version, see [9]) by way of an oxidative carbon–carbon sigma bond cleavage pathway. More than a decade later, Suggs [10] discovered that decarbonylation of the Rh(III)ketoacetylide intermediate was prevented by stabilizing the metal center within a chelate through a process known as cyclometalation [11]. As a result, the two metal–carbon bonds would have the potential to be further functionalized with control.

The cyclometalation reaction was achieved by treating 8-quinolinyl *tert*-butylacetylenyl ketone **1a** with $RhCl(PPh_3)_3$ in $CH_2Cl_2$ for 10 min at 40°C. The resulting stable yellow solid **2a** displayed a ketone IR band shifted from 1,645 to 1,670 $cm^{-1}$ and a $^{31}P$ NMR doublet at $-19.35$ ppm, $J_{Rh–P} = 110$ Hz (Scheme 1a) [10]. Upon treatment with anhydrous acid, the Rh(III) compound **2a** underwent protonation to provide *tert*-butylacetylene **3** in high yield. An analogous C–C bond insertion reaction was successful with $RhCl(PPh_3)_3$ and 8-quinolinyl phenylacetylenyl ketone **1b**.

Given that $sp^2$–sp C–C bonds of ynones can be readily cleaved by strong reducing agents and/or nucleophiles owing to the low $pK_a$ of acetylenes, it was questioned whether the less reactive $sp^2$–$sp^2$ C–C bond of an enone could be activated in a similar manner. Attempts to add $RhCl(PPh_3)_3$ oxidatively across the $sp^2$–$sp^2$ carbon–carbon bond of 8-quinolinyl styrenyl ketone **4** did not provide complex **5**, even after extended reflux in benzene (Scheme 1b) [10]. The $^1H$ NMR spectrum of the product showed an olefin multiplet at 4.4 ppm, and the 2-quinolinyl signal shifted downfield at 8.45 ppm, which was consistent with rhodium bound by both quinoline nitrogen and olefin as depicted in **6**. No reports indicated whether other Rh(I) complexes were evaluated on this system.

## 2.2 *Directed Cleavage of $sp^2$–$sp^3$ Carbon–Carbon Bonds*

### 2.2.1 Substrates Without β-Hydrogens

Since the $sp^2$–$sp^2$ α-ketone C–C bond is less reactive than the sp–$sp^2$ α-ketone C–C bond toward nucleophilic cleavage, one might expect the $sp^2$–$sp^3$ α-ketone C–C bond to be even less reactive. Perhaps owing to this notion, no reaction of 8-quinolinyl alkylketones **7a–c** was observed when treated with $RhCl(PPh_3)_3$ [12]. However, reacting **7a–c** with $[RhCl(C_2H_4)_2]_2$ in benzene at room temperature provided an insoluble material, which was presumed to be the chlorine-bridged oligomer **8a–c** (Scheme 2). The oligomeric complexes were solubilized with excess pyridine and crystallized from ether to give the six-coordinate dipyridyl Rh(III)–acyl complexes **9a–c**. The benzyl **9a** [12] and ethyl **9b** [13] complexes were confirmed by X-ray crystallography and **9b** was independently synthesized via a hydroacylation reaction of 8-quinolinecarboxaldehyde and ethylene [13, 14]. Whether oxidative addition occurred prior to the addition of pyridine was probed by treatment of both **8** and **9** with bromine. Each reaction gave the corresponding alkyl bromide **10** which indicated that pyridine was not necessary for carbon–carbon bond activation to occur.

In solution, the five-coordinate mono-pyridyl rhodium complex predominates (not shown). As with other five-coordinate Rh(III) structures [15], it is likely square pyramidal with the alkyl group at the apex. With the Rh–$N_{pyridine}$ bond *trans* to the benzyl group 0.13 Å longer than the Rh–$N_{pyridine}$ bond *trans* to quinoline in complex **9**, it is likely that this pyridine is lost in solution. The Rh(III)–acyl bond length in **9a** was measured at a short distance of 1.949 Å [12] (1.938 Å [13] for **9b**) relative to other Rh(III)–acyl bonds (1.971–2.062 Å) [16]. These bond length values mirror those found for rhodium carbenes (1.968 Å) [17]. The inherent strength of these shortened bonds could reflect the thermodynamic driving force of the reaction. The strong *trans* effect of the acyl group, in turn, lengthens the Rh–Cl bonds by 10% in comparison to typical values [18]. This lengthened Rh–Cl bond may play a significant role in the transmetalation step shown in the cross-coupling

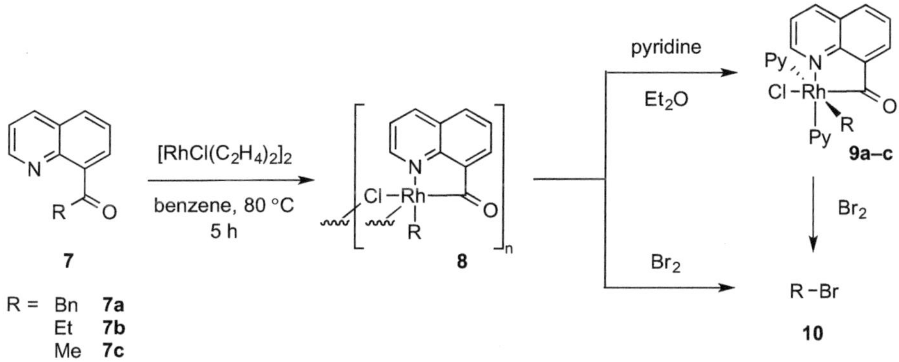


**Scheme 2** Trapping of the $sp^2$–$sp^3$ C–C bond activation adducts and derivatization

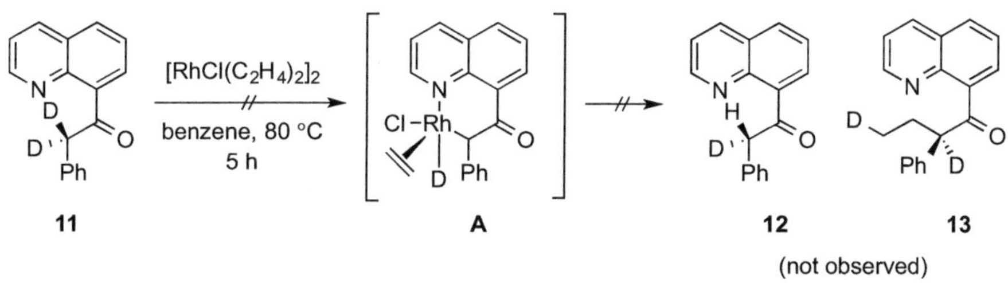


**Scheme 3** Hypothetical C–H bond activation via a six-membered metallacycle

reaction described in Sect. 3.3. In corroboration with X-ray crystallography, the IR $\upsilon_{CO}$ value of 1,633 $cm^{-1}$ portrays **9** as the Rh–acyl.

The proximity of the benzylic protons in 8-quinolinyl benzylketone **7a** makes it plausible that carbon–hydrogen activation could precede C–C bond activation and thus be a potential competing pathway (Scheme 3). In order to probe this, deuterated 8-quinolinyl benzylketone **11** was subjected to the reaction conditions. Should a C–D bond insertion occur via the six-membered metallacycle **A**, the resulting Rh(III)–D would presumably undergo migratory insertion onto a bound ethylene ligand. A subsequent β-hydride elimination would generate an Rh(III)–H that could reductively eliminate to give **12**. No H/D scrambling was detected and no evidence of the ethylation product **13** (see Sect. 2.2.2) was observed. This led Suggs and Jun to conclude that C–H bond activation was not occurring. The propensity toward C–C bond activation in this case could be explained by the enhanced stability of five-membered metallacycles in cyclometalation reactions [11].

The octahedral dipyridyl Rh(III) complexes **9a–c** (Scheme 2) were shown to be stable in solution for extended periods of time. Attempts to promote the back reaction (reductive elimination) by the addition of hard ligands such as triethylamine, dimethyl sulfoxide, or *N,N*-dimethylaminopyridine were unsuccessful. This observation was rationalized in terms of hard/soft acid base theory. The resulting soft Rh(I) species would be unstable when coordinated to hard ligands.

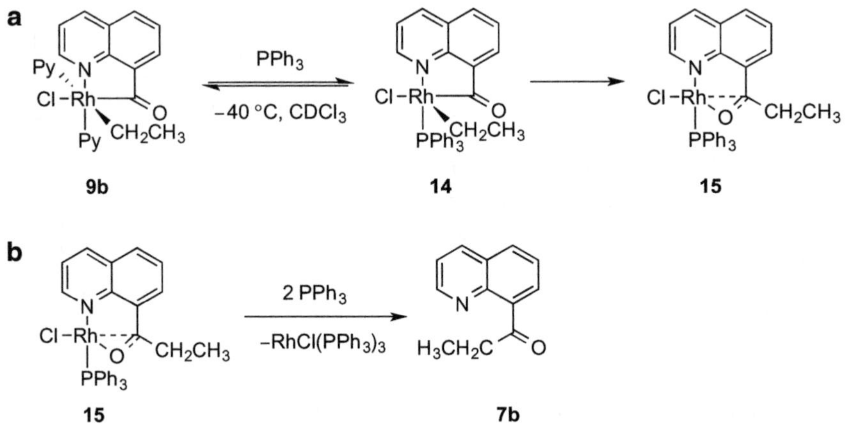


**Scheme 4** (**a**) Phosphine-promoted reductive elimination. (**b**) Excess phosphine-promoted ligand exchange

The addition of soft ligands (i.e., phosphines, phosphites, or CO) promoted reductive elimination via Rh(I) stabilization. Alternatively, the lower oxidation state may be stabilized by π-acidic ligands. Treating compound **9b** with $PPh_3$ at −40°C in $CDCl_3$ allowed for the rapid formation of a five-coordinate monophosphine Rh(III) complex **14** characterized by the $^{31}$P NMR doublet at 8.3 ppm, $J_{Rh\text{–}P} = 65$ Hz, and the $CH_2$ carbon as a doublet of doublets, $J_{P\text{–}C} = 82$ Hz, and $J_{Rh\text{–}C} = 35$ Hz in the $^{13}$C NMR spectrum (Scheme 4a) [13]. An equilibrium ratio greater than six to one favoring compound **14** was established within minutes. Over time, the appearance of a second $^{31}$P signal at 35.6 ppm with a large coupling constant ($J_{Rh\text{–}P} = 188$ Hz), and the absence of Rh–C and P–C coupling to the $CH_2$ in the $^{13}$C NMR spectrum indicated the formation of an Rh(I) species via reductive elimination. However, the persisting nonequivalence of the $CH_2$ protons and appearance of a $^{13}$C NMR carbonyl doublet at 99.9 ppm, $J_{Rh\text{–}C} = 15$ Hz suggested the $\eta^2$-ketone complex **15**. In the absence of excess $PPh_3$, the bound $\eta^2$-ketone either dissociates or is displaced by the pyridine in solution at temperatures above −10°C. Kinetic measurements demonstrated a first-order dependence on phosphine, which implied only one phosphine was required to promote reductive elimination. In the presence of excess $PPh_3$, complex **15** was readily converted to the 8-quinolinyl ethylketone **7b** and $RhCl(PPh_3)_3$ (Scheme 4b). This process was followed by the variations in the coupling constants between rhodium and phosphorus.

With phosphine ligands shown to promote reductive elimination, and phosphine-containing complexes oxidatively unreactive toward substrates **7a–c**, Suggs investigated whether removing phosphine from the metal center would encourage oxidative addition. A $CDCl_3$ solution of **15** was treated with $[Rh(cod)Cl]_2$ which served as a phosphine sponge [19] (Scheme 5). Although the $^{31}$P resonance of **15** rapidly disappeared, no indication of the rhodium insertion product **9b** was observed, even upon warming to 25°C.

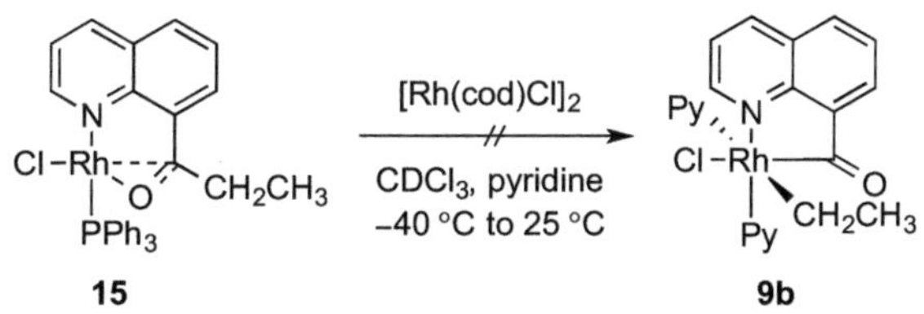


**Scheme 5** Failure to promote oxidative addition by $PPh_3$ adsorption onto $[Rh(cod)Cl]_2$

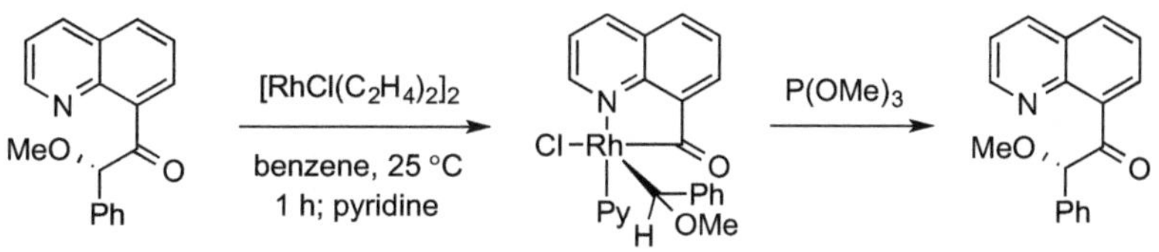


**Scheme 6** Stereochemical outcome of oxidative addition and reductive elimination

In order to gain insight into the reaction mechanism of carbon–carbon bond activation, the stereochemical outcome of oxidative addition and reductive elimination was probed. Combining $[RhCl(C_2H_4)_2]_2$ and (*S*)-8-quinolinyl α-methoxybenzyl ketone derivative **16** ($[\alpha]_D = -117°$) in benzene at room temperature provided the Rh(III)–alkyl complex **17** as a single diastereomer (Scheme 6) [20]. Phosphite-promoted reductive elimination regenerated **16** ($[\alpha]_D = -111°$). With reductive elimination known to proceed with retention of configuration in other systems [21], it is likely that both steps took place with retention rather than inversion of configuration.

In the absence of phosphite ligand, benzaldehyde was detected upon thermal decomposition of complex **17** at 90°C. With α-alkoxy radicals known to undergo fragmentation to carbonyl compounds [22], the formation of benzaldehyde could readily be explained by homolysis of the rhodium–alkyl bond. In the presence of $CCl_4$ the putative methyl radical was trapped as the methyl chloride under the thermolysis conditions.

Thermolysis in the presence of $CCl_4$ trapped the methyl radical as methyl chloride. Lowering the temperature to 45–60°C prevented radical fragmentation, but was sufficient for homolysis as indicated by the racemization of **16** upon reductive elimination. A cage-escape crossover experiment, in which complex **17** and the methylated ethoxy derivative **18** were heated together at 45°C, yielded the corresponding racemates of **16** and **19** along with racemic crossover products **20** and **21** (Scheme 7). A control experiment in which complexes **17** and **18** were immediately treated with trimethyl phosphite did not yield crossover products. These observations were interpreted as a result of the radical fragmentation and recombination mechanistic process.

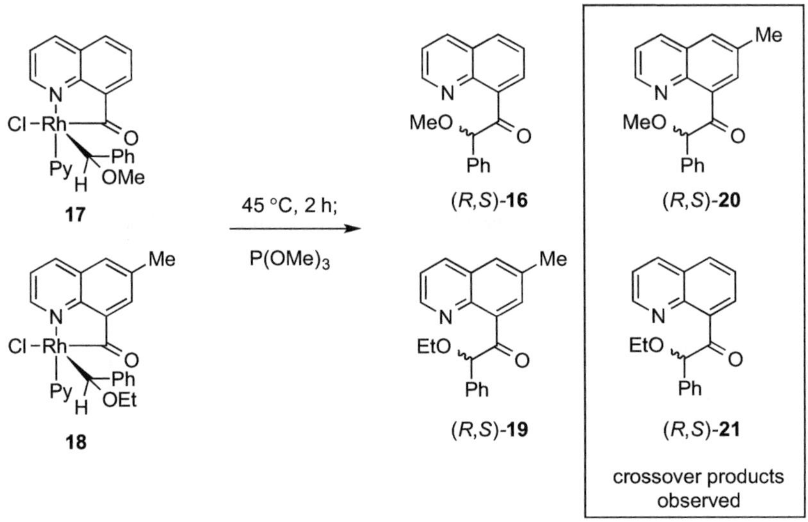


**Scheme 7** Crossover experiment: evidence for an Rh–C homolysis-recombination mechanism

The rates of racemization of the carbon center over the temperature range of 37–52°C provided an enthalpic value of $\Delta H^{\ddagger}_{(\mathrm{racemization})} = 32.5 \pm 1.5$ kcal/mol. Assuming the carbon radical has a very low racemization barrier [23], the calculated $\Delta H^{\ddagger}_{(\mathrm{racemization})}$ should reflect the activation enthalpy for homolysis of the Rh–C bond. Should the radical recombination barrier for Rh(II) mirror the value reported for Co(II) systems (ca. 2 kcal/mol) [24, 25], an Rh–C bond dissociation energy (BDE) of approximately 31 kcal/mol can be estimated.

### 2.2.2 Substrates with β-Hydrogens

Subjecting 8-quinolinyl butylketone **7e** to reaction with $[RhCl(C_2H_4)_2]_2$ and pyridine did not provide the analogous bipyridyl Rh(III)–butyl complex **22** as expected (Scheme 8a) [26]. Instead, the ethyl complex **9b** was isolated in >90% yield and 1-butene was observed in the $^1$H NMR spectrum. The observation of 1-butene suggested that Rh(III)–butyl complex **B** underwent β-hydride elimination to give the Rh(III)–H complex **C** (Scheme 8b). Migratory insertion of the metal–hydride across an ethylene ligand (**D**) (ethylation) would generate the corresponding Rh(III)–ethyl complex that is subsequently trapped by pyridine (**9b**). Without the detection of 1-hexene, it appears that β-hydride elimination of **B** has a lower kinetic barrier than migratory insertion of *n*-butyl. Other ketones containing β-hydrogens (**7f–h**) were also shown to form ethyl complex **9b** exclusively, with exception of a cyclopropyl derivative that underwent rearrangement to a π-allyl system (not shown) [27]. Heating **9b** at 100°C or treatment with excess $PPh_3$ afforded 8-quinolinyl ethyl ketone **7b** via reductive elimination (Scheme 8c).

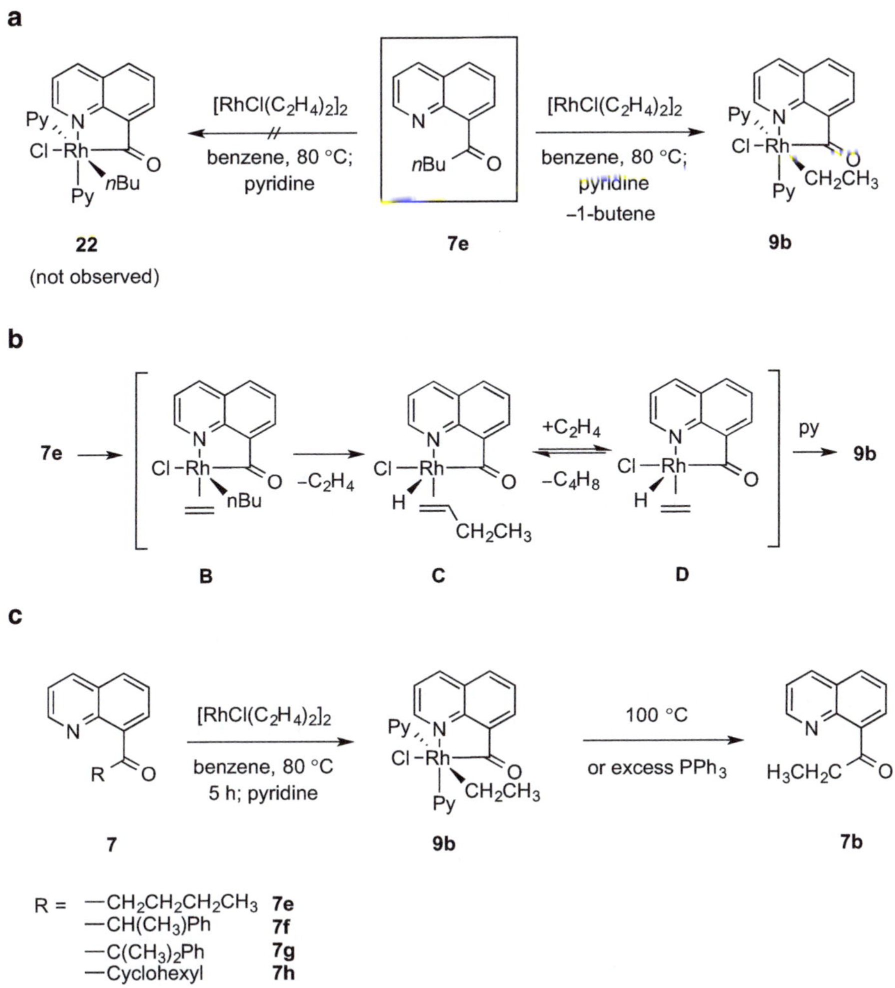


**Scheme 8** Alkyl exchange to form ethyl ketone **7b**: (**a**) unexpected complex formation; (**b**) ethylation mechanism; (**c**) additional alkyl exchange reactions

# 3 Catalytic Carbon–Carbon Bond Activation: Fragmentations

## *3.1 $Sp^2$–$sp^3$ Carbon–Carbon Bonds: Hydroacylation*

Reactions with compounds **7e–h** were made catalytic using higher temperature (100°C), longer reaction time (48 h), and 6 atm of ethylene pressure (Scheme 9) [26]. Substrate **7e** yielded 8-quinolinyl ethyl ketone **7b** in 61% yield with the remaining material being unreacted starting material. The insufficient conversion

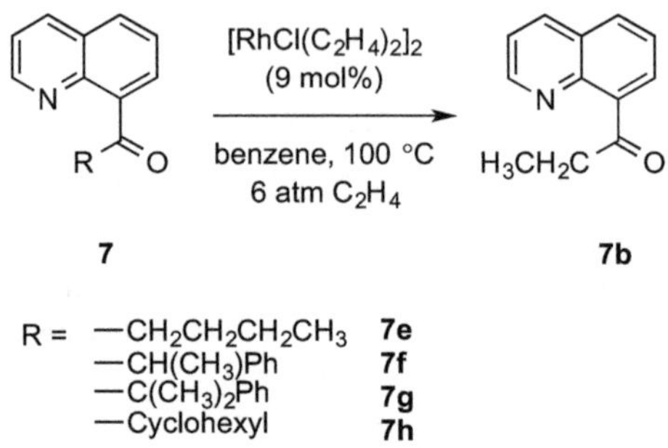

Scheme 9 Catalytic conversion of $sp^2$–$sp^3$ substrates with β-hydrogens to ethyl ketone **7b**

was attributed to catalyst deactivation rather than full equilibration considering the large excess of ethylene employed. Attempted reactions with other alkenes were mentioned, but the outcomes of these experiments were not clear. The authors simply stated that "the exchange reaction with alkenes other than ethylene was not efficient," and that "β-hydride elimination is too fast to compete with reductive elimination except for ethylene." It is plausible that the binding of more sterically hindered alkenes promotes reductive elimination. It was found that catalytic C–C bond activation reactions were successful under conditions utilizing $[Rh(cod)Cl]_2$, $[Ir(cod)Cl]_2$, and even $RhCl(PPh_3)_3$, which had been shown to be inactive in related stoichiometric reactions. Complexes that did not catalyze the exchange reaction included $Pd(PPh_3)_4$, $Pd(OAc)_2$, $Pt(PPh_3)_4$, $RuCl_2(PPh_3)_3$, and $Rh(C_5H_5)(C_2H_4)_2$.

## 3.2 *$Sp^2$–$sp^2$ Carbon–Carbon Bonds: Hydroacylation*

Although oxidative addition into the $sp^2$–$sp^2$ C–C bond of enone **4** did not occur with $RhCl(PPh_3)_3$ (Scheme 1), catalytic conversion of 8-quinolinyl phenyl ketone **7i** to ethyl ketone **7b** with $[RhCl(C_2H_4)_2]_2$ proceeded in quantitative yield (Scheme 10) [26]. The formation of 1 equiv. styrene (**23**) suggested that the migratory aptitude of phenyl is greater than for alkyl analogs since such ethylation products were not observed in the latter cases (see Scheme 9). This difference in reactivity was attributed to the ability of the resulting homobenzylic moiety to coordinate to rhodium through the phenyl π-bond, thus maintaining the electron-count around the metal center post insertion.

A plausible mechanism for the exchange reaction is illustrated in Scheme 11. Following oxidative addition into the acyl C–C bond, the resulting Rh(III)–phenyl complex **E** undergoes migratory insertion across ethylene. β-Hydride elimination of the homobenzylic intermediate **F** and migratory insertion of the subsequent Rh(III)–H **D** across another ethylene unit generates Rh(III)–ethyl species **G**. Reductive elimination delivers **7b** to complete the net hydroacylation process.

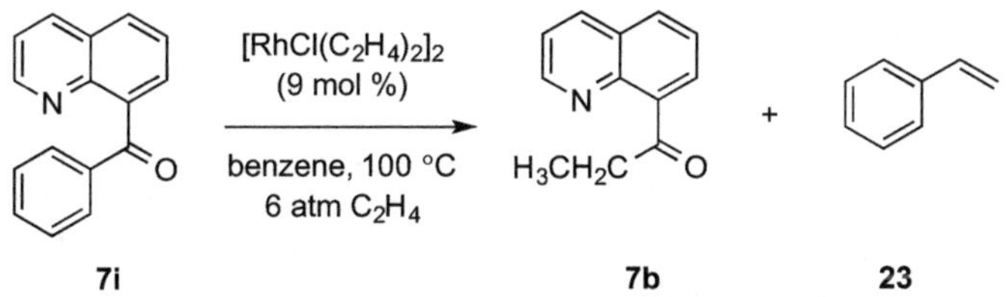


**Scheme 10** Catalytic conversion of sp$^2$–sp$^2$ phenyl substrate into ethyl ketone **7b** and styrene

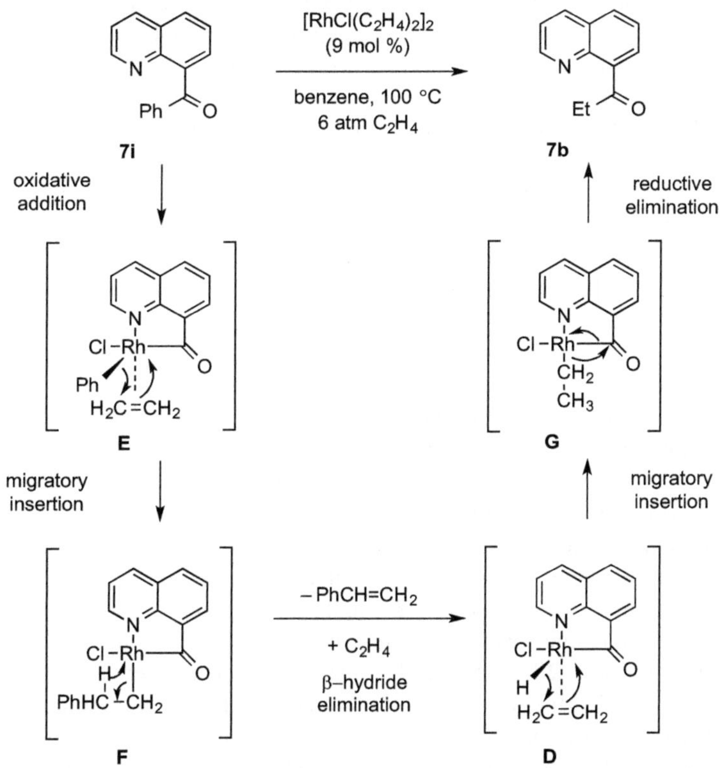


**Scheme 11** Mechanistic proposal for the exchange of phenyl for ethyl

The potential for competitive *ortho* C–H activation was explored through deuterium-label studies. Reaction with penta-deutero phenyl **7i** quantitatively yielded **7b** with complete deuterium retention in the styrene by-product. While this experiment suggested that *ortho* C–H bond activation was not in competition with C–C bond activation in 8-acylquinolines systems, subsequent work by Douglas [28] has identified this as a challenge (see Sect. 4.1).

## 3.3 Cross-Coupling Reactions

Although many organometallic intermediates have been shown to undergo a wide variety of subsequent functionalization reactions, carbon–carbon bond activation reactions have been relatively limited to additions across π-bonds. Wang [29] and co-workers embarked on merging the activation of 8-acylquinoline C–C bonds with other known C–C bond-forming reactions, such as the Suzuki–Miyaura coupling. With the exception of carbon–nitrile (C–CN) bond activation ([30] and references therein), and β-carbon elimination processes [31], examples of direct C–C bond cleavage reactions with successive cross-coupling functionalization are rare.

Methyl ketone **7c** was allowed to react with 2 equiv. phenylboronic acid (R=H) **24** in the presence of CuI, $O_2$, $K_2CO_3$, and catalytic $RhCl(PPh_3)_3$ in xylene at 130°C for 18 h. The aryl-exchanged product **25** was isolated in 93% yield, along with toluene (R=H) **26** as a by-product (Scheme 12) [29]. Molecular oxygen as a terminal oxidant was critical for the reaction, as no product was formed under anaerobic conditions. Under 1 atm of $O_2$, the reaction proceeded in less than 12 h, providing **25** in 72% yield along with unidentified side-products. The increase in reaction rate led to the speculation that $O_2$ must be involved in the rate-limiting step of catalysis.

The stereoelectronic effects of the reaction were explored with various arylboronic acids containing electron-donating or electron-withdrawing groups. Although requiring longer reaction times (36 or 48 h), arylboronic acids possessing *meta* and *para* electron-donating groups were well tolerated with yields ranging from 74% to 90%, whereas those with *ortho* substitution failed to undergo the reaction. Aryl boron species with electron-withdrawing groups requiring extended reaction times, provided products in lower yields (35–57% yields) with the exception of 4-chlorophenylboronic acid which proceeded in 83% yield.

The 8-quinolinyl aryl ketones **27** and arylboronic acids **24** successfully underwent the exchange reaction to yield the corresponding aryl ketones **25** and biphenyl by-products **28** (Scheme 13). The synergistic stereoelectronic effect between coupling partners significantly impacted the reaction outcomes. Aryl ketone **27** with $R^1$=H reacted with various electron-rich boronic acids with yields ranging from 45% to 71%. The combination of electron-donating groups on both **27** and **24** gave similar results. Optimal conditions employed electron-deficient aryl ketones with electron-rich boronic acids (61–93%). In all cases, boronic acids with electron-withdrawing groups failed to undergo the reaction.

The authors proposed a catalytic cycle that involves oxidative addition into the acyl C–C bond to form the five-membered Rh(III) metallacycle **H** (Scheme 14). Transmetalation with boronic acid generates intermediate **I** that undergoes phosphine-promoted reductive elimination. Oxidation of the resulting Rh(I) chelate **J** with $O_2$ in the presence of CuI gives an Rh(III) species **K** that participates in a second transmetalation. Reductive elimination of complex **L** yields product **7i**.

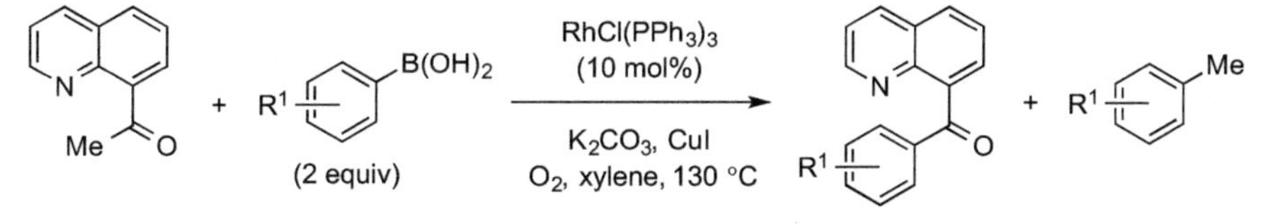


**Scheme 12** Methyl-for-phenyl exchange reaction

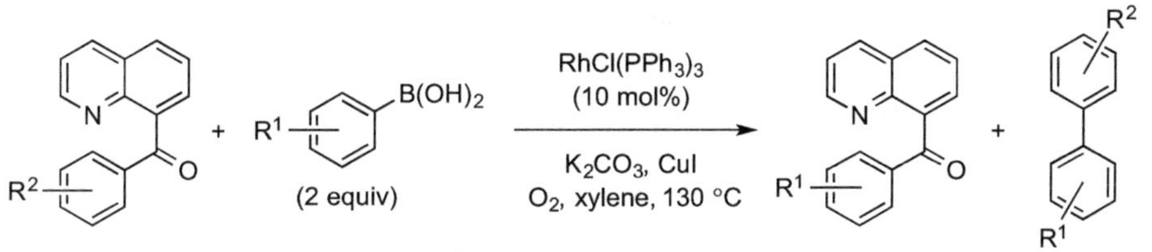


**Scheme 13** Aryl-for-aryl exchange reaction

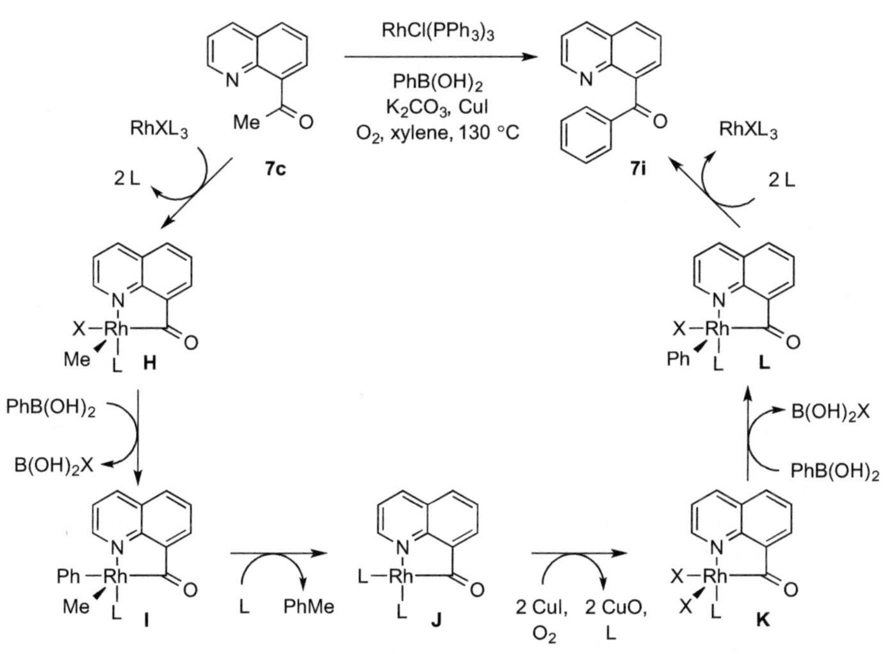


**Scheme 14** Proposed cross-coupling mechanism

With the oxidation of **J** suggested to be rate-limiting, it is unclear how the rate is affected by the electronic nature of the boronic acid, and thus further investigation is warranted.

# 4 Catalytic C–C Bond Activation Reactions: Carboacylation

## *4.1 Intermolecular Carboacylation*

Limited to reactions with ethylene, quinoline-directed carbon–carbon bond activation catalysis went underdeveloped. After more than 20 years had passed since Suggs and Jun's alkyl exchange reactions were developed, Douglas [28] and co-workers designed a catalytic system that would allow for C–C bond activation to forgo the β-hydride elimination process that leads to fragmentation. In theory, employing strained [2.2.1]bicycloheptenes would produce intermediates void of accessible *syn* β-hydrogens and thus allow for a complexity-building pathway.

Heating equimolar amounts of 8-quinolinyl phenyl ketone **7i** and norbornene (**29**) with several Rh(I) catalysts in a range of solvents and temperatures gave varying ratios of C–C bond activation (carboacylation) product **30** and *ortho* C–H bond activation (hydroarylation) product **31** (Table 1). Although Suggs had not reported a competitive C–H bond insertion pathway (see Sect. 3.2) [26], advances in C–H activation chemistry has shown *ortho*-metalation to occur with carbonyl-like directing groups via five-membered chelates [32]. The relief of ring strain for norbornene (ca. 25 kcal/mol) may lower the barrier for this hydroarylation process. Since migratory insertion is known to proceed in a *syn* fashion, the *anti* stereorelationship in **30** likely occurs via epimerization post carboacylation, presumably through an inter- or intramolecular deprotonation by the basic quinoline nitrogen. The possibility for β-hydride elimination to remain a competing pathway through the reverse reaction (metal insertion into the *anti* product) exists.

Reaction between 8-quinolinyl phenyl ketone **7i**, norbornene **29**, and $[RhCl(C_2H_4)_2]_2$ in toluene at 130°C for 24 h exclusively gave the hydroarylation product **31** in 79% isolated yield (entry 1). Cationic rhodium complexes $[Rh(cod)_2]BF_4$ and $[Rh(cod)_2]OTf$ allowed for C–C activation to compete with C–H activation, giving a 1:6 (38%, entry 2) and 4:5 (56%, entry 3) product ratio (**30**:**31**), favoring C–H activation. The remaining mass in these reactions consisted of unreacted starting material and unidentified side-products. Increasing the alkene loading tenfold did not improve the conversion. Switching to a more polar solvent and reducing the temperature to 100°C reversed the chemoselectivity to favor C–C activation, thereby providing products in a 5:3 ratio in acetonitrile (41%, entry 4) and 1:0 in tetrahydrofuran (50%, entry 5). Reaction with $RhCl(PPh_3)_3$ was relatively ineffective (<10%, entry 6) even though it had been shown to be efficient in other catalytic systems (see Sects. 3.1, 3.3, and 4.2) [26].

**Table 1** Intermolecular carboacylation and hydroarylation

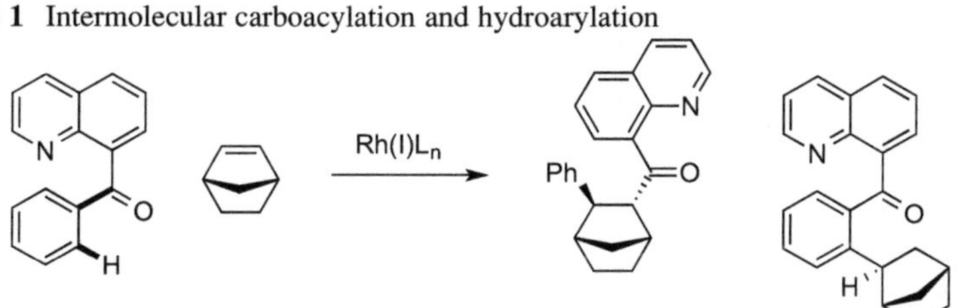


7i 29 30 31

C-C activation (carboacylation) C-H activation (hydroarylation)

| Entry | Catalyst | Solvent | Temperature (°C) | Yield (%) | **30:31** |
|---|---|---|---|---|---|
| 1 | $[RhCl(C_2H_4)_2]_2$ | PhMe | 130 | 79 | 0:1 |
| 2 | $[Rh(cod)_2]BF_4$ | PhMe | 130 | 38 | 1:6 |
| 3 | $[Rh(cod)_2]OTf$ | PhMe | 130 | 56 | 4:5 |
| 4 | $[Rh(cod)_2]OTf$ | MeCN | 100 | 41 | 5:3 |
| 5 | $[Rh(cod)_2]OTf$ | THF | 100 | 50 | 1:0 |
| 6 | $RhCl(PPh_3)_3$ | PhMe | 130 | <10 | – |

The nature of the rhodium counterion appears to influence the chemoselectivity of the reaction. Assuming that intermediates **M** and **N** are in equilibrium, it is possible that the coordinating chloride ligand raises the barrier ($\Delta G^{\ddagger}$) to migratory insertion in **M** and thus favors the consumption of **N** (Curtin–Hammett kinetics) (Scheme 15). The cationic rhodium complex with the less-coordinating triflate ligand in a stabilizing polar solvent could make the metal center more accessible to olefin complexation and thus lower the barrier ($\Delta G^{\ddagger}$) to migratory insertion for intermediate **M**.

Alternatively, should there be an equilibrium between **30** and epimer **O**, and between **O** and **M** (via β-aryl elimination), it is conceivable that **30** and **31** would also be in equilibrium. In order to examine this, **30** was subjected to the reaction conditions with $[RhCl(C_2H_4)_2]_2$ in toluene without detection of **31** by $^1$H NMR spectroscopy. Although this proves that **30** and **31** do not equilibrate, it does not infer about the relative energy levels of **M** and **N**.

The addition of a *para*-$CH_3$ group on the substrate phenyl ring negatively impacted the reactivity and chemoselectivity of the reaction, producing a 1:1 product mixture in 30% yield (66% brsm) under the optimized reaction conditions ($[Rh(cod)_2]OTf$, THF, 100°C) with norbornene. The addition of an electron-withdrawing *para*-$CF_3$ substituent effectively suppressed the C–H insertion pathway to give the carboacylation product in 24% yield. Exchanging the phenyl group for a methyl group (**7c**, Schemes 2 and 12), where C–H activation is not applicable, gave the carboacylation product in 39% yield (60% brsm) under $[RhCl(C_2H_4)_2]_2$ catalysis.

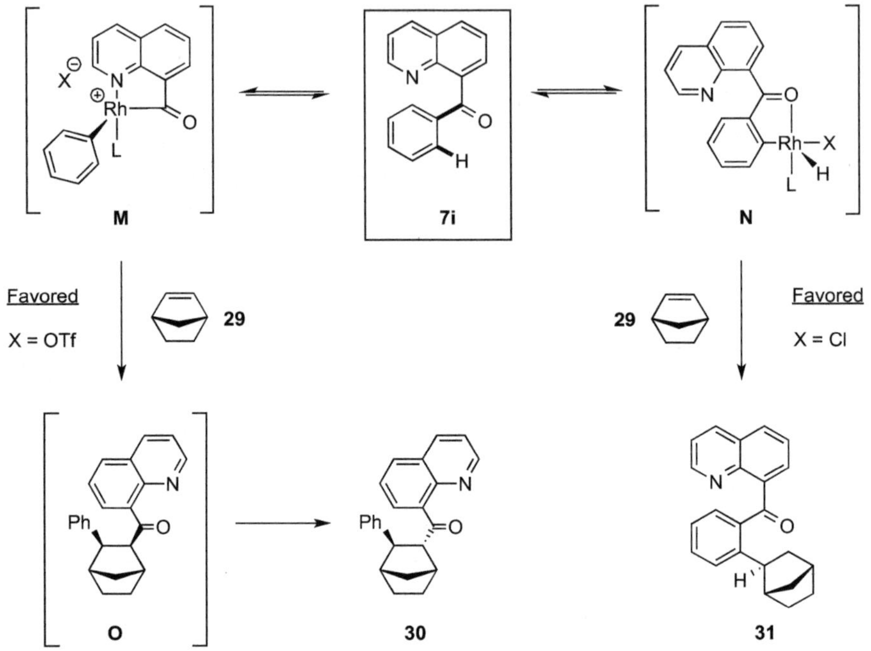


**Scheme 15** Carboacylation favored with cationic rhodium complexes in a polar solvent, and hydroarylation favored with coordinating chloride ligand in a non-polar solvent

## *4.2 Intramolecular Carboacylation*

In order to be an effective synthetic strategy, the activation of a C–C bond should result in metal–carbon bonds that can be functionalized to produce a more complex product. Concurrent with their intermolecular carboacylation efforts, Douglas et al. envisioned that an intramolecular reaction had the potential to overcome the kinetic limitations associated with olefin insertion reactions with ethylene (see Sects. 3.1 and 3.2) and norbornene (Sect. 4.1).

Substrate (**i**) was designed to force the appended alkene into proximity of the reactive metal center (**ii**) and thus allow for cyclization to form two new C–C bonds (Scheme 16). The 1,1-disubstitution on the alkene served to avoid a potential β-hydride elimination pathway and to provide a scaffold that could access the synthetically elusive all-carbon quaternary stereogenic center [33–36], as seen in (**iii**) (Scheme 16).

Heating allylic ether substrate **32** with various Rh(I) catalysts in toluene at 130°C for 48 h under nitrogen gave the desired dihydrobenzofuran product **33** in good to excellent yields (Table 2) [37]. Although phosphine ligands were shown to promote reductive elimination [13] (see Scheme 4) and be rather ineffective in facilitating intermolecular carboacylation reactions (see Table 1, entry 6), reaction of **32** with $RhCl(PPh_3)_3$ provided the desired product **33** in 96% isolated yield

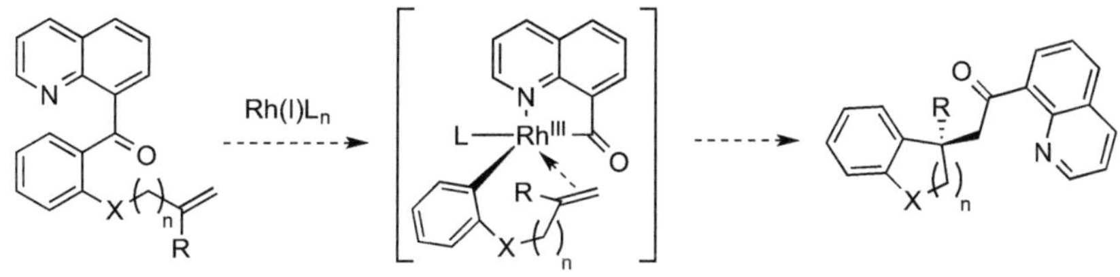


**Scheme 16** Proposed intramolecular carboacylation

**Table 2** Optimization of intramolecular carboacylation

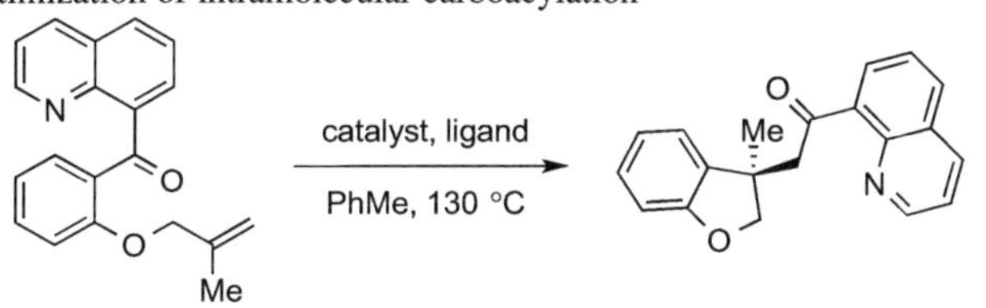


| Entry | Catalyst | mol% | Ligand | Yield **33** (%) |
|---|---|---|---|---|
| 1 | $[RhCl(C_2H_4)_2]_2$ | 5 | None | 95 |
| 2 | $RhCl(PPh_3)_3$ | 10 | None | 96 |
| 3 | $RhCl(PPh_3)_3$ | 2 | None | 90 |
| 4 | $[Rh(cod)_2]OTf$ | 5 | None | 62 |
| 5 | $[Rh(cod)_2]BF_4$ | 10 | None | 54 |
| 6 | $[RhCl(C_2H_4)_2]_2$ | 5 | $PCy_3$ | 62 |
| 7 | $[RhCl(C_2H_4)_2]_2$ | 5 | $PMe_3$ | 53 |
| 8 | $[RhCl(C_2H_4)_2]_2$ | 5 | P(*t*Bu)$_3$ | 54 |
| 9 | $[Rh(cod)_2]OTf$ | 5 | $PMe_3$ | 72 |
| 10 | $[Rh(cod)_2]OTf$ | 5 | BINAP | <5 |

(Table 2, entries 2 and 3)! With asymmetric catalysis predominantly achieved with the use of chiral ligands, bidentate phosphine BINAP was investigated. Unfortunately, the reaction favored alkene isomerization to the corresponding enol ether along with deallylation to give the resulting phenol (entry 10). Although palladium and nickel catalysts have been shown to participate in other C–C bond activation reactions, they were unsuccessful in achieving carboacylation.

The scope of the reaction was investigated by altering the alkene substituent, tethering atom, length of the olefin appendage, and nature of the aryl ring (Scheme 17). Dihydrobenzofuran **36** (R=Ph) was obtained in a 94% isolated yield, despite the sterically encumbered diaryl all-carbon quaternary center. Although β-hydride elimination was expected to complicate the formation of dihydrobenzofuran **37** (R=H), cleavage of the allyl ether was the predominant decomposition pathway. The less nucleophilic $[Rh(cod)_2]OTf$ complex served to limit deallylation, providing **37** in 25% yield.

Although the lack of reactivity with bromine substitution (**38**) was attributed to the inability of the electron-deficient alkene to bind to the metal center,

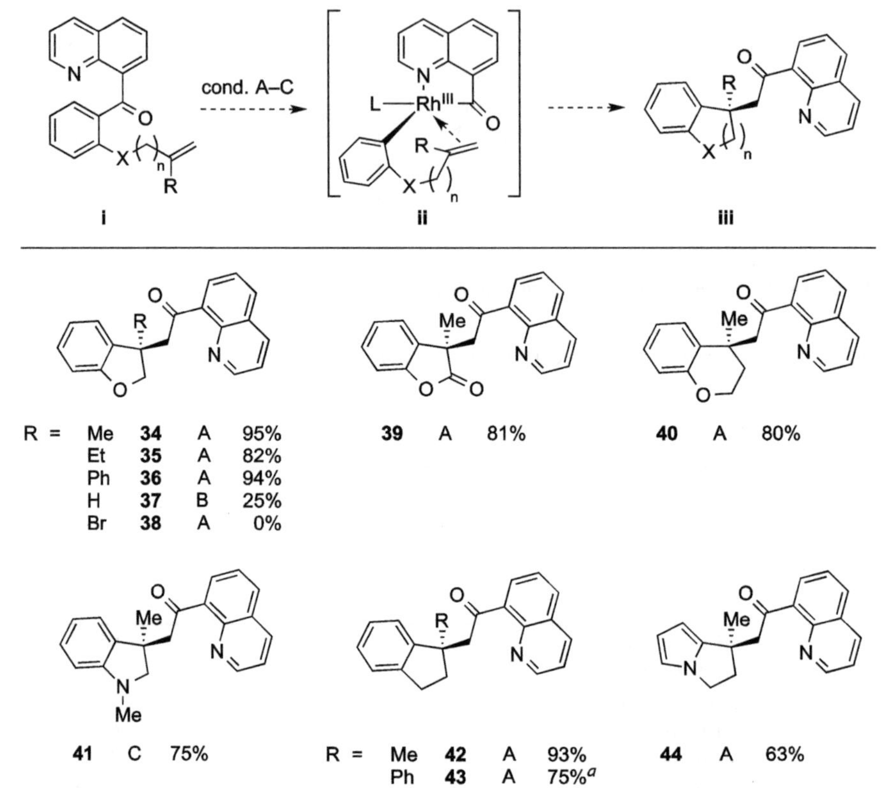


**Scheme 17** Isolated yields of intramolecular carboacylation products. *Condition A*: 5 mol% $[RhCl(C_2H_4)_2]_2$, PhMe, 130°C, 48 h. *Condition B*: 5 mol% $[Rh(cod)_2]OTf$, PhMe, 130°C, 48 h. *Condition C*: 10 mol% $RhCl(PPh_3)_3$, PhMe, 130°C, 48 h. [a]Conversion by $^1H$ NMR spectroscopy

carboacylation with a methacrylate ester successfully gave benzofuranone **39** in 81% yield. Extending the tether length allowed for the synthesis of dihydrobenzopyran **40** from the corresponding homoallylic ether substrate. Exchanging the tethering heteroatom for nitrogen significantly impeded the reaction. Dihydroindole **41** was formed in 10% conversion upon reaction with $[RhCl(C_2H_4)_2]_2$ after 48 h. Running the reaction in the presence of **32**, which underwent complete conversion to **33**, disproved the hypothesis that the anthranilic ketone had sequestered the catalyst through non-productive binding. Increasing the nucleophilicity of the catalyst ($RhCl(PPh_3)_3$) allowed for dihydroindole **41** to be obtained in 75% yield, which suggested that electron donation from the 2-amino group simply renders the carbonyl less electrophilic and thus slower in undergoing oxidative addition. Removing the tethering heteroatom allowed for the preparation of dihydroindenes **42** and **43**. Comparable to the analogous cyclization of **32**, dihydroindene **42** (R=Me) was isolated in 93% yield. In contrast, incomplete conversion of the phenyl-substituted derivative was observed under all conditions, providing dihydroindene **43** (R=Ph) in 75% with $[RhCl(C_2H_4)_2]_2$, 65% with $[Rh(cod)_2]OTf$, and <10% conversion with $RhCl(PPh_3)_3$. The difference in reactivity upon forming dihydrobenzofuran **36** and dihydroindene **43** is unclear. The phenyl ring was replaced with a pyrrole, allowing for the isolation of dihydropyrrolizine **51** in 63% yield.

## *4.3 Mechanistic Considerations*

Johnson et al. [38] reported mechanistic investigations of the intramolecular carboacylation developed by Douglas. The rate law, $^{12}C/^{13}C$ kinetic isotope effects, and activation parameters were discovered for both $RhCl(PPh_3)_3$ and $[RhCl(C_2H_4)_2]_2$ complexes to determine the rate-limiting step of catalysis. The rate laws differed between catalyst systems, and the reaction mechanism was shown to be substrate dependent. In addition, several other substrate derivatives were identified to undergo the reaction successfully.

### 4.3.1 Investigations with $RhCl(PPh_3)_3$

Traditional kinetic methods revealed a linear consumption of starting material (concentration vs time) through at least 80% conversion when substrate **32** was treated with $RhCl(PPh_3)_3$ at 130°C. This data indicated a zero-order dependence upon substrate (saturation kinetics) and a first-order dependence upon $RhCl(PPh_3)_3$. The overall first-order rate law was determined where $k = 4.98 \times 10^{-4}\ s^{-1}$. Exogenous $PPh_3$ ligand was introduced to the reaction in concentrations ranging from 0 M to 0.108 M. Under these conditions, a change from zero- to first-order in substrate dependence was observed, and an inverse first-order dependence upon $PPh_3$ indicated reaction inhibition. From these data, it was apparent that $RhCl(PPh_3)_3$ must lose 1 equiv. $PPh_3$ prior to reaction and that $PPh_3$ disrupts a pre-equilibrium where catalyst and substrate are bound in the resting state. The activation parameters were determined in which the relatively neutral change in entropy suggested minimal molecular reorganization in the rate-limiting step ($\Delta S^{\ddagger} = -4.3 \pm 2.4$ eu) and the enthalpic change quantifies the energy required to cleave a C–C bond under transition-metal catalysis ($\Delta H^{\ddagger} = 27.8 \pm 1.0$ kcal/mol).

Without the ability to probe the relative energies regarding oxidative addition, migratory insertion, or reductive elimination for intramolecular reactions by traditional kinetic methods, analysis of $^{12}C/^{13}C$ kinetic isotope effects by the Singleton method [39, 40] gave insight into the rate-limiting step of the reaction. Negligible isotope effects were seen at both alkene carbons (less than $1.003 \pm 0.005$) suggesting that neither migratory insertion nor reductive elimination are rate limiting. Conversely, the ketone ($1.027 \pm 0.005$) and α-aryl carbon ($1.028 \pm 0.004$) showed significant isotope effects, which support carbon–carbon bond activation as the catalytic slow step.

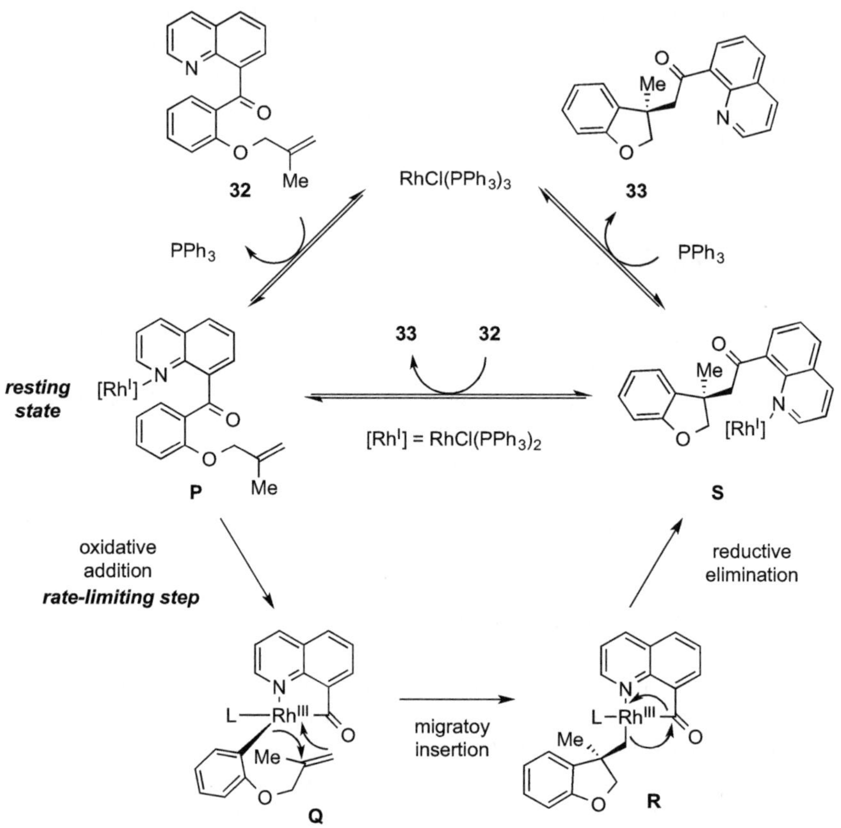


**Scheme 18** Mechanistic proposal for intramolecular carboacylation with $RhCl(PPh_3)_3$

The combination of kinetic isotope effects, activation parameters, and established rate law are consistent with the mechanism proposed in Scheme 18. The nature of intermediates **R** and **S** are unable to be probed since they occur after the rate-limiting step. Intermediates **P** and **Q** are inferred from the inhibitory effects of product and $PPh_3$ on the rate of reaction. The relatively neutral activation entropy is in agreement with intermediate **P** as a resting state and oxidative addition being the rate-limiting step in that little change in the overall molecular organization would be expected.

The search for additional insight into the reaction mechanism and the nature of the rate-limiting step continued with a linear free-energy correlation study. Several analogs were prepared containing both electron-withdrawing and electron-donating substituents on the aryl ring (Table 3) and the rates of reaction were determined with $RhCl(PPh_3)_3$ as catalyst at 130°C. Electron-donating groups (entries 2–5) accelerated the reaction whereas electron-withdrawing groups (entries 6 and 7) had an inhibitory effect. This counterintuitive observation could be explained with the acknowledgement that increased electron density would have a stabilizing

**Table 3** Linear free-energy relationships with $RhCl(PPh_3)_3$

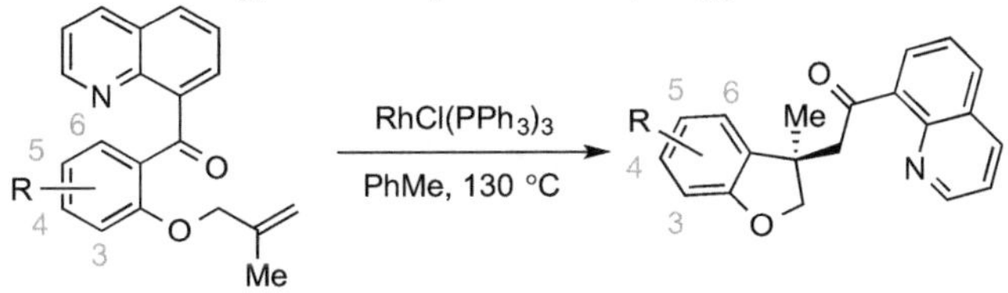


| Entry | R, substituent | Yield (%) | $k$ ($s^{-1} \times 10^{-4}$) |
|---|---|---|---|
| 1 | H **32** | 97 | 4.98 |
| 2 | 3-OMe **45** | 95 | 4.14 |
| 3 | 4-OMe **46** | 93 | 8.35 |
| 4 | 5-OMe **47** | 97 | 5.13 |
| 5 | 4-$NEt_2$ **48** | 85 | 12.5 |
| 6 | 5-$NO_2$ **49** | 91 | 2.32 |
| 7 | 5-Cl **50** | 94 | 3.77 |
| 8 | 3,5-(*t*Bu)$_2$ **51** | 0 | n/a |

effect on the metallacycle intermediate by strengthening the Rh–$C_{aryl}$ bond rather than rendering the ketone less electrophilic. Substrate **48** with a 4-diethylamino group underwent the reaction more than twice as fast as the parent substrate **32**. This data seemingly conflicts with the observation Douglas made concerning the formation of dihydroindole **41** (see Scheme 17). Perhaps the amino heteroatom is either not in conjugation with the aryl ring because the alkene is bound to the metal center during oxidative addition, or that oxidative addition is not the rate-limiting step when an amino group is at the 2-position. It was also observed that the reaction did not tolerate the sterically hindered *tert*-butyl groups (entry 8). Overall, the reaction was shown to be tolerant of several functional groups not examined in the initial report [37], including aryl chlorides and nitro groups.

With oxidative addition established as the rate-limiting step for $RhCl(PPh_3)_3$ catalyzed intramolecular carboacylation of allyl ether substrate **32**, it is reasonable to assume that modifications made to the alkene substituent would have no effect on the rate. However, the remaining allylic and homoallylic ether derivatives (see Scheme 17), reported by Douglas to have cyclized with $[RhCl(C_2H_4)_2]_2$, did not provide product with $RhCl(PPh_3)_3$. Additional substrates with 1,2-disubstituted alkenes also failed to undergo the reaction with $RhCl(PPh_3)_3$. These results imply a change in mechanism, which will be discussed in Sect. 4.3.2.

Suggs and Jun had determined that a homolysis-recombination mechanism was operative in the thermolysis of (*S*)-8-quinolinyl α-methoxybenzyl ketone **17** (see Scheme 7). An analogous cage-escape crossover experiment was performed on 3-methoxy-aryl derivative **45** with 6-methyl-quinolinyl derivative **52** in order to probe the nature of the intermediates (Scheme 19). Both substrates underwent complete conversion to their respective product (**53** and **54**) without detection of crossover products **55** or **33**. It is therefore unlikely that this transformation proceeds through a radical mechanism.

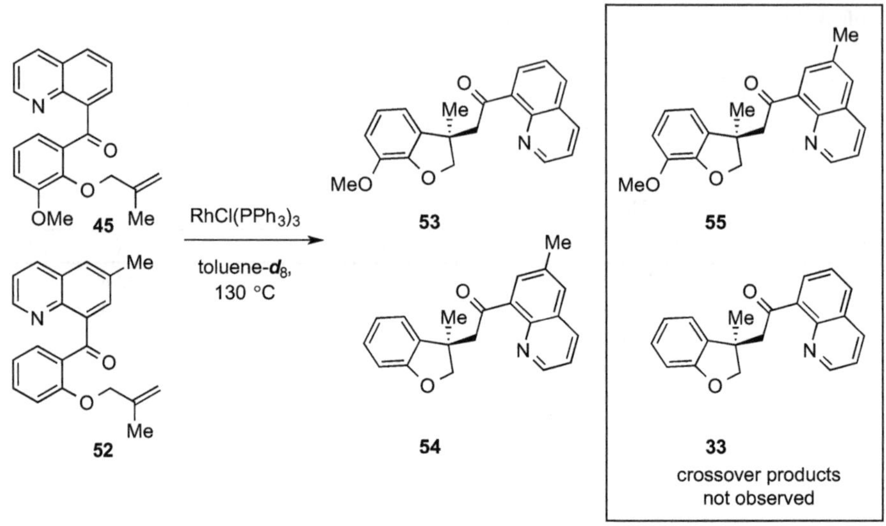


**Scheme 19** Crossover experiment

### 4.3.2 Investigations with $[RhCl(C_2H_4)_2]_2$

The selective reactivity of $RhCl(PPh_3)_3$ prompted Johnson to investigate the mechanism associated with the more active $[RhCl(C_2H_4)_2]_2$ catalyst [41]. A first-order dependence on both reactants led to an overall second-order rate law with a rate constant of $k = 7.59 \times 10^{-4}\ M^{-1}\ s^{-1}$. In contrast to the first-order rate law observed with $RhCl(PPh_3)_3$, the second-order rate law for catalysis with phosphine-free $[RhCl(C_2H_4)_2]_2$ suggested a lack of a pre-equilibrium coordination of substrate to metal. Additionally, the reaction rate was not subject to product inhibition. The observed rates from reactions at varying temperatures provided a comparable enthalpic value of $\Delta H^{\ddagger} = 28.4 \pm 1.3$ kcal/mol (in comparison to $RhCl(PPh_3)_3$, $\Delta H^{\ddagger} = 27.8 \pm 1.0$ kcal/mol) and a relatively large entropic value of $\Delta S^{\ddagger} = -26.4 \pm 2.6$ eu (in comparison to $RhCl(PPh_3)_3$, $\Delta S^{\ddagger} = -4.3 \pm 2.4$ eu) that supports a bimolecular rate-limiting step which is consistent with $[RhCl(C_2H_4)_2]_2$ as the resting state of catalysis.

The Singleton $^{12}C/^{13}C$ kinetic isotope effects were determined for carboacylation catalysis with $[RhCl(C_2H_4)_2]_2$ and parent substrate **32**. Similar to catalysis with $RhCl(PPh_3)_3$, negligible isotope effects at the alkene carbons (less than 1.002) and significant isotope effects at the ketone carbon (1.1016 ± 0.0005) and the α-aryl carbon (1.1012 ± 0.0004) were discovered for catalysis with $[RhCl(C_2H_4)_2]_2$, thus maintaining oxidative addition as the likely rate-limiting step.

Additional support for oxidative addition as the rate-limiting step for $[RhCl(C_2H_4)_2]_2$ catalysis was demonstrated through linear free-energy correlation studies. As with $RhCl(PPh_3)_3$, the rate of reaction was enhanced with the addition of electron-donating groups on the central aryl ring (Table 4). The magnitude of rate enhancement, however, was substantially less for $[RhCl(C_2H_4)_2]_2$. For example,

**Table 4** Linear free-energy relationships with $[RhCl(C_2H_4)_2]_2$

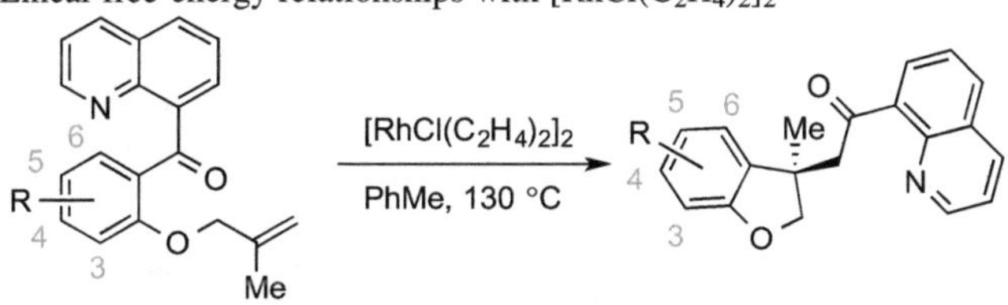

| Entry | R, substituent | Yield (%) | $k$ ($M^{-1}$ $s^{-1}$ × $10^{-4}$) |
|---|---|---|---|
| 1 | H **32** | 96 | 7.59 |
| 2 | 3-OMe **45** | 91 | 7.11 |
| 3 | 4-OMe **46** | 87 | 9.69 |
| 4 | 4-$NEt_2$ **48** | 93 | 11.4 |
| 5 | 5-$NO_2$ **49** | 89 | 6.39 |
| 6 | 5-Cl **50** | 94 | 6.72 |

**Table 5** Alkene substitution effects on the rates of reaction

| Entry | $R^1$ | $R^2$ | $R^3$ | $n$ | Yield (%) | $k$ ($M^{-1}$ $s^{-1}$ × $10^{-4}$) |
|---|---|---|---|---|---|---|
| 1, **32** | Me | H | H | 1 | 96 | 7.59 |
| 2, **56** | Me | H | Me | 1 | 94 | 7.90 |
| 3, **57** | Ph | H | H | 1 | 91 | 2.62 |
| 4, **58** | Me | H | H | 2 | 89 | 1.20 |
| 5, **59** | H | Me | H | 1 | <5 | n/a |
| 6, **60** | H | Ph | H | 1 | <5 | n/a |

4-diethylamino substrate **48** ($k = 11.4 \times 10^{-4}$ $M^{-1}$ $s^{-1}$) underwent the reaction 1.5 times faster than parent compound **32** ($k = 7.59 \times 10^{-4}$ $M^{-1}$ $s^{-1}$) with $[RhCl(C_2H_4)_2]_2$, whereas the rate was enhanced by a factor of 2.5 for $RhCl(PPh_3)_3$ (see Table 3).

Mechanistically, $[RhCl(C_2H_4)_2]_2$ and $RhCl(PPh_3)_3$ differ only in the resting state of catalysis. The nature of the diminished reactivity of $RhCl(PPh_3)_3$ was therefore not apparent. Continued kinetic examination with regard to alkene substitution showed significant rate dependence. Substitution on the quinolinyl 6-position (Table 5, entry 2, **56**) with a mild electron-donating group gave rise to a slight increase in reaction rate. The added electron density presumably stabilizes the Rh(III) metallacycle intermediate by enhancing the coordinating ability of the quinoline nitrogen. This observation is in accordance with an oxidative addition rate-limiting step. Replacing the vinyl methyl group in parent substrate **32** (entry 1) with a phenyl group (entry 3, **57**) consequently decreased the rate threefold.

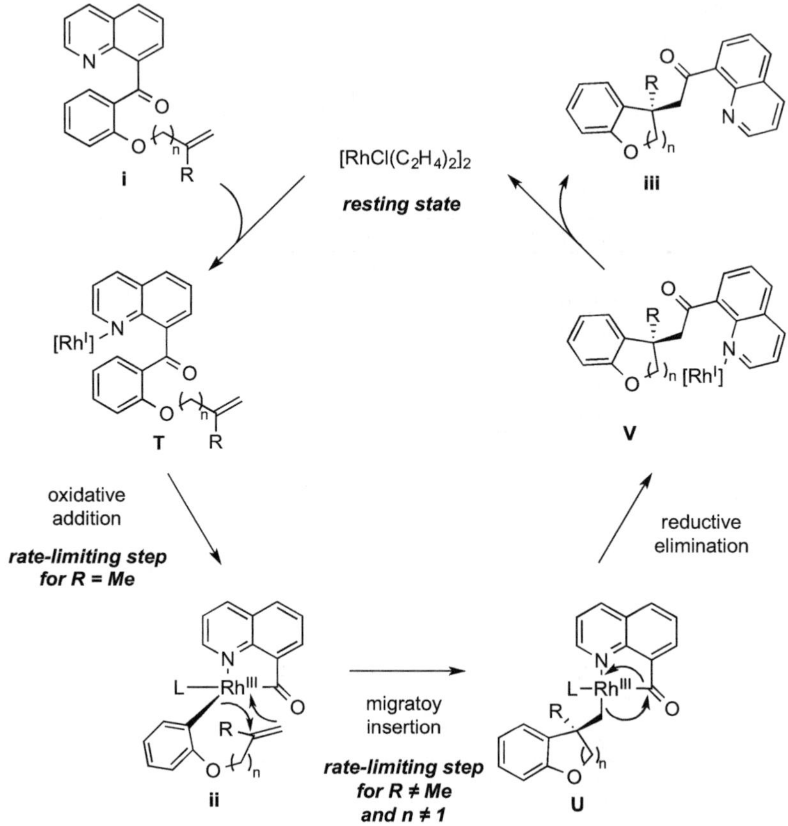


**Scheme 20** Intramolecular carboacylation under $[RhCl(C_2H_4)_2)]_2$ catalysis

With the literature suggesting that reductive elimination for carbon–carbon bonds should be fast relative to oxidative addition and migratory insertion [42–45], and the assumption that the alkene plays no role in oxidative addition, it was concluded that migratory insertion became rate-limiting for the phenyl substrate. With the general notion that five-membered ring cyclizations are kinetically more facile than six-membered ring cyclizations, the slower rate of homoallylic substrate **58** (entry 4) relative to the allylic substrate **32** (entry 1) is consistent with migratory insertion as rate-limiting for larger ring-forming reactions.

With 1,1-disubstituted alkenes shown to be optimal, Johnson considered 1,2-disubstitued alkenyl substrates (Table 5, entries 5 and 6) to target products containing two vicinal stereogenic centers. As mentioned before, these substrates failed to undergo reaction with $RhCl(PPh_3)_3$. Under $[RhCl(C_2H_4)_2]_2$ catalysis, substrates **59** and **60** underwent less than 5% conversion to product, returning mostly starting material as *cis*:*trans* isomers. The added steric environment (secondary vs primary Rh-alkyl) post migratory insertion would undoubtedly make the resulting intermediate less stable. These results also imply migratory insertion as the highest energy step of the reaction.

The combination of data including the overall second-order rate law, activation parameters, and Hammett relationships for the intramolecular carboacylation of 2-methyl allylic substrate **32** with $[RhCl(C_2H_4)_2)]_2$ indicate a free-catalyst resting state, and oxidative addition as the rate-limiting step to catalysis as shown in Scheme 20. The substantial differences in rate based upon alterations made to the alkene appendage suggest a change in mechanism based on the assumption that the alkene is not involved in the oxidative addition step for these substrates. For the 2-phenyl allylic substrate **57** and the 2-methyl homoallylic substrate **58**, migratory insertion likely becomes rate limiting. Unfortunately $^{12}C/^{13}C$ kinetic isotope effects were not determined for these derivatives. Observing isotope effects on both alkene carbons would provide further evidence suggesting migratory insertion as rate limiting, whereas isotope effects on the terminal alkene carbon and ketone carbon would indicate reductive elimination is the slow step.

# 5 Conclusion

Carbon–carbon bond activation is just beginning to develop its potential in organic synthesis. With the exception of carbon–nitrile (C–CN) functionalization [46], the majority of examples have been plagued by the use of highly specific substrates, the necessity for directing groups, and limited to fragmentation reactions. The intramolecular carboacylation reaction described herein, though requiring a directing group, is one of the few examples of C–C bond activation reactions that build molecular complexity without the need for high-energy starting materials. We envision that more processes based on the activation of unstrained C–C bonds will be developed based on the groundwork summarized in this chapter. The foundations for unconventional creativity have been laid for those willing to take up this challenge.

**Acknowledgement** The authors thank the National Science Foundation (CHE-115157), the donors of the Petroleum Research Fund (47565-G1), and Research Corporation for Science Advancement (Cottrell Scholar Award to CJD, 19985) for support of the work in this chapter carried out at Minnesota.

# References

1. Crabtree RH (1985) Chem Rev 85:245
2. Rybtchinski B, Milstein D (1999) Angew Chem Int Ed 38:870
3. Jun CH (2004) Chem Soc Rev 33:610
4. Murakami M, Matsuda T (2011) Chem Commun 47:1100
5. Korotvička A, Nečas D, Kotora M (2012) Cur Org Chem 16:1170
6. Ruhland K (2012) Eur J Org Chem 2683

7. Murakami M, Ito Y (1999) In: Murai S (ed) Activation of unreactive bonds and organic synthesis. Springer, New York, pp 97–129
8. Müller E, Segnitz A, Langer E (1969) Tetrahedron Lett 10:1129
9. Dermenci A, Whittaker RE, Dong G (2013) Org Lett 15:2242
10. Suggs JW, Cox SD (1981) J Organomet Chem 221:199
11. Bruce MI (1977) Angew Chem Int Ed 16:73
12. Suggs JW, Jun CH (1984) J Am Chem Soc 106:3054
13. Suggs JW, Wovkulich MJ, Cox SD (1985) Organometallics 4:1101
14. Suggs JW (1978) J Am Chem Soc 100:640
15. Cheng CH, Spivack BD, Eisenberg R (1977) J Am Chem Soc 99:3003
16. Adamson GW, Daly JJ, Forester D (1974) J Organomet Chem 71:C17
17. Hitchcock PB, Lappert MF, McLaughlin GM (1974) J Chem Soc Dalton Trans 68
18. Bombieri G, Graviani R, Panattoni C, Volponi L (1967) J Chem Soc Chem Commun 977a
19. Milstein D (1982) Organometallics 1:1549
20. Suggs JW, Jun CH (1986) J Am Chem Soc 108:4679
21. Flood TC (1981) Top Stereochem 12:37
22. Steenken S, Schushmann HP, von Sonntag C (1975) J Phys Chem 79:763
23. Greene FD (1959) J Am Chem Soc 81:2688
24. Ng FTT, Rempel GL, Halpern J (1982) J Am Chem Soc 104:621
25. Finke RG, Hay BP (1984) Inorg Chem 23:3041
26. Suggs JW, Jun CH (1985) J Chem Soc Chem Commun 92
27. Lee DY, Jun CH (2003) Bull Kor Chem Soc 24:1059
28. Wentzel MT, Reddy VJ, Hyster TK, Douglas CJ (2009) Angew Chem Int Ed 48:6121
29. Wang J, Chen W, Zuo S, Liu L, Zhang X, Wang J (2012) Angew Chem Int Ed 51:12334
30. Tobisu M, Kinuta H, Kita Y, Remond E, Chatani N (2012) J Am Chem Soc 134:115
31. Gribov DV, Pastine SJ, Schnürch M, Sames D (2007) J Am Chem Soc 129:11750
32. Murai S, Kakiuchi F, Sekine S, Tanaka Y, Kamatani A, Sonoda M, Chatani N (1993) Nature 366:529
33. Trost BM, Jiang C (2006) Synthesis 369
34. Christoffers J, Baro A (eds) (2005) Quaternary stereocenters. Wiley, Weinheim
35. Douglas CJ, Overman LE (2004) Proc Natl Acad Sci U S A 101:5363
36. Denissova I, Barriault L (2003) Tetrahedron 59:10105
37. Dreis AM, Douglas CJ (2009) J Am Chem Soc 131:412
38. Rathbun CM, Johnson JB (2011) J Am Chem Soc 133:2031
39. Singleton DA, Thomas AA (1995) J Am Chem Soc 117:9357
40. Frantz DE, Singleton DA, Snyder JP (1997) J Am Chem Soc 119:3383
41. Lutz PJ, Rathbun CM, Stevenson SM, Powell BM, Boman TS, Baxter CE, Zona JM, Johnson JB (2012) J Am Chem Soc 134:715
42. Brown JM, Cooley NA (1988) Chem Rev 88:1031
43. Espinet P, Echavarren AM (2004) Angew Chem Int Ed 43:4704
44. Denmark SE, Sweis RF (2004) J Am Chem Soc 126:4876
45. Jones GD, Martin JL, McFarland C, Allen OR, Hall RE, Haley AD, Brandon RJ, Konovalova T, Desrochers RJ, Pulay P, Vicic DA (2006) J Am Chem Soc 128:13175
46. Nakao Y (2012) Bull Chem Soc Jpn 85:731

Top Curr Chem (2014) 346: 111–162
DOI: 10.1007/128_2013_521
© Springer-Verlag Berlin Heidelberg 2014
Published online: 27 April 2014

# Catalysis of Diazoalkane–Carbonyl Homologation. How New Developments in Hydrazone Oxidation Enable the Carbon Insertion Strategy for Synthesis

**David C. Moebius, Victor L. Rendina, and Jason S. Kingsbury**

**Abstract** Diazo compounds continue both to challenge and to fascinate practitioners of chemical synthesis. The most strategically powerful and unique type of reactivity observed with these reagents is a formal insertion of the donor-acceptor carbon into C–C or C–H bonds alpha to carbonyl groups. Although the reaction does not involve discrete carbon–metal bonds, it can be catalyzed by metal-based Lewis acids. This chapter investigates both classical and modern developments in diazoalkyl carbon insertion with a special emphasis on nonstabilized nucleophiles.

**Keywords** α-Functionalization · C–C bond activation · Carbon insertion · Diazo compounds · Scandium

## Contents



D.C. Moebius
Onyx Pharmaceuticals, Inc., 249 E. Grand Avenue, South San Francisco, CA 94080

V.L. Rendina
Assistant Professor of Pharmaceutical Sciences, Massachusetts College of Pharmacy and Health Sciences, 179 Longwood Avenue, Boston, MA 02115

J.S. Kingsbury (✉)
Assistant Professor of Chemistry, California Lutheran University, 60 West Olsen Road, #3700, Thousand Oaks, CA 91360
e-mail: jkingsbu@callutheran.edu

# 1 Introduction

Synthesis of the first diazo compound dates back over 100 years and began with a preparation of ethyl diazoacetate (EDA) by Curtius [1]. This was followed by a synthesis of diazomethane by Pechmann [2]. Almost immediately, the power and value of these fascinating donor-acceptor molecules were acknowledged. The diazo reactive function is one that is *amphoteric*: able to react as a nucleophile in basic and neutral media or, alternatively, as an electrophile in acid. Among the first applications of a diazo compound to be discovered was a ketone synthesis under mild, neutral conditions [3, 4]. Many early reports focused on the reaction of diazo esters derived from glycine. As shown in Scheme 1, aldehyde electrophiles undergo formal carbon insertion into the C–H bond. The mechanism, though not known at the time, involves nucleophilic addition of the diazoalkyl carbon to furnish a tetrahedral alkoxide (the diazonium betaine) and subsequent 1,2-migration with concerted loss of nitrogen. It is widely believed that steric factors orient the diazoalkane during addition to the carbonyl [3, 4]. The preferred migrating group – in this case hydride – is that which is smallest and situated anticoplanar to the leaving group. A general and efficient Sn-catalyzed variation of this approach to β-keto esters is now known as the Roskamp synthesis [5, 6].

The nucleophilicity, and thus reactivity, of diazoalkanes is highly dependent upon substitution, with a potential for further delocalization of the lone electron pair. Careful kinetic experiments by Mayr and coworkers have established a series of relative diazoalkane nucleophilicity parameters (Fig. 1) [7, 8]. At the more reactive end of the spectrum, the nucleophilicity of diazomethane was found to be comparable to an enamine functional group. At the opposite end of the spectrum, diethyl 2-diazomalonate was found to have a nucleophilicity similar to styrene.

With this scale serving as a general guideline, diazoalkanes may be classified into two primary categories. Those referred to as "stabilized" diazoalkanes contain an adjacent carbonyl, phosphoryl, or sulfonyl moiety ($N < 5$). Several of these compounds have become commercialized (Fig. 2), and their use has been extensively reviewed (for lead references, see [9–11]). With the exception of (trimethylsilyl)diazomethane (TMSD, **1**), all commercially available diazoalkanes are flanked by either one or two electron-withdrawing carbonyl groups. This

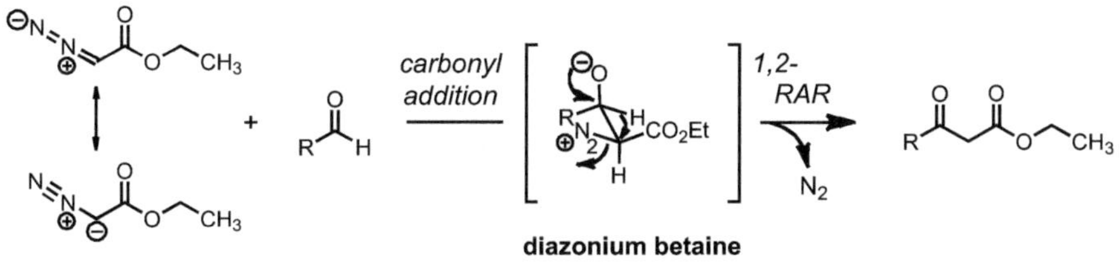


**Scheme 1** Diazoalkyl carbon atoms undergo net insertion into the α C–H bond of aldehydes in two stages through addition and 1,2-rearrangement

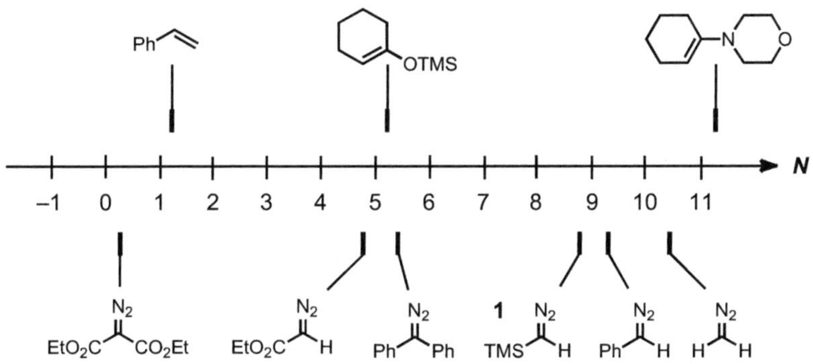


**Fig. 1** Some common diazo compounds overlayed on the nucleophilicity scale ($N$) defined by Mayr and coworkers

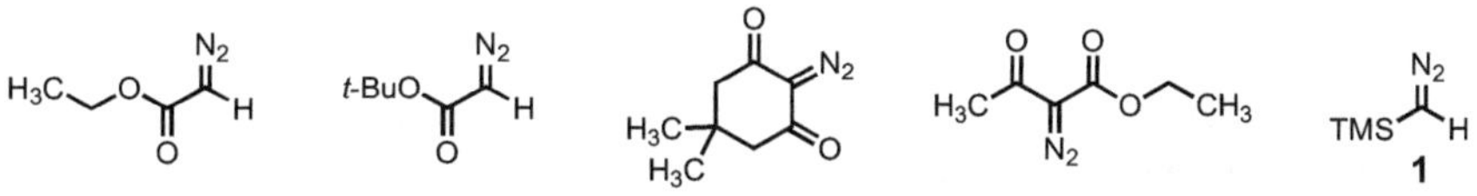


**Fig. 2** Commercially available diazoalkanes, save one exception, are stabilized

distinguishes them from their more reactive, noncarbonyl-stabilized counterparts. As in many domains of preparative chemistry, a trade-off exists. "Non-stabilized" diazo compounds are comparably underexploited as carbon nucleophiles in synthesis because of greater toxicity and instability. Nowhere is the danger more explicit than in the case of diazomethane, whose low molecular weight and high volatility create an explosion hazard (informative details are available in Aldrich Technical Bulletin #AL-180: *Diazald®, MNNG, and Diazomethane Generators*. This material is freely available under the "Related Information" tab of the Search command at http://www.sigma-aldrich.com. Explosion of a supersaturated, crystallizing solution was once observed [12, 13]). Regrettably, and in spite of studies published for both ethyl diazoacetate (EDA) [14, 15] and diazomethane [16], thermal decomposition data is unavailable for a wide range of other diazo compounds, most of which would be considered more synthetically useful.

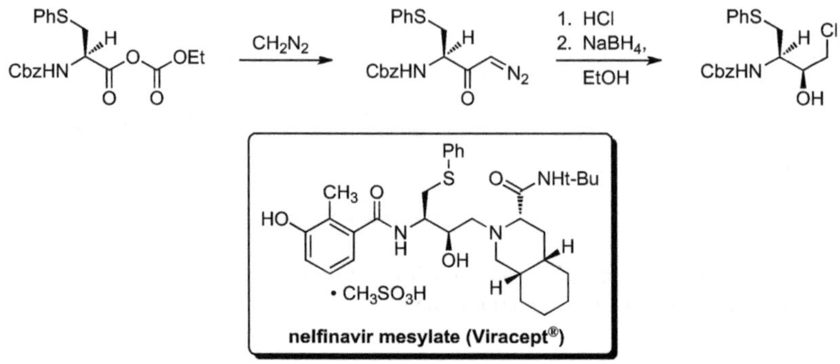


**Scheme 2** Hydrochlorination of a stereochemically labile α-chiral diazoketone made by acyl substitution in the production-scale synthesis of nelfinavir mesylate

Fortunately, however, pervasive concern over stability and toxicity has done little to stunt a universal interest in diazo compounds within the community of synthetic chemists. Beyond the carbon insertion concept introduced in Scheme 1, non-stabilized diazoalkanes smoothly react with carboxylic O–H bonds, making them atom-economical reagents for esterification [17, 18]. They are also capable of etherification with alcohols [19–30] and phenols [2, 31]. The diazoalkyl 1,3-dipole will participate in cyclization reactions with both alkynes [32–40] and electron-deficient alkenes [3, 41]. In addition, diazoalkanes are the precursors used in many reactions that generate metallocarbenes and carbenoids [42, 43], such as sulfur-mediated aziridination [44, 45], epoxidation [46], and cyclopropanation [45]. Despite the wide and demonstrated benefit of these neutral donor-acceptor carbon sources, the chemical and pharmaceutical industries have been slow to apply the technologies on the large scale. Recently, however, patent filings [47–49] show a willingness to design and implement novel practices for synthesis and consumption of diazoalkane reagents. Continuous processes [49] allow for the manufacture of multi-ton quantities over extended periods of time while real-time inventories are maintained at safe, secure levels. As an example, Phoenix Chemicals, Ltd. has been able to execute a critical homologation/chlorination/reduction sequence toward the anti-HIV protease inhibitor Viracept® (nelfinavir mesylate) at a level of 200 metric tons of diazomethane per year [16] (Scheme 2). Although, in principle, a myriad of other strategies could have achieved the required structural outcome, in reality, none surpassed the efficiency and neutrality ensured by the illustrated use of diazomethane.

Alongside shifting attitudes in the private sector have come a number of dramatic changes in the way chemists prepare non-stabilized diazoalkanes (for a comprehensive review on the synthesis of diazo compounds in general, see [50, 51]). Foremost among these are new developments in hydrazone dehydrogenation [17, 52–54]. Improved oxidants and solvent media now define methods capable of delivering pure solutions of diazoalkanes for titration and manipulation under controlled conditions. The result has been a revival of interest in methodologies based on simple aryl- and alkyl-substituted diazomethanes. Time, together with a relentless drive in favor of “ideal” syntheses [55], will tell whether challenges in handling these innately unstable reagents will be muted by their power and versatility, especially in the generation of molecular complexity.

This review chapter begins with a historical account of the most significant developments in non-stabilized diazoalkane chemistry, with specific emphasis on carbonyl homologation in the context of ring enlargement. Its classification with other forms of C–C bond activation is unmistakable the moment the reader appreciates the simplicity and brevity of this approach to α-substituted carbonyls. From proton-promoted to Lewis acid-mediated reactions, a foundation for the recent discovery of catalytic turnover in diazoalkyl addition/rearrangement will be presented. The discussion then turns to the core diazotization procedures that have made the methodological advances possible, culminating in a survey of modern carbon insertion processes that are highly efficient and stereocontrolled. Literature coverage will focus primarily on reactions that rely on noncarbonyl-stabilized diazo compounds, but, whenever possible, parallel developments derived from the application of α-diazocarbonyl compounds will be included.

## 2 History of Diazoalkane Ring Expansion and Chain Extension Reactions

The reaction of diazoalkanes with carbonyl-containing compounds dates back to observations made by Buchner and Curtius as early as 1885 [56]. Although others examined this novel reactivity pattern [57, 58], Schlotterbeck is largely credited with uncovering the merger of aldehydes and diazoalkanes in 1907 [59, 60]. Schlotterbeck was able to confirm by careful experimentation that various aliphatic aldehydes afforded the corresponding methyl ketones when treated with diazomethane. The reaction of aldehydes with diazomethane to form methyl ketones (Scheme 3) later became known as the Buchner–Curtius–Schlotterbeck reaction [61]. Application of this chain elongation to ketones and eventually cycloalkanones did not come until several decades later and required a critical new discovery.

### *2.1 Protic Solvent-Promoted Reactions*

In 1928, Meerwein recorded one of the first reactions of diazomethane with a ketone promoted by the presence of a hydroxylic solvent [62, 63]. When acetone was treated with diazomethane, no reaction was observed. However, in the presence of water or alcohols, dimethylethylene oxide and 2-butanone were readily produced (Scheme 4). This important new discovery could be rationalized by invoking a model based on general acid catalysis. Since the mechanism involves initial rate-limiting addition of diazomethane to the carbonyl group, methanol facilitates the reaction by hydrogen bonding to the incipient alkoxide, thereby enhancing the electrophilicity of the carbonyl acceptor. Under these conditions, the homologation

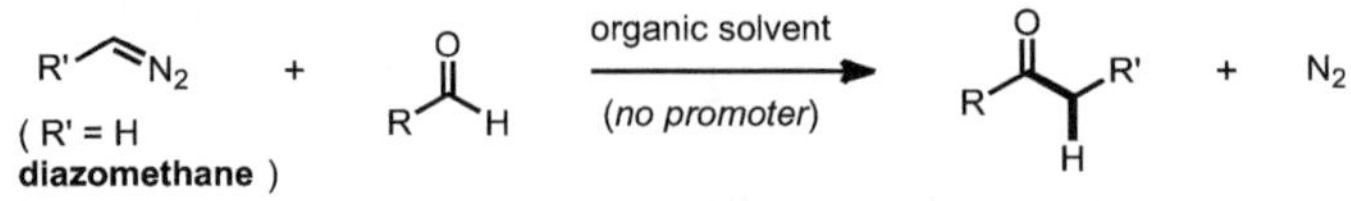


**Scheme 3** The Buchner–Curtius–Schlotterbeck reaction (methyl ketone synthesis)

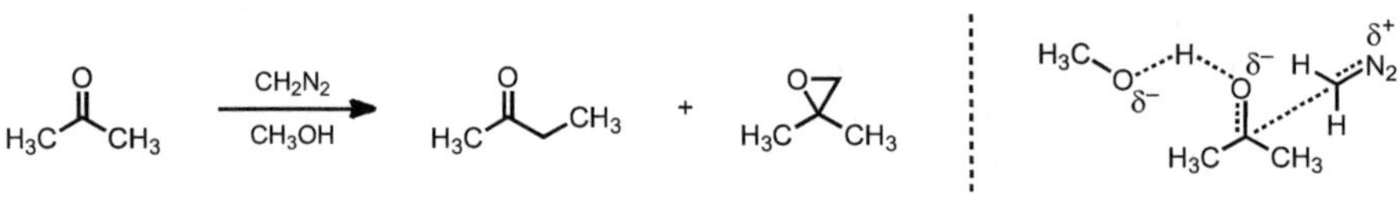


**Scheme 4** Discovery of rate acceleration by hydroxylic solvents

reaction can intercept the same hydroxy diazonium intermediate formed upon treatment of β-amino alcohols with nitrous acid in the related Tiffeneau–Demjanov rearrangement (for a review, see [64]. For a representative total synthesis application, see [65]). It is of interest to consider whether epoxide byproducts become a general, yield-limiting drawback to the use of protic solvents, since they derive from internal, 3-*exo*-tet displacement within the tetrahedral betaine intermediate as opposed to the typical 1,2-migratory shift.

After Meerwein's critical discovery of proton-based rate acceleration, Mosettig [66] reported the first carbocyclic ring expansions (Heller had observed the production of dihydroxyquinoline from isatin several years prior to Mosettig's work [67, 68]). Cyclohexanone, when treated with excess diazomethane in ethereal solvents, was totally unreactive (as evidence of greater electron density at the diazoalkyl carbon, a subsequent report indicated that cyclohexanone would undergo ring expansion with diazoethane in the absence of methanol to furnish 2-methylcycloheptanone as the primary product [69]). Upon the introduction of methanol, nitrogen gas evolved vigorously and the production of cycloheptanone, cyclooctanone, and an epoxide isomeric with cycloheptanone was observed (Scheme 5). When the same reaction was performed with cyclopentanone ($n = 0$) as the substrate, cycloheptanone and cyclooctanone were again the primary products. Residual cyclopentanone and cyclohexanone could not be detected, thus indicating complete consumption of the starting material and subsequent homologation of the intermediate cyclohexanone. 1,2-Addition of diazomethane to cyclopentanone increases torsional strain by introducing an additional $sp^3$ hybridized center within the confined ring system. Cyclohexanone is universally regarded as being more reactive due to the staggered nature of its bonds in the low energy chair conformation [3, 4]. This early example serves to illustrate three fundamental challenges associated with the diazoalkane–carbonyl homologation reaction: (1) controlling the ring size is difficult when the products are more reactive than the starting materials – the products formed possess an identical functional group poised for further reaction, (2) formation of oxirane byproducts can be unavoidable, and (3) an excess of diazomethane is typically employed to offset its background decomposition via a formal O–H insertion with the Brønsted acidic promoter (typically present as a cosolvent in excess).

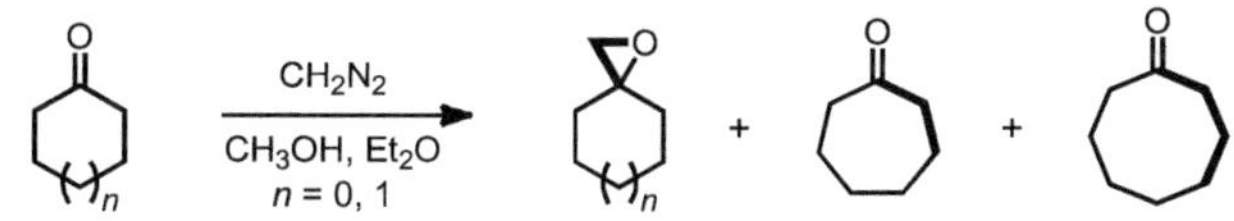


**Scheme 5** First example of carbocyclic ring expansion using diazomethane

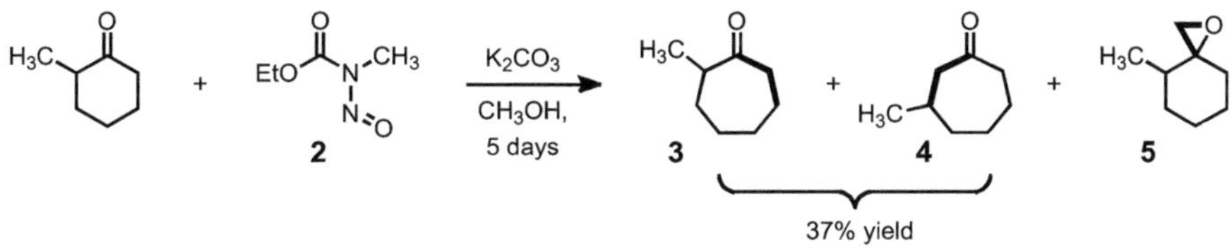


**Scheme 6** First ring expansion of a 2-substituted cycloalkanone

Like Mosettig's first disclosure, a subsequent investigation (a number of medium ring cyclic ketones were prepared on kilogram scale according to Meerwein's protocol [63]; see also [70]) was limited in scope to symmetrical cycloalkanones. It was nearly a decade later when Adamson and Kenner reported homologation of 2-methylcyclohexanone with diazomethane (Scheme 6) [71]. Generation of the nucleophile in situ from *N*-nitrosomethyl-urethane **2** in the presence of 2-methylcyclohexanone afforded both possible regioisomers of ring expanded product (**3** and **4**) in a combined 37% yield along with an equal amount of epoxide **5**. The 2- and 3-substituted cycloheptanones were separated and positively identified after selective formation of a crystalline bisulfite adduct; however, the regioisomeric ratio was not clearly reported [71].

In 1949, Gutsche began to examine carefully the regiochemical outcome when various 2-aryl cyclohexanones were exposed to diazomethane [72, 73]. As in the study by Adamson and Kenner, three outcomes were possible: formal methylene insertion into the more substituted α C–C bond, insertion into the less substituted α C–C bond, or epoxidation via addition and $S_N2$ closure. Gutsche hypothesized that upon introduction of electronic perturbations at the substituted α carbon, the more electron-rich group would migrate preferentially. The results of his findings, along with the corresponding Hammett $\rho$ values [74], are summarized in Table 1.

It was anticipated that entry 5 (G = *p*-Cl) would show the highest levels of regioselectivity, with preferential migration of the less substituted (α′) carbon. Entry 4 (G = 2,3,4-$OCH_3$) was expected to show the lowest levels of regiocontrol, or potentially an inversion of selectivity, favoring migration of the aryl substituted carbon. Unfortunately, the data were inconclusive and attempts were made to rationalize the results. The highest level of regioselectivity was observed for entry 1 (G = H), not entry 5 (G = *p*-Cl). The lowest level of selectivity was observed in entry 4 as expected, but regardless, there appeared to be little difference between the values in each entry. Gutsche proposed that three factors were significant in determining which C–C bond would migrate during breakdown of the intermediate diazonium betaine: (1) the relative electron-releasing ability of the two respective α carbon atoms, (2) the strain involved in the transition state for 1,2-rearrangement,

**Table 1** An early regiochemical investigation by Gutsche and coworkers

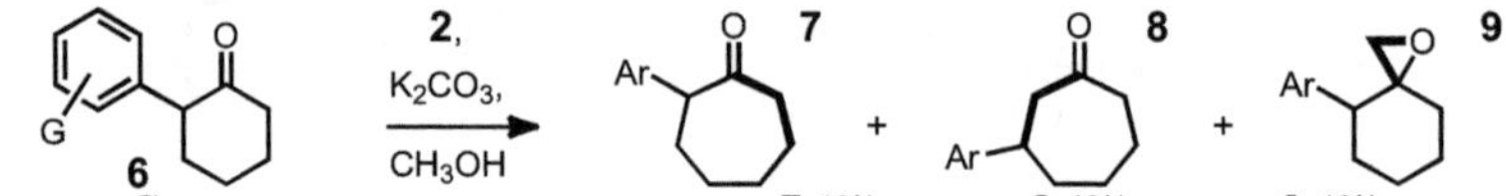

| Entry | G | $\rho$ | **7** (%) | **8** (%) | **9** (%) | rr (**7**:**8**) |
|---|---|---|---|---|---|---|
| 1 | H | 0 | 59 | 14 | 21 | 4.2:1 |
| 2 | *p*-$CH_3$ | −0.170 | 55 | 20 | 21 | 2.8:1 |
| 3 | *p*-$OCH_3$ | −0.268 | 57 | 21 | 14 | 2.7:1 |
| 4 | 2,3,4-$OCH_3$ | NA | 40 | 28 | 18 | 1.4:1 |
| 5 | *p*-Cl | +0.227 | 45 | 20 | 26 | 2.2:1 |

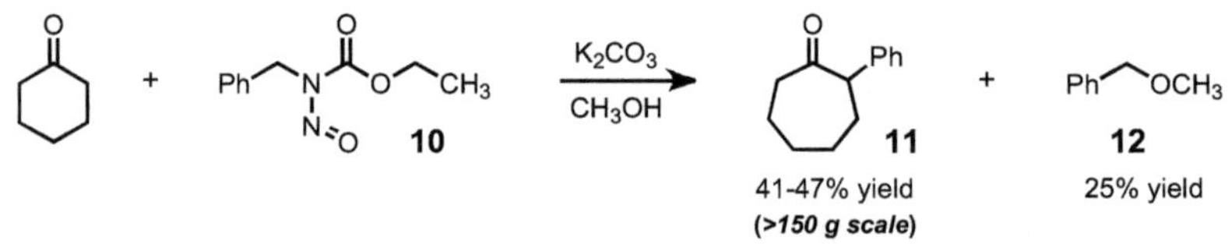

**Scheme 7** Large-scale preparation of 2-phenylcycloheptanone in a basic medium

and (3) the steric environment around the diazonium cation after carbonyl addition. Gutsche concluded that the reactions were largely insensitive to electronic perturbations at the aromatic ring and that the observed selectivities were the result of counterbalancing each of these factors. In general, however, there was a strong intrinsic regiochemical preference for migration of the less substituted (smaller) ketone substituent, irrespective of electronic perturbations (Baeyer–Villiger and Criegee oxidations, which are pertinent as *oxygen* insertion reactions, typically display the opposite regiochemical preference with unsymmetrical ketones – selective migration of the more substituted α carbon atom; see [75–77]).

Gutsche was also a pioneer with respect to diversification of the nucleophile, reporting some of the first examples of protic solvent-mediated reactions with monosubstituted diazomethanes (Scheme 7) [78–80]. Although a number of examples were disclosed, the most striking case was a large-scale preparation of 2-phenylcycloheptanone (**11**) by the in situ generation of phenyldiazomethane from ethyl *N*-nitroso-*N*-benzylcarbamate (**10**). The yield was only moderate, but over 150 g of product were obtained in a single run. In addition to the desired carbon insertion product, methyl benzyl ether (**12**) was recovered in 25% yield, highlighting one of the serious problems associated with the use of protic solvents to enhance the efficiency of carbonyl homologation.

Building upon Gutsche's work aimed at the elucidation of migratory aptitudes, Greene later found that α,α-dichlorocyclobutanones furnished products resulting from preferential migration of the more electron-rich C–C bond (Scheme 8) [81]. The common epoxide byproduct was not observed, perhaps owing to ring strain involved in constructing a [2.3] spirocyclic system (Jaz made a similar observation during the ring expansion of cyclobutanone; see [82]). Greene also

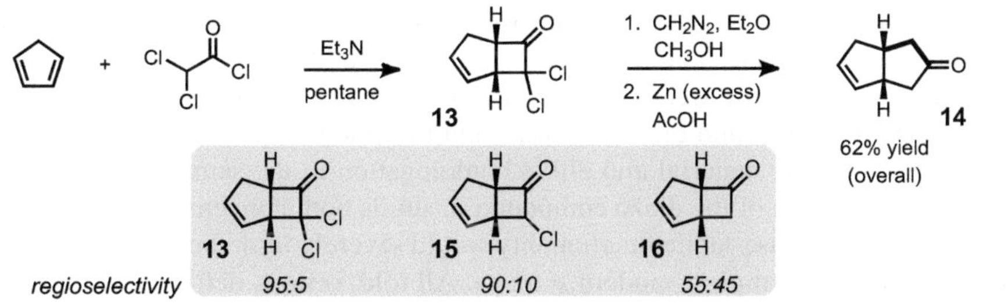


**Scheme 8** High levels of regiochemical control with α,α-dichlorocyclobutanones

observed significant rate acceleration for the electron deficient cyclobutanones, consistent with carbonyl addition being rate-limiting. The rate enhancement was attributed to carbonyl-π electron donation into the adjacent C–Cl σ* orbital and increased polarization of the C–O bond through inductive effects. In this system, the electronics of the cyclobutanone had a remarkable impact on the regioselectivity. The *des*-chloro cyclobutanone **16** provided a 55:45 mixture of regioisomers, slightly favoring the production of **14**.[1] With a single chlorine atom (**15**), a 90:10 ratio was observed. The best selectivity was observed with **13**, which gave predominantly the β-substituted ketone **14** in a 95:5 regioisomeric ratio after reductive dehalogenation. A more recent investigation in the Hegedus laboratories [84] has led to excellent regioisomeric ratios (up to >98:2) favoring methylene migration in the case of α-methoxy cyclobutanone electrophiles that also bear an electron-withdrawing substituent at the C3 position. Noteworthy is that an alcohol additive was not required to ensure smooth conversion with these substrates. Presumably, the inherent ring strain and lack of α′ substitution was the root cause of higher reactivity relative to the [3.2.0]-bicycloheptanone ring expansions explored by Greene.

## 2.2 *Lewis Acid-Promoted Reactions*

Recourse to an alcohol cosolvent was the premier means of accelerating diazoalkane-based ring expansion reactions for more than half a century. Alkyl diazomethanes were prepared by base-catalyzed decomposition (for further

[1] Interestingly, Tiffeneau–Demjanov rearrangement provides mostly the α-ketone product in an 85:15 ratio of regioisomers (determined by IR spectroscopy); see [83])

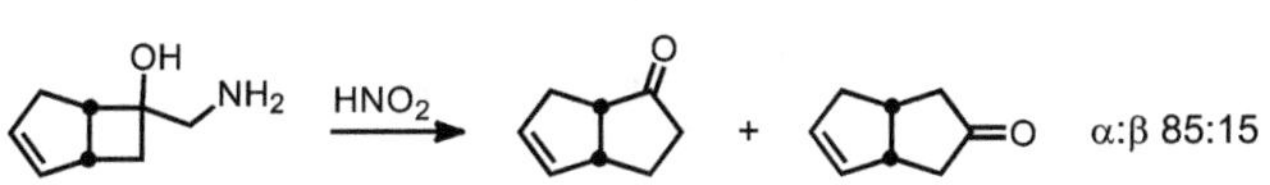

discussion, see [3, 4, 85, 86]) of *N*-nitrosoalkylureas, -urethanes, or -sulfonamides, and aryl diazomethanes could be accessed by the Bamford–Stevens reaction [87]. Methanol and ethanol were typically the best choice for solubilizing both the diazalkane precursor and the alkali base, and this encouraged many simply to add the desired starting material and effect homologation in the same medium. While direct consumption of the diazo compound in situ is both convenient and practical, an intolerance of base-labile functionality would severely limit the suitability of this strategy in more complex, modern settings. All told, serious deficiencies detracted from the preparative value of proton-promoted carbon insertion. As discussed above, the reactions suffered from sluggish reaction rates, O–H insertion, epoxide byproducts, multiple homologations, regiochemical issues, and poor efficiencies with more sterically crowded internal (doubly substituted) diazoalkanes. Early mechanistic data pointed to the carbonyl addition step that forms the diazonium betaine as rate-limiting. In principle, a stronger proton source would serve as an even better activator, but stronger Brønsted acids were long known to decompose diazoalkanes rapidly [3, 4]. Further development of the transformation required discovery of a new class of promoter.

Knowing that alcohol solvents were problematic, and aware of the accepted mechanism, Müller and coworkers screened a variety of Lewis acids and showed the halogen salts of Al, Zn, Ti, and Zr to be marginally effective promoters [88]. It was House, however, who developed the first broadly applicable reactions of diazomethane with ketones [89]. A prior report had indicated that diazomethane would undergo rapid decomposition to form polymethylene and fluoromethyl boron difluoride when treated with boron trifluoride.[2] Despite this precedent, by premixing $BF_3 \cdot Et_2O$ and a solution of the appropriate ketone prior to the addition of nucleophile, House recorded dramatic improvements in reaction efficiency relative to protic solvent-based reactions (Table 2). Products that had previously taken days to form with methanol as the promoter were now forming in minutes. The reaction of diazomethane with the neopentyl acceptor pinacolone was completely unsuccessful in methanol (entry 9), yet it proceeded smoothly with stoichiometric quantities of $BF_3 \cdot Et_2O$ in diethyl ether (entry 10). Epoxy byproducts were not found in the unpurified reaction mixture for any run. However, coproduction of aldehydes from said epoxides through a Lewis acid-mediated rearrangement pathway was detected in two cases for the more hindered ketones (entries 8 and 10). House undertook a careful study of the regiochemical outcome and compared that directly with data obtained from methanol-promoted reactions. For acyclic ketones, a moderate preference was observed for carbon insertion on the

[2] See [90].

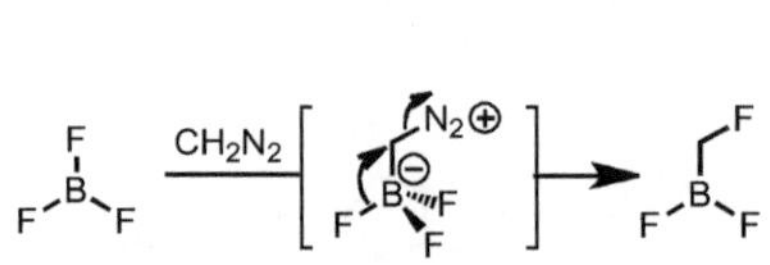

**Table 2** Regiochemical studies by House and the onset of Lewis acid promotion

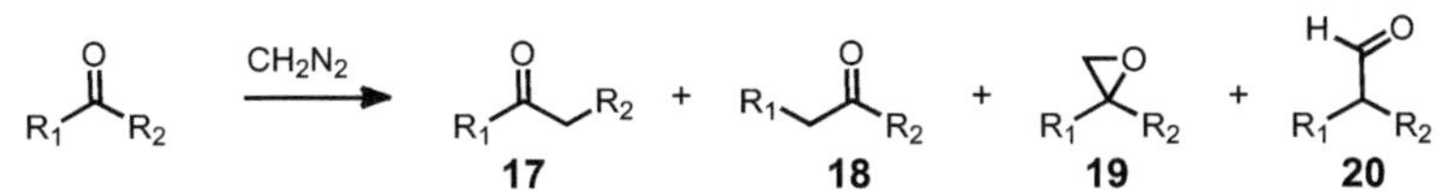


| Entry[a] | $R_1$ | $R_2$ | Time | Promoter | % Conv.[b] | **17**:**18**:**19**:**20**[c] |
|---|---|---|---|---|---|---|
| 1 | Ph | $CH_3$ | 4 day | $CH_3OH$ | 55.8 | 4:69:27:0 |
| 2 | Ph | $CH_3$ | 2 min | $BF_3 \cdot Et_2O$ | 36.3 | 22:78:0:0 |
| 3 | Bn | $CH_3$ | 3 day | $CH_3OH$ | 65.4 | 32.5:20.5:47:0 |
| 4 | Bn | $CH_3$ | 2 min | $BF_3 \cdot Et_2O$ | 36.5 | 78.5:21.5:0:0 |
| 5 | Pr | $CH_3$ | 3 day | $CH_3OH$ | 25.0 | 33:34:33:0 |
| 6 | Pr | $CH_3$ | 4 min | $BF_3 \cdot Et_2O$ | 19.0 | 50.5:49.5:0:0 |
| 7 | *i*-Pr | $CH_3$ | 1 day | $CH_3OH$ | 4.9 | 65.5:34.5:0:0 |
| 8 | *i*-Pr | $CH_3$ | 2 min | $BF_3 \cdot Et_2O$ | 6.8 | 46:22.5:0:31.5 |
| 9 | *t*-Bu | $CH_3$ | – | $CH_3OH$ | 0 | nd |
| 10 | *t*-Bu | $CH_3$ | 2 min | $BF_3 \cdot Et_2O$ | 0.8 | 44:15.5:0:40.5 |

[a]Conditions: run in $CH_3OH$ or with diethyl ether as solvent with 1.0 equiv. $BF_3 \cdot Et_2O$
[b]Determined from recovered starting material
[c]Determined by gas chromatography

less sterically demanding side of the carbonyl. These observations were consistent with Gutsche's regiochemical studies discussed above for α-aryl-substituted cycloalkanones [72, 73]. In House's studies, reactions were run to a low level of conversion to avoid complications arising from multiple homologation. Regardless of that limitation, the notable enhancement in reaction rate opened the door to further investigations and an expanded substrate scope. The use of Lewis acids also paved the way for ring expansion reactions with the less nucleophilic carbonyl-stabilized diazoalkanes, allowing facile access to ring-expanded β-keto ester products [91].

Another major advance in diazoalkane-based ring expansion chemistry came about with Shioiri's introduction of (trimethylsilyl)diazomethane (**1**) in 1980 [92]. With prior reactions often plagued by the problem of overhomologation, the new reagent served to mitigate the issue by allowing for Lewis acid-promoted Brook rearrangement (reviews [93, 94]. For the reaction of acyl silanes with diazomethane, see [95]. For a retro-Brook rearrangement that is synthetically valuable, see [96]) of the expected α-trimethylsilyl ketone product. 1,3-Migration of silicon from carbon to oxygen converts the keto silane to a trimethylsilyl enol ether, which lacks a carbonyl function and cannot undergo further reaction. As an example of this, consider the illustrated transformation of fluorenone (**21**) to phenanthrenyl silyl ether **22** under optimized conditions (Scheme 9). Desilylation occurs readily upon quenching with water, providing the same product one would obtain using diazomethane, albeit with a potential for higher regioselectivity if the ketone were dissymmetric. Aside from mediating internal Brook rearrangement as a traceless form of protection, the Lewis acid serves to counteract the lower

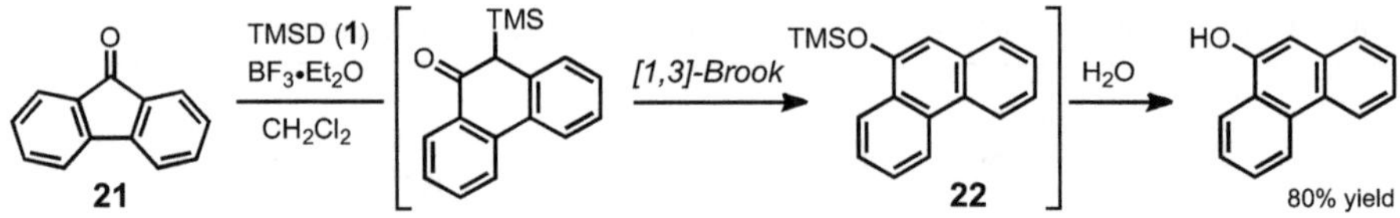

**Scheme 9** Facile 1,3-Brook reorganization of the anticipated α-keto silane adduct

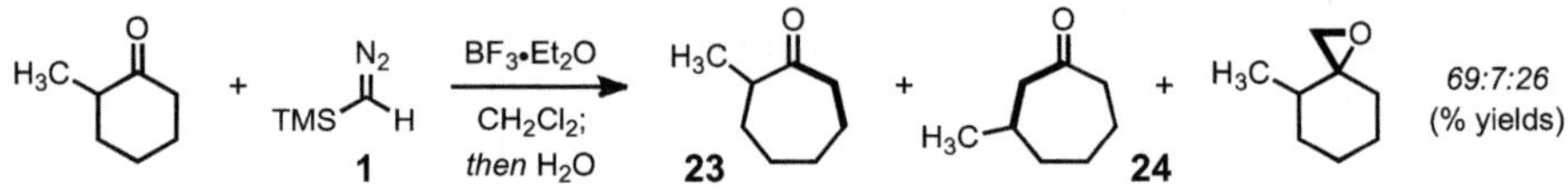

**Scheme 10** Use of (trimethylsilyl)diazomethane as a substitute for diazomethane

nucleophilicity of TMSD relative to diazomethane. Shioiri found that the highest efficiencies were achieved when $BF_3 \cdot Et_2O$, as formerly described by House, was used in the non-coordinating solvent dichloromethane. Attempts to use ethereal solvents gave lower chemical yields of the target compounds.

A second representative example of Shioiri's homologation protocol is shown in Scheme 10. When 2-methylcyclohexanone was treated with 1.5 equiv. $BF_3 \cdot Et_2O$ and 1.5 equiv. TMSD (**1**) in dichloromethane for 4 h at −15° C, 2- and 3-methylcycloheptanone were formed in a 10:1 regioisomeric ratio (→**23** and **24**). The 2-methyl regioisomer **24**, derived from 1,2-shift of the unsubstituted carbon, was isolated in 69% yield. This represents a marked improvement over Adamson and Kenner's previous effort, which netted only a 37% combined yield of 2- and 3-methylcyclohexanone after 5 days with methanol as the promoter. The regiochemical outcome is in accord with previous reports in the literature, underscoring an intrinsic preference for migration of the smaller α carbon atom, regardless of the chosen promoter or diazoalkane. At the time TMSD was introduced, it was praised for its greater stability and safety profile relative to diazomethane. While it is true that TMSD is thermally stable in inert solvents and has since become commercialized, it should be regarded as being equally toxic, and great care must be exercised to avoid inadvertent protodesilylation of the reagent (for a note on the safety of TMSD, see [97]). At least two chemists have died as a result of lung failure following exposure to TMSD in the absence of appropriate ventilation [98].

Although the introduction of TMSD presented advantages over the traditional diazomethane-based homologations, there was still ample room to improve product distributions and to uncover more effective, mild promoters. Yamamoto and coworkers made remarkable strides by evaluating various aluminum-based Lewis acids [99, 100] (an earlier report by Müller and Bauer had discussed the effect of $AlCl_3$ [101]). When cyclopentanone was treated with TMSD (**1**) under Shioiri's optimal conditions, a modest 35% net yield was obtained with a poor product distribution (64% cyclohexanone, 23% cycloheptanone, 10% cyclooctanone, and 3% epoxide). As illustrated in Scheme 11, a switch to trimethylaluminum as

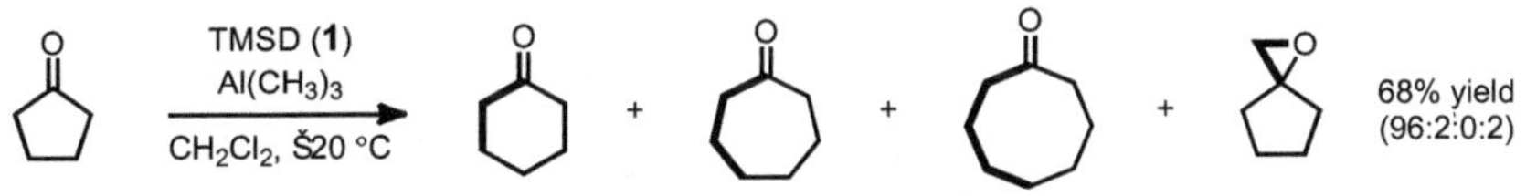


**Scheme 11** Improved product distributions with aluminum-based Lewis acids

promoter provides a much higher 68% overall yield with a greatly improved product distribution (96% cyclohexanone). In a manner analogous to the use of boron-based Lewis acids, trialkylaluminums were previously reported to furnish decomposition products when treated with diazomethane.[3] Yamamoto found it essential to pre-mix cyclopentanone and the aluminum reagent in order to ensure a productive merger with the diazo compound. One detail that should not be overlooked is the minor but nonetheless detectable amount (2% mass balance) of the doubly homologous cycloheptanone in this example. It can only be formed if the predominant cyclohexenyl TMS ether undergoes Lewis acid-mediated hydrolysis or protodesilylation before complete consumption of starting material. It is possible that trace amounts of adventitious moisture were present under the reaction conditions, or, alternatively, that any unreacted diazoalkane was not effectively quenched prior to the aqueous workup. Cha and coworkers have also employed trimethylaluminum to great effect in 4→5 carbon ring expansion, but here again no deliberate attempt was made to validate the intermediacy of the silyl enol ether prior to its routine hydrolysis to the corresponding ketone [103].

Although trimethylaluminum has proven to be an effective promoter with TMSD, reactions with diazomethane have afforded less desirable product distributions. To improve further both efficiency and generality, Yamamoto began modifying the steric and electronic environment around the aluminum center. When the bulky reagent methyl aluminum bis(2,6-di-*tert*-butyl-4-methyl-phenoxide (MAD)[4] was introduced as a promoter, excellent yields with minimal byproducts derived from overhomologation or epoxidation were achieved [99]. Specifically, the union of 4-*tert*-butyl cyclohexanone and diazomethane took place cleanly with MAD at low temperature, affording results that were far superior to those obtained with either trimethyl- or triisobutylaluminum. As shown in entry 4 of Table 3, a 95% yield of all products was obtained with the desired singly homologated cycloheptanone **25** accounting for 84% of the recovered material.

In a conscious effort to broaden the scope of Al-promoted carbon insertion, Yamamoto and coworkers also tested their MAD reagent using a small number of

[3] See [102].

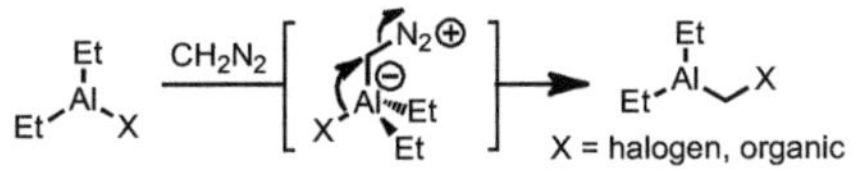


[4] Readily prepared by treating trimethylaluminum with two equivalents of BHT; see [99].

**Table 3** Highly selective result with the bulky, oxophilic methylaluminum MAD

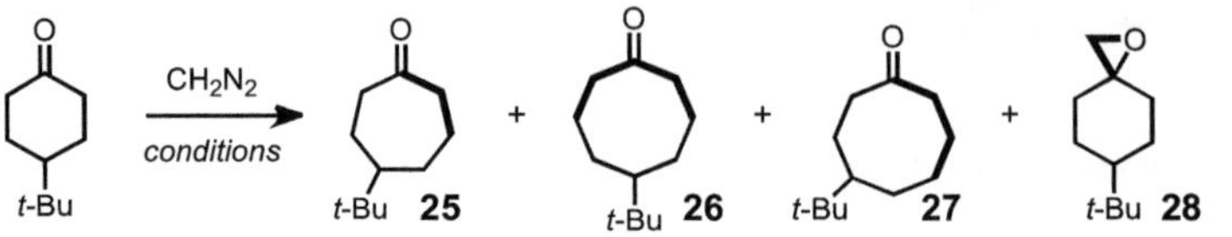


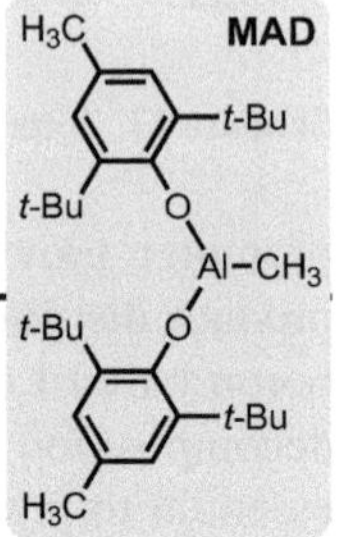


| Entry | Promoter | Solvent | Temperature (°C) | Yield (%) | **25:26:27:28** |
|---|---|---|---|---|---|
| 1 | $CH_3OH$ | $Et_2O$ | 0 | 63 | 50:25:25:0 |
| 2 | $i$-$Bu_3Al$ | $CH_2Cl_2$ | −78 | 68 | 54:22:22:2 |
| 3 | $(CH_3)_3Al$ | $CH_2Cl_2$ | −78 | 70 | 66:15:15:4 |
| 4 | MAD | $CH_2Cl_2$ | −78 | 95 | 84:3:3:10 |

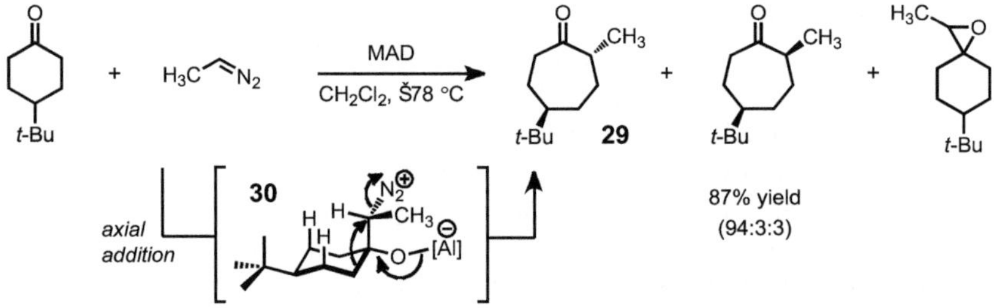


**Scheme 12** Desymmetrization via bond-selective insertion of diazoethane into 4-*tert*-butylcyclohexanone

substituted diazoalkanes. With diazoethane and the former crystalline "workhorse" substrate 4-*tert*-butyl cyclohexanone, which contains a prochiral (pro-stereogenic) center, Yamamoto recorded the first stereoselective diazoalkyl insertion reactions by desymmetrization [99]. As shown in Scheme 12, use of 1.2 equiv. MAD led predominantly to the *trans*-cycloheptanone **29** in 82% yield (87% combined) and with >30:1 diastereomeric ratio (dr).[5] The high level of diastereoselectivity can be explained by a model involving axial approach of diazoethane in an orientation that situates the small diazoalkyl proton out over the six-membered ring (see **30**). A least motion collapse of the antiperiplanar C–C bond, assuming no free rotation once the nucleophile has added, correctly predicts the major *trans* stereoisomer. Applying the same analysis with equatorial approach of the diazoalkane leads to the minor *cis* diastereomer.

[5] The *cis*/*trans* configuration of 2-methyl-5-*tert*-butylcycloheptanone was established following equilibration in methanolic $NaOCH_3$.

## 2.3 Lewis Acid-Catalyzed Reactions

The aforementioned work of House [89] and Shioiri [92] convincingly shows that diazoalkyl carbon insertion reactions are effectively promoted by stoichiometric quantities of $BF_3 \cdot Et_2O$. However, background decomposition of the nucleophile is a competing problem when using this agent. Indeed, nearly all known examples of carbocyclic ring expansion are superstoichiometric with regard to both diazoalkane and promoter. In the realm of understudied α,β-unsaturated ketone substrates, Johnson *et al.* were the first to achieve turnover with substoichiometric amounts of fluoroboric acid or boron trifluoride. A threefold to fivefold excess of diazomethane was still needed in order to counteract decomposition (for details, see [104]). In Yamamoto's later work with aluminum-based Lewis acids, catalytic turnover was never observed, perhaps due to the high oxophilicity of aluminum [99]. For over a decade, Yamamoto's work remained state of the art. Notwithstanding the fact that true catalysis was elusive, Yamamoto's thorough study set a benchmark, showcasing some of the most selective diazoalkane ring expansions ever reported.

In 2007, experimentation in the Kingsbury group began with a search for a broadly applicable and tunable catalyst for diazoalkane–carbonyl ring expansion [105]. A range of modified aluminum- and boron-based Lewis acids were evaluated in keeping with the literature precedents, but catalytic turnover was not observed for the merger of phenyldiazomethane and 4-*tert*-butylcyclohexanone [106]. A survey of chiral proton sources was also carried out, with still less than optimal results. A readily available TADDOL system was the best of the H-bond promoters tested, but even at 40–50 mol% loadings, catalytic turnover was still in question (due to the presence of two hydroxyl groups), and long reaction times (4–5 days) were needed for the formal benzyl insertion event to reach completion [105]. At last, a cursory screen of commercial rare earth metal complexes was conducted and led to a gratifying discovery: the tris(triflate) salts of scandium, yttrium, lanthanum, samarium, europium, and ytterbium all gave clean and rapid conversion to *trans*-2-phenyl-5-*tert*-butylcycloheptanone in less than 20 min (10 mol % loadings). $Sc(OTf)_3$ was selected for further experimentation because of its proven utility in asymmetric catalysis, especially in the context of readily available chiral tridentate ligands (for a lead reference, see [107]; for a review, see [108]). Scheme 13 shows a preliminary reaction scope for aryl, alkyl, and mixed aryl-alkyl diazomethanes in the context of 4→5 carbon ring expansion. Cyclobutanone was an ideal substrate for initial reaction development and optimization, since the shift in the C=O stretch as well as decay of diazo absorptions were easily discernable with direct, ReactIR monitoring of reactions. This new catalytic ring expansion method does not produce any byproducts derived from epoxidation or overhomologation. At the time, no precautions were taken to dry the commercial $Sc(OTf)_3$ – sold as a polyhydrate – and thus a control reaction was conducted to rule out protic catalysis. Exposure of cyclobutanone and phenyldiazomethane to 1 mol% triflic acid (toluene, 23° C) gave none of the homologation product and only stilbene (~1:1 *E*/*Z*). Stilbene was also produced, together with 1,4-diphenylazine, when phenyldiazomethane was treated

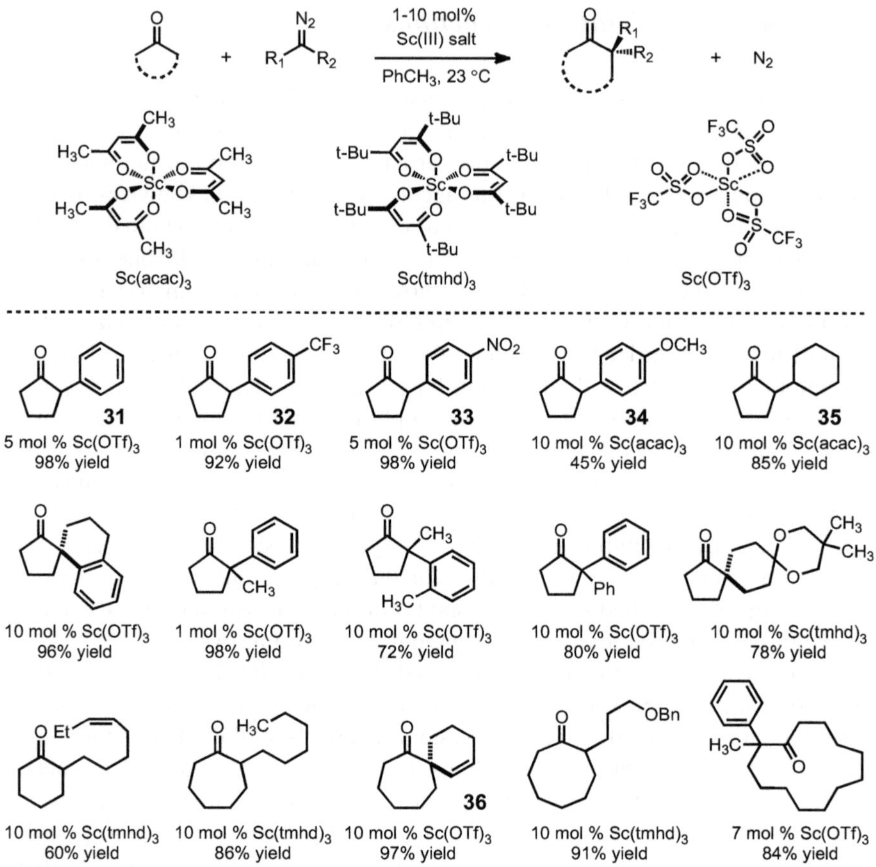


**Scheme 13** Efficient catalysis of diazoalkyl carbon insertion with Sc(III) salts

with 10 mol % $Sc(OTf)_3$ in the absence of electrophile (for earlier work on Rh-catalyzed dimerization of aryl diazomethanes, see [109]).

The illustrated (Scheme 13) α-aryl cyclopentanone syntheses are noteworthy for their tolerance of both steric and electronic modifications in the diazoalkane precursor. Switching to electron-poor diazoalkanes (*p*-$CF_3$ or $NO_2$) had no effect on isolated yields (→**31**–**33**, >92% yields). The electron-rich *p*-(methoxy)phenyl-diazomethane required the less Lewis acidic $Sc(acac)_3$ and afforded a diminished 45% yield of **34** under conditions equimolar in cyclobutanone.[6] The low yield could

[6] The less electron-poor, more hindered Lewis acids ($Sc(acac)_3$ and $Sc(tmhd)_3$) were substituted for $Sc(OTf)_3$ in reactions with more Lewis basic diazoalkanes in order to maximize the yield of product (tmhd = 2,2,6,6-tetramethyl-3,5-heptanedianato, or *tert*-butyl(acac)). See [99] and citations therein for more details on the Lewis-acid mediated destruction of diazoalkanes.

be improved by adding excess nucleophile.[7] Aliphatic diazo compounds could be employed, and reaction efficiency did not suffer with β-branching (→**35**, 85% yield). In the initial report, reactions were effected with 5–10 mol % catalyst, but additional studies showed that lower loadings (1 mol %) were tolerated [110]. The additional examples further underscore a remarkable facility with which sterically congested, all-carbon quaternary carbon centers can be installed in a single step from ketone-derived, "internal" diazoalkanes. Good to excellent yields were obtained for a number of different ring sizes and substitution patterns, and the method was shown to be tolerant of common functional groups, such as halogens, alkenes, benzyl ethers, and ketals. The direct elaboration of spirocycles is an especially attractive feature of the ring expansion strategy. Bicyclic α-vinyl cycloheptanone **36** – a structure not readily accessed by classic metalloenamine or enolate alkylation chemistry – was synthesized in high yield on a 5 mmol scale from over 0.5 g of cyclohexanone.

The first substrates tested under catalytic conditions were all symmetrical, cyclic ketones. Given the longstanding challenges with regiochemical control in carbonyl homologation, a parallel investigation was commenced targeting a single carbon ring expansion of dissymmetric, α-quaternary substrates [111]. TMSD (**1**) was chosen as the one-carbon source, with its promise of greater stability (for commentary on a possible stereoelectronic basis for this trend, see [112]) and regioselectivity [113] than diazomethane and the potential for access to versatile organosilicon products. Identification of a catalyst subject to facile modification creates the opportunity for altered reactivity and selectivity in any synthetic transformation, and this follow-up study was no exception.

As shown in Scheme 14, exposure of α,α-disubstituted cyclobutanone **37** to **1** in the presence of 10 mol % $Sc(OTf)_3$ delivered silyl enol ether **38** in 85% yield as a single compound (9:1 regioselectivity on the basis of $^1H$ NMR analysis of the crude reaction mixture). ReactIR analysis of the event revealed a dual role for $Sc(OTf)_3$, catalyzing first a rapid (trimethylsilyl)methine insertion to give **39** and then a more gradual conversion to the observed **38** by Brook isomerization [93]. In contrast to previously discussed methods, the enol silane could be purified by silica gel chromatography in an eluant pre-treated with ammonium hydroxide. This allowed access to its useful functional handle; alternatively, a deprotection in dilute acid gave the homologous cyclopentanone (**40**) in high yield. The research team was quick to deduce that a less Lewis acidic Sc trication might terminate the reaction at the α-silyl ketone **39**, especially since the kinetics of addition and 1,2-migration appeared to be much faster than 1,3-Brook rearrangement. Ultimately, $Sc(hfac)_3$ (hfac = (hexafluoro)acetylacetonate) was identified as a commercially available catalyst that allowed for recovery of pure **39** in 76% yield with an 8:1 dr and 12:1 regioisomeric ratio (rr) for the process overall. As expected, acidic hydrolysis of keto silane **39** likewise confers access to ketone **40**. However, it was further

[7] *p*-(Methoxy)phenyldiazomethane is a potent Lewis base and has been reported to decompose at temperatures as low as –80° C; see [125].

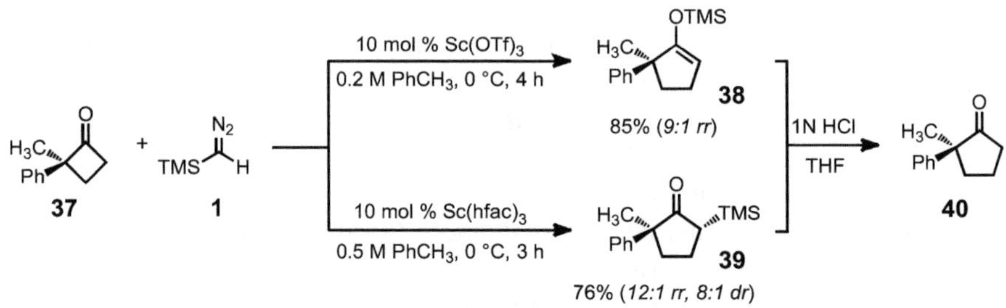


**Scheme 14** Divergent access to enol silane or keto silane ring expansion products as a function of Sc(III) counterions

demonstrated that for an α-dimethylphenylsilyl substituent introduced by ring enlargement with synthetic $PhMe_2SiCHN_2$ (prepared by the procedure used for TMSD from (chloromethyl)dimethylphenylsilane [114]), a simple sequence involving diastereoselective reduction, protection, and stereoretentive Fleming oxidation [115] could deliver a complex cyclopentane *syn*-diol in 54% overall yield [111].

The seminal report from the Kingsbury group in 2009 [105] disclosed (1) the first catalytic examples of non-stabilized diazoalkane–carbonyl homologation and, as far as we can tell, (2) the first construction of α-quaternary carbon atoms by cycloalkanone ring expansion. Subsequent work showed the inexpensive Sc catalysts to be equally effective with α-(trialkylsilyl)diazomethanes, not only for a regioselective 1-C homologation to higher homologs but also for the simultaneous installation of useful functionality. Additional reports from the authors' laboratory concern strategies for linear ketone synthesis by aldehyde homologation [116, 117] and enantioselective catalysis of α-arylation [118]. The new methods confer significant advantages with short reaction times, high chemical yields, and, in certain cases, outstanding levels of regio- and stereochemical control. Although these advances have renewed interest in carbonyl homologation as a core transform in synthesis and spurred progress in related areas, none would have come to fruition without modern methods capable of delivering pure solutions of diazoalkanes for storage and usage under controlled conditions. Such enabling procedures now warrant the reader's continued interest, representative of the frequently synergistic relation between synthetic methodology development and the small molecule reagents that make the science possible.

## 3 Modern Methods in Non-stabilized Diazoalkane Synthesis

Much of the development of chemistry based on the diazoalkyl reactive function has been hindered by a lack of convenient synthetic methods with broad substrate scope. To combat the lack of generality, synthetic chemists have devised multiple

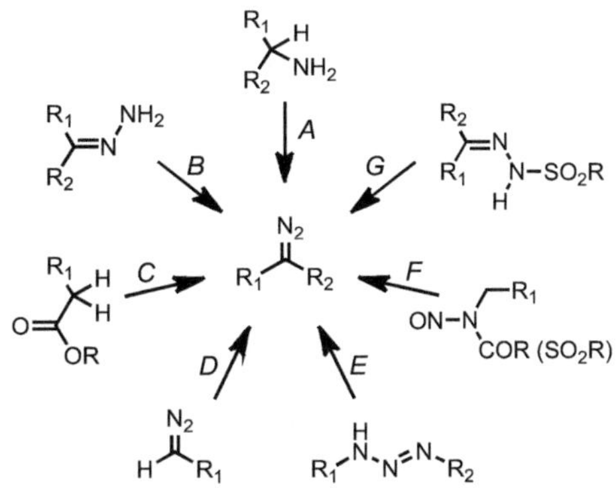


**Fig. 3** Diazoalkanes and a number of potentially viable precursors

routes (Fig. 3) for the preparation of diazoalkanes by (A) nitrosation of alkyl amines bearing an electron accepting group, (B) dehydrogenation of hydrazones, (C) diazo transfer to activated methylene compounds, (D) substitution within a preexisting diazoalkane, (E) triazine decomposition, (F) base-induced hydrolysis of *N*-nitroso carbamates, amides, or sulfonamides, and (G) base-mediated decomposition of sulfonyl-substituted hydrazones. The discussion below tracks new and mainstay procedures exclusive to the realm of non-carbonyl-stabilized diazoalkanes. By being both deliberate and restrictive in this task, such methods are brought together – for summary and scrutiny – in what may be the first time. For details on preparative entries to diazocarbonyls and other stabilized species, the reader is directed to other articles with a thorough coverage of this area [50, 51].

## *3.1 Base-Mediated Hydrolysis of* N*-Nitrosylamides, -Carbamates, and -Ureas*

The pioneering synthesis of diazomethane by von Pechmann more than 100 years ago [2] occurred by the reaction of base with an acetylated *N*-nitroso methylamine derivative (1). Throughout the twentieth century, experimentation has given way to various preferred or optimized procedures (for syntheses of nitrosoamine precursors and diazomethane, see [79, 119–121]). This strategy is of singular relevance in the preparation of less common diazoalkanes such as the doubly functionalized compounds **41** and **42** [122–124]. The reaction medium cannot be described as mild, but the simplicity and dependability ensures a robust entry to non-stabilized diazoalkanes lacking base-labile functional groups. Worth noting is the fact that von Pechmann's original strategy for producing diazomethane still receives mainstream usage today, in both the academic and industrial sectors, on scales both small and large [12, 13, 16].

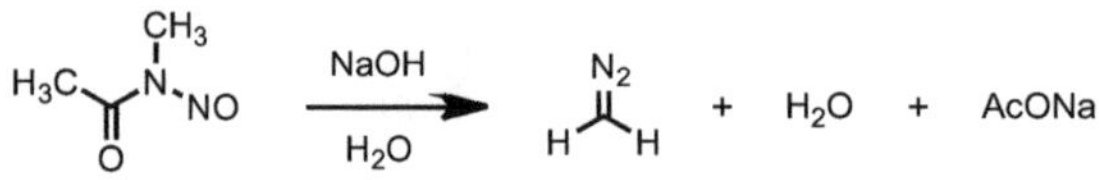

**41** (n = 1)
**42** (n = 2)

(1)

### 3.2 Diazotization of Alkylamines

Diazotization of alkylamines represents the most direct conceivable route to diazoalkanes (2). However, isolation of non-stabilized diazoalkanes by this strategy is not feasible; they fail to survive the strongly acidic conditions of the synthesis. If diazotization is performed on 1,2-amino alcohols with sodium nitrite at low pH, Tiffeneau–Demjanov rearrangement (TDR) [64, 65] occurs through a cationic (vs zwitterionic) intermediate to give a rearranged ketone (3).

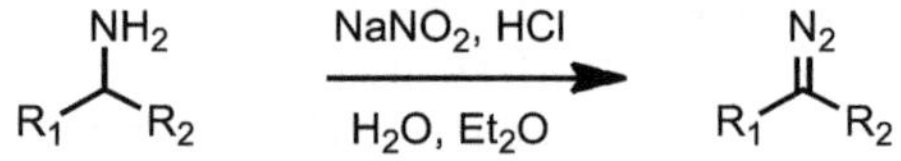

(2)

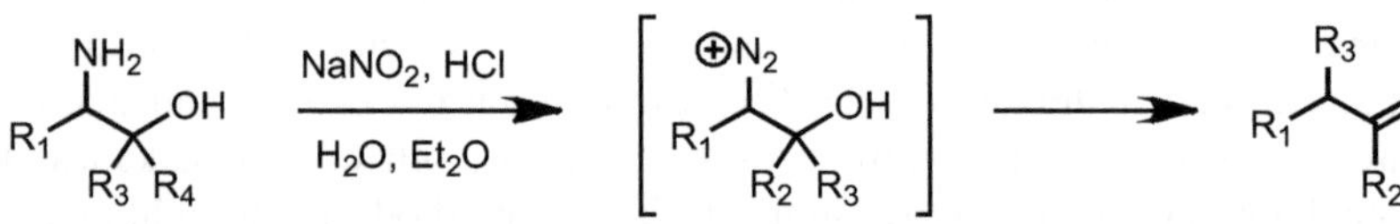

(3)

The TDR strategy has seen use in complex molecule synthesis, though a lack of regioselectivity will on occasion serve to limit its efficiency. In an early, highly successful example, the synthesis of the longipinenes was explored through ring expansion of bicyclic amino carbinol **43** to isomers **44** and **45** (Scheme 15) [125].

TDR was also applied toward the diversity-oriented synthesis of analogs of the antimicrobial agent spectinomycin by researchers at Upjohn in 1988 (4) [126]. Although homospectinomycin **46** was isolated in very low yield (byproducts of epoxidation and subsequent ring opening comprised the mass balance), it is important to consider that 8-*epi*-spectinomycin failed to yield any ring-enlarged ketone products; only the epoxide and its corresponding diol could be recovered.

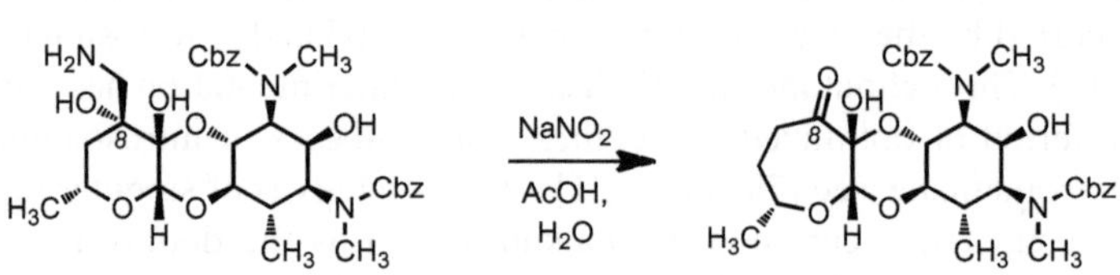

spectinomycin **46** (*23% yield*)

(4)

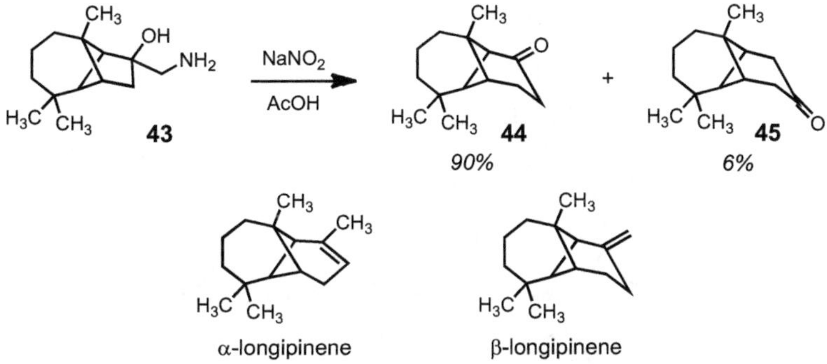


**Scheme 15** Elaboration of the bridged longipinene carbon framework by TDR

## *3.3 Bamford–Stevens Reaction and the Rise of In Situ Methodologies*

Though analogous with respect to the reaction medium, the decomposition of *p*-(toluenesulfonyl)hydrazones gives distinct outcomes relative to that of *N*-nitroso-amides/carbamates/ureas. When (tosylhydrazonyl)alkane substrates were first treated with alkali metals, only the olefinic products of elimination or Wagner–Meerwein rearrangement were observed (5, 6). Conversely, when treated under identical conditions, tosylhydrazones of aromatic carbonyls (e.g., benzaldehyde, acetophenone, benzophenone, and fluorenone) gave rise to isolable diazoalkanes (e.g., phenyl-, methyl phenyl-, and diphenyldiazomethane, and diazofluorene, respectively) [87]. Since this original finding by Bamford and Stevens, their procedure has been widely used in conjunction with a host of chemical transformations.

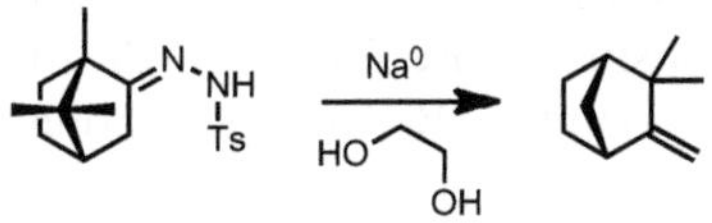


(5)

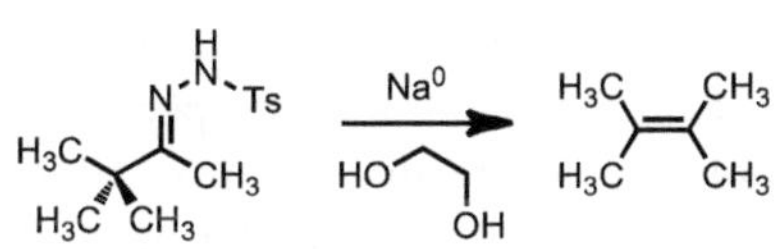


(6)

### 3.3.1 Reactions that Consume Aryldiazomethanes Without Isolation

The facile synthesis of sulfonylhydrazones from aromatic aldehydes and ketones, not to mention their inherent stability (and frequent crystallinity), makes these

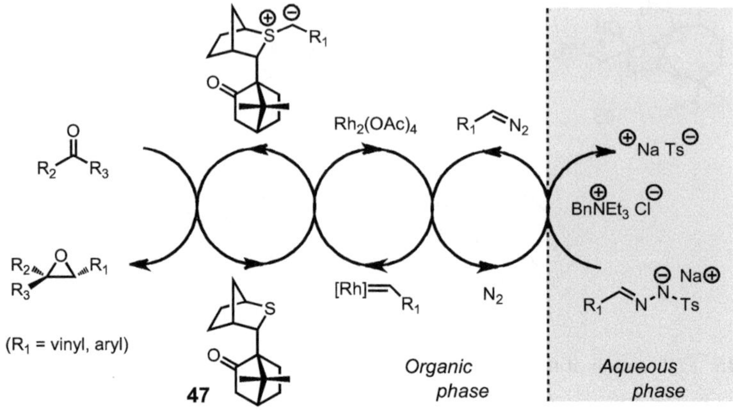


**Scheme 16** A cascade of catalysis involving diazoalkyl carbon transfer in situ

starting materials particularly attractive as diazoalkane precursors. Aggarwal and coworkers have exploited the use of phase-transfer conditions for generation and immediate consumption of resonance-delocalized diazo compounds (Scheme 16) [46, 127]. Benzyltriethylammonium chloride serves to shuttle a tosylhydrazone into solution, whereby decomposition by mild heating affords the diazoalkane. At this point, the nucleophile is converted into a rhodium carbenoid through catalysis by $Rh_2(OAc)_4$. By linking this reaction to that of a previously reported asymmetric ketone epoxidation – itself catalyzed by chiral sulfide **47** [128, 129], good to excellent yields of enantioenriched epoxides can be achieved for a variety of vinyl and aryl-substituted tosylhydrazones without any need for manipulation of the reactive diazoalkanes. Success in orchestrating this complex and interdependent process, which proceeds by the intermediacy of no less than three donor-acceptor carbon sources, is very impressive. Perhaps the only pitfall associated with this strategy is its reliance on a strong base for tosylhydrazone breakdown, which can limit its applicability with complex "$R_1$" groups bearing epimerizable stereogenic centers.

The above in situ strategy was also applied to Schiff base substrates (in lieu of aldehydes or ketones), affording aziridination products in fair yield, moderate dr, and outstanding enantioselectivity [44, 45]. The specific examples of benzylidene transfer illustrated in Table 4 confirm that electron-poor (entry 1), electron-rich (entry 4), aliphatic (entry 2), and α,β-unsaturated groups are tolerated in the *N*-SES imine electrophiles. As shown in Scheme 17, the utility of the method was showcased in an unconventional preparation of paclitaxel side-chain **49** that ultimately begins with 3-furfural. The optically pure heterocyclic aziridine **48** was obtained as an inconsequential diastereomeric mixture by the trio of coordinated phase transfer, achiral rhodium, and chiral sulfide catalyses, albeit in the antipodal series (see *ent*-**47**); subsequent conversion to the target amido ester was by way of standard manipulations [130].

**Table 4** Imine acceptors for benzylidene transfer from phenyldiazomethane

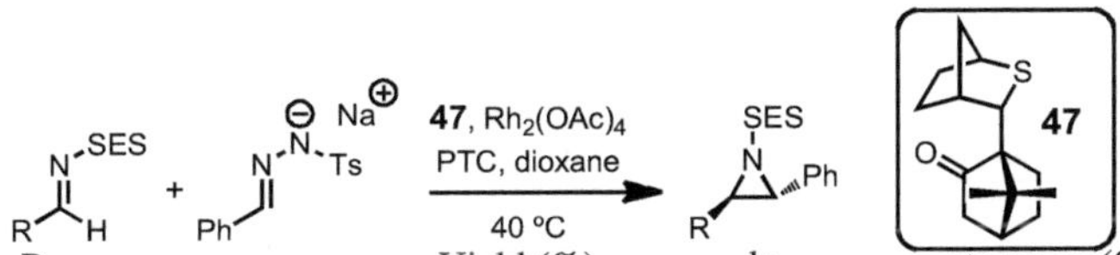


| Entry | R | Yield (%) | dr | *trans ee* (%):*cis ee* (%) |
|---|---|---|---|---|
| 1 | 4-ClC$_6$H$_4$ | 82 | 2:1 | 98:81 |
| 2 | Cy | 50 | 2.5:1 | 98:89 |
| 3 | *trans*-PhCH=CH | 59 | 8:1 | 94 |
| 4 | 4-MeOC$_6$H$_4$ | 60 | 2.5:1 | 92:78 |

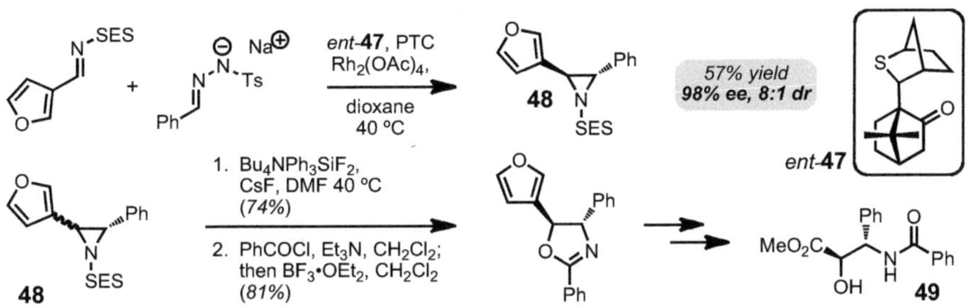


**Scheme 17** A ring-opening approach to a crucial taxol fragment based on formal carbene transfer from phenyldiazomethane

By employing alkenes within the framework of simultaneous diazoalkane generation and consumption, functionalized, stereodefined cyclopropanes can be prepared by carbene transfer as well. The use of metal carbenoids to effect cyclopropanation is now commonplace, and chemists routinely turn to stabilized diazo compounds as the carbene progenitors. Less common, however, is the exploitation of non-stabilized diazoalkanes due to their susceptibility to homodimerization in the presence of metal catalysts [109]. The maintenance of very minute concentrations of the diazoalkane has again turned out to be a valid solution to the problem in the context of a sulfur ylide-based process. The Aggarwal lab has synthesized a range of biologically relevant cyclopropane amino acids by net diazoalkyl carbon transfer to electron-poor alkenes, including **50**, anti-Parkinson's compound **51**, and the natural product (±)-coronamic acid (**52**) [45, 131, 132].

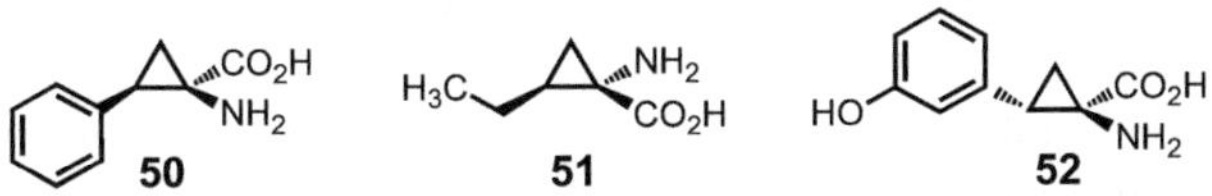


Finally, as if to come full circle and again ponder carbonyl electrophiles that served so admirably in asymmetric epoxidation, the well-known Schlotterbeck chain extension of aldehydes by formal C–H insertion with diazoalkanes has been

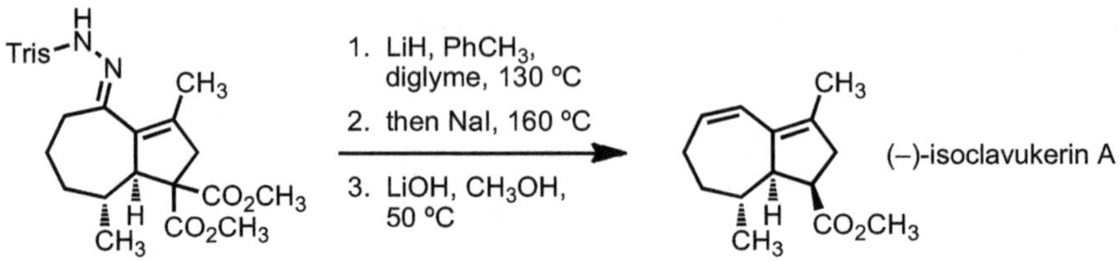


**Scheme 18** Bamford–Stevens olefination in a total synthesis of isoclavukerin A

explored. Angle and coworkers have investigated Bamford–Stevens degradation of tosylhydrazones in the presence of benzaldehydes, whereby a protic medium serves a dual role as solvent for the strong base and hydrogen bond promoter for homologation [133]. The Aggarwal lab has further expanded the chain elongation reaction through a reliance on bench-stable tosylhydrazone sodium salts [134] and the use of water to ensure homogeneity. Though water, with its uninterrupted network of hydrogen bonds, may reign supreme in its capacity for proton promotion, it is interesting to note that superstoichiometric amounts of lithium bromide were also present in these reactions to foster a highly ionic medium. It is difficult to address the confluence of both water and lithium cations since they may act together to promote carbonyl homologation. Nonetheless, group I metal halides could be viewed as yet another inexpensive tactic for lowering the activation barrier to rate-limiting diazoalkyl–carbonyl 1,2-addition.

### 3.3.2 Examples in Total Synthesis

A critical ramification of the examples presented thus far is that Bamford–Stevens reaction is subject to restrictions on the identity of carbon groups incorporated into the sulfonylhydrazone. Aryl and α,β-unsaturated derivatives give rise to the corresponding diazoalkanes with lifetimes sufficient to allow chemists to capitalize safely on the reactive function, whereas saturated starting materials lead only to alkenes. This trend pervades the total synthesis literature as well. The majority of target-based applications of Bamford–Stevens degradation involve generation of an alkene by elimination of nitrogen (commonly referred to as the Bamford–Stevens olefination; see [135]), such as in Trost's synthesis of (–)-isoclavukerin A (Scheme 18). Still, this is not to say that some level of complexity can't be tolerated. Employing the phase-transfer strategy for diazoalkane genesis in the direct presence of vinylphthalimide, the Che group achieved an efficient *cis* selective cyclopropanation catalyzed by ruthenium porphyrin **53** en route to the potent HIV-1 reverse transcriptase inhibitor **54** (Scheme 19) [136]. The target has deactivating chloro and fluoro substituents that did not interfere in the event in spite of their positioning *ortho* to the donor-acceptor function. Similar strategies by Aggarwal [128] and Doyle [137] toward **54** have also proven successful.

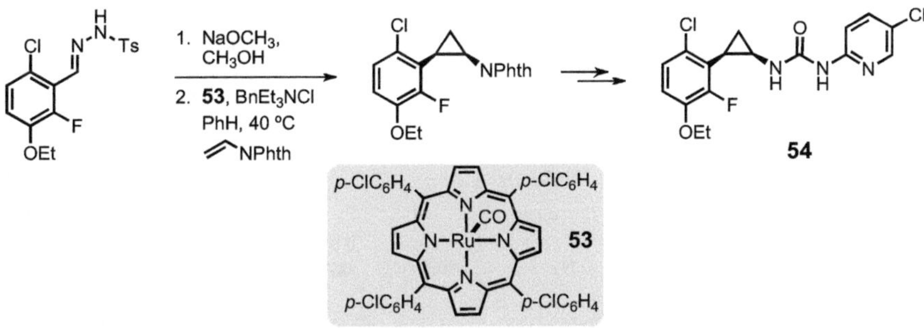


**Scheme 19** Double *ortho* halogens tolerated in an Ru-catalyzed approach to HIV-1 reverse transcriptase inhibitor **54**

## *3.4 Dehydrogenation of Hydrazones and Diazoalkane Stock Solutions*

Slowly but surely, hydrazone oxidation has evolved to become the premier preparative route for a broad array of more functionalized diazoalkanes. Hydrazones are easily obtained from simple starting materials [138], and numerous reagent systems have been identified that commence two-electron oxidation by accepting the terminal nitrogen lone pair. Historically, heterogeneous agents such as mercury oxide, silver(I) oxide, and manganese dioxide were preferred oxidants owing to their facile removal by filtration. Unfortunately, inconveniently sluggish reaction rates have detracted from their use, especially for hindered hydrazones or more reactive diazoalkyl products that can be prone to bimolecular degradation in solution (dimerization to azines) (bimolecular decomposition paths to azine or olefin impurities are common for noncarbonyl-stabilized diazoalkanes [139–141]). Through both rational design and serendipity, new oxidation strategies have appeared which allow clean and rapid conversion at low temperature and frequently maintain a purification advantage. These enabling developments, together with selected examples of their utility, will be highlighted in the sections that follow.

### 3.4.1 Lead(IV) Oxidation of Hydrazones

Optimization of a protocol for $Pb(OAc)_4$ oxidation of hydrazones by Shechter and coworkers has made previous heavy metal oxidations obsolete [52]. The use of $Pb(OAc)_4$ as a powerful, homogeneous oxidant is well known for many functional group interconversions [142], yet as a means to oxidize hydrazones it had received only passing attention because of a propensity for the oxidation byproducts (and even lead(IV) itself) to react with the nascent diazoalkane product. Prior to 1995, the limited cases that were reported comprised highly stabilized products such as bis(trifluoromethyl)diazomethane [143], aryldiazoacetates [144], aryldiazoacetonitriles [145], tetrabromodiazocyclopentadiene [146], and dicyanodiazomethane [147]. Shechter's oxidation system had to overcome the strongly acidic reaction

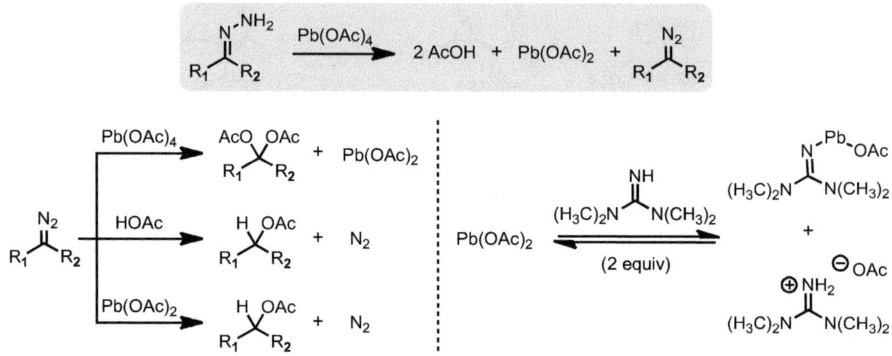


**Scheme 20** Shechter's reliable lead(IV) oxidation system overcame a number of challenges within the reaction medium in order to promote diazoalkane longevity

conditions created by $Pb(OAc)_4$, the $Pb(OAc)_2$ coproduct, and the 2 equiv. acetic acid generated during the course of dehydrogenation [52]. To combat these factors, a solvent mixture containing DMF and 1,1,3,3-tetramethylguanidine (TMG) (or alternatively, chloroform and triethylamine) was implemented. The benefit of this mixture is twofold. The large excess of base serves to quench the acetic acid that would quickly esterify the nascent diazoalkane. At the same time, the amine attenuates the Lewis acidity of $Pb(OAc)_4$ and $Pb(OAc)_2$, preventing deleterious interaction with the diazoalkane (Scheme 20).

Perhaps most importantly, the selection of DMF as the solvent allows facile, biphasic extraction of relatively non-polar diazoalkanes from the reaction mixture with hydrocarbon washes involving pentane, hexanes, or pentane/ether mixtures. Such incremental yet deliberate painstaking improvements have greatly enhanced the efficiency, substrate scope, and purity of access to isolable diazoalkanes. It should not be underemphasized that the extraction technique fosters recovery of products as stock solutions under cryogenic conditions that greatly reduce the risk of violent thermal decompositions that were documented in the realm of prior isolation techniques [80, 148]. Although the method has augmented the chemical yield for several previously known conjugated internal diazoalkanes (Fig. 4), it is unique in its provision for formerly unknown aliphatic derivatives, such as those derived from pinacolone and methyl cyclopropyl ketone.[8] If one had to claim a drawback associated with this approach, it would only concern the environmental implications of disposing the heavy metal-laden DMF layer after siphoning off the organic layers. Arguably, this only becomes an impediment on large scales, and as we will see momentarily, other oxidants can be substituted for $Pb(OAc)_4$ within this basic overall strategy.

[8] Yields in Fig. 4 were determined by manometric titration, as described in [51].

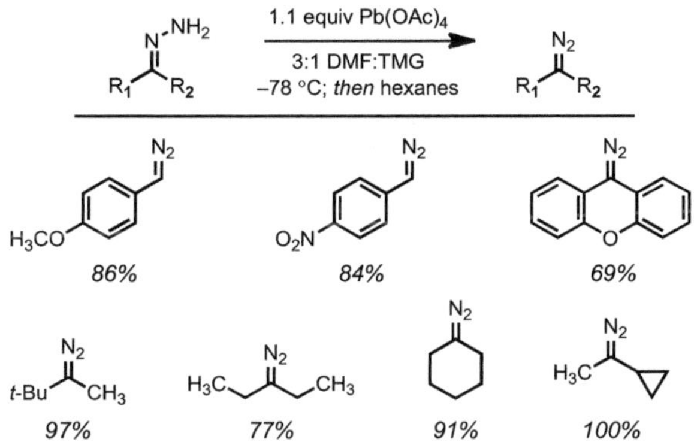

**Fig. 4** Demonstrative diazoalkane syntheses from Shechter's lead(IV) protocol

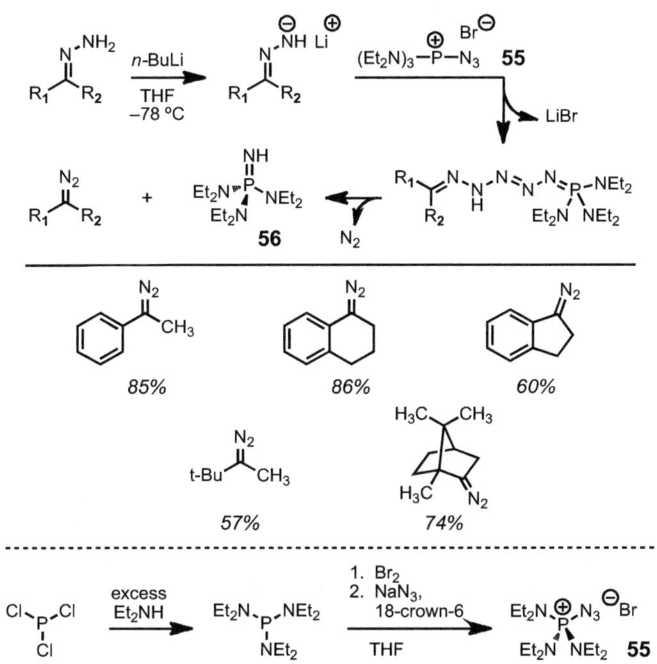

**Scheme 21** A metal-free oxidation of *N*-lithio hydrazones with $(Et_3N)_3PN_3{}^+Br^-$

### 3.4.2 Azidotris(diethylamino)phosphonium Bromide with Lithium Hydrazinides

Abolishing the need for heavy metal oxidants, Shechter and coworkers introduced a novel method for hydrazone dehydrogenation by reacting lithium hydrazonides with phosphonium azide **55** (Scheme 21) [53]. The inherently basic reaction

**Table 5** An industry-relevant synthesis of aryl or biaryl diazomethanes, catalytic in iodine and stoichiometric in peracid

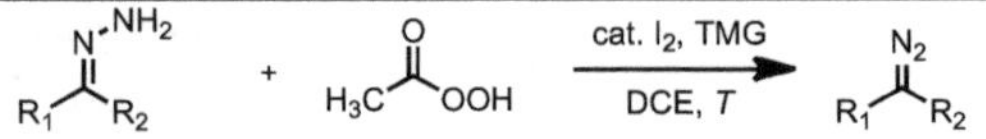


| Entry | $R_1$ | $R_2$ | *T* (°C) | Yield (%) |
|---|---|---|---|---|
| 1 | *o*-Tolyl | Ph | 0 | 99 |
| 2 | 2-Thienyl | Ph | 0 | 83 |
| 3 | H | 2-Thienyl | −5 | 79 |
| 4 | H | Ph | 0 | 89 |
| 5 | H | 4-OMePh | −20 | 77 |
| 6 | Me | Ph | −5 | 96 |
| 7 | H | 2-Furyl | −20 | 85 |

conditions obviate the need for exogenous base. A simplified workup procedure can be undertaken, and diazoalkanes are recovered as cold solutions after ethylene glycol extraction. If an allowance is made for the solutions to sit over dry ice, subsequent filtration removes any traces of the phosphorimine byproduct (**56**). In spite of a reliance on distinct reagents, this alternative method compares favorably to the aforementioned lead(IV)-based protocol and also includes high yielding syntheses of quite hindered, fully saturated diazo compounds, such as the example involving camphor hydrazone. Importantly, as indicated at the bottom of Scheme 21, azidophosphonium bromide **55** is readily produced on scale in just three steps from commercially available starting materials [149].

### 3.4.3 Oxidation by Peracids and Catalytic Iodine

Barton and coworkers have shown that treatment of ketohydrazones with iodine in the presence of base yields vinyl iodides, whereas aldohydrazones furnish *gem*-diiodides [136, 150]. Although this transformation invokes reactive diazoalkanes as intermediates, the potential for isolation is precluded by their rapid addition to the molecular iodine responsible for oxidation. Nonetheless, another metal-free and highly practical hydrazone dehydrogenation has been reported by a process group at the former Glaxo Labs [151–153] consisting of catalytic iodine in the presence of peracetic acid as the stoichiometric oxidant. As illustrated in Table 5, when various aryl and heteroaryl hydrazones were treated with catalytic iodine (0.01 mol %), peracetic acid, and TMG in dichloroethane, good to excellent yields of the corresponding diazoalkanes were obtained.

### 3.4.4 Oxidation of *N*-(*tert*-Butyldimethylsilyl)hydrazones with Hypervalent Iodine

Hydrazone oxidation by hypervalent iodine finds precedent in the literature [154–157], but until now has not enjoyed much in the way of generality. Recently, Myers and Furrow [17] reported the use of *N-tert*-butyldimethylsilylhydrazones (TBSHs)

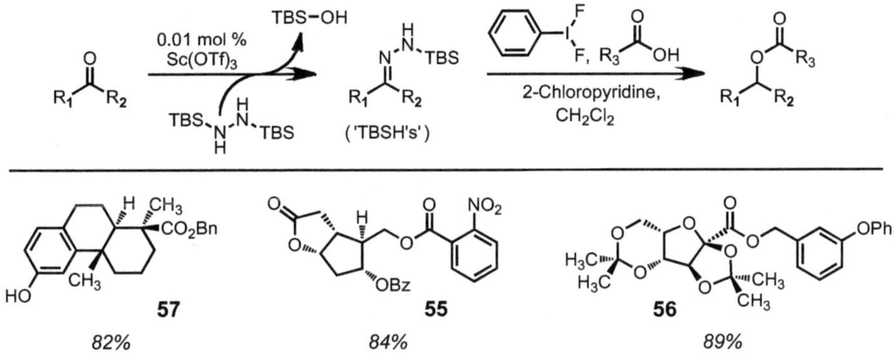


**Scheme 22** Myers's (difluoroiodo)benzene oxidation of *N*-silylhydrazones allows for esterification events that are wide in scope

[158] along with oxidation by iodine(III) to effect complex, functionally tolerant esterification reactions in situ (Scheme 22). For TBSH preparation, low catalytic loadings of $Sc(OTf)_3$ were used for Lewis activation of the carbonyl precursors in the presence of bis(TBS)hydrazine. Under these conditions, TBSOH is formed as a coproduct that can be removed by overnight heating under vacuum. While the requirement for vacuum distillation of silanol precludes the use of low molecular weight starting materials, the strategy gives efficient, reliable access to sterically deactivated hydrazones. To effect oxidation and esterification, the silylhydrazones are treated with (difluoroiodo)benzene in the presence of a carboxylic acid and a 2-chloropyridine buffer. This is a clever design given that the fluoride counterion gives rise to the free hydrazone by desilylation, and the products of esterification were isolated in high yield without a need for handling of the potentially unstable diazoalkanes. In fact, real time NMR analyses for several representative esters showed that the diazoalkane concentrations failed to ever reach detectable levels (<2%). The conditions were tolerant of very sensitive functionality, such as that seen in **55** and **56**, and without competitive O–H insertion for phenols like **57**. These formal carboxylic O–H insertions are unrivaled examples in which the in situ strategy for diazoalkane genesis and capture has extended to unconjugated substrates and occurs outside the confines of the harshly basic reaction conditions associated with Bamford–Stevens decomposition.

### 3.4.5 Oxidation of Hydrazones by "Activated" DMSO

Chlorodimethylsulfonium chloride (Swern reagent) represents another metal-free alternative for hydrazone oxidation [54]. The strategy was discovered by chance during research on a new conversion of free hydrazones to alkyl chlorides [159]. When benzophenone hydrazone was exposed to Swern reagent in the absence of an amine base, Brewer and Javed observed localized yet transient red coloration of the reaction mixture. Diligent experimentation then led to the

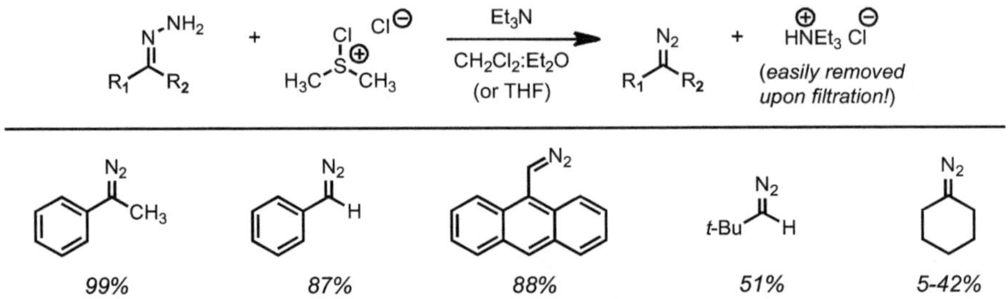


**Scheme 23** Brewer's simple and reliable Swern-based approach to diazoalkanes

identification of a preferred solvent to favor diazoalkane formation over competitive hydrochlorination in the presence of $Et_3N$. In practice, the oxidant is pre-formed at low temperature in THF and treated with a solution of $Et_3N$ and hydrazone to effect dehydrogenation. The procedure is extremely convenient and dependable for the synthesis of monoaryl, diaryl, and mixed aryl-alkyl diazomethanes (Scheme 23). However, more reactive alkyl diazoalkanes, despite forming efficiently under the Swern conditions, appear to experience finite lifetimes. Instability could result from deleterious interactions at the surface of the heterogeneous, yet nonetheless acidic, triethylammonium chloride precipitate. Alternatively, aliphatic diazo compounds may be labile to dimethylsulfide, the nucleophilic coproduct of Swern oxidation. In any case, like Shechter's lithium hydrazonide method, Brewer's discovery fosters a simple, filtration-based workup and the opportunity to isolate pure diazoalkane solutions, all the while maintaining the species at cryogenic temperatures and promoting safety. The use of an appropriate solvent (THF or a dichloromethane/ether mixture) is vital to strike a balance between solubility of the starting materials and solubility of triethylamine hydrochloride. Segregation of the diazoalkane and the conjugate acid are the key to achieving high efficiency.

### 3.4.6 Hybrid Protocols and Modern Titration Methods

As part of their campaign to discover the scandium(III) salts that now effectively catalyze carbonyl homologation with both cycloalkanone and aldehyde acceptors, Kingsbury and coworkers favored a blended approach to diazoalkane synthesis by borrowing heavily from several of the key innovations just discussed, specifically (1) Myer's transient *N*-silyl protection that discourages hydrazone dimerization to unwanted azines upon storage and (2) Shechter's DMF/TMG solvent combination which allows for diazoalkane purification by low temperature extraction into an immiscible hydrocarbon layer. This plan was ideal for preparing aliphatic mono- and dialkyldiazomethanes that have been difficult to make by Brewer's method due to the lack of resonance stabilization imparted by an aromatic substituent.

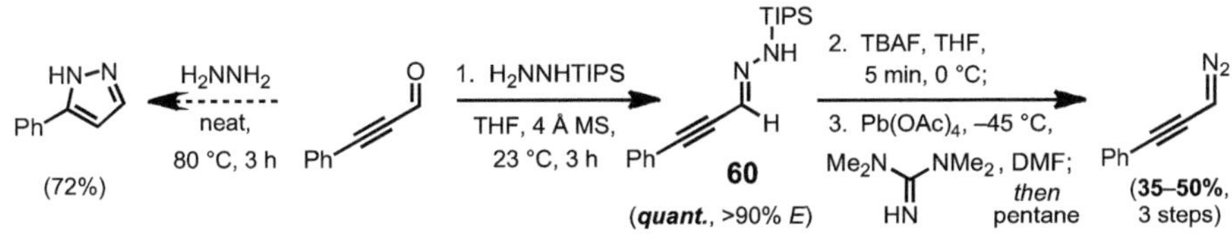


**Scheme 24** Fast and convenient access to alkyl and α,β-unsaturated diazoalkane stock solutions by a modified strategy involving *N*-TIPS hydrazone intermediates

Triisopropylsilylhydrazine [160], readily distilled on scale upon mixing TIPSCl with anhydrous hydrazine, was deliberately chosen as a reagent for condensation with the carbonyl compound. Compared to bis(*tert*-butyldimethylsilyl)hydrazine [17, 158], it removes a need to catalyze the condensation for sterically accessible carbonyls and also prevents coproduction of a high molecular weight silanol that can be cumbersome to remove from hydrazone adducts, even under high vacuum. Scheme 24 shows representative results obtained for a model propargylic aldehyde, a challenging test case given its susceptibility to Michael addition and subsequent heterocyclization. Exposing 3-phenylpropionaldehyde to a traditional procedure for hydrazone formation (hydrazine hydrate as solvent, 80° C), led exclusively to 5-phenyl-1*H*-pyrazole. By contrast, mild treatment with $TIPSNHNH_2$ and powdered 4 Å molecular sieves (THF, 0° C) delivered TIPS-hydrazone **60** in quantitative yield as an inconsequential 10:1 mix of geometric isomers [117]. The blocking group was then cleaved by exposure to a commercial solution of TBAF, and after a solvent switch in the same vessel, the free hydrazone was treated with an oxidant[9] and purified by Shechter's extraction/siphoning techniques [52].

The yields of fully saturated and α,β-unsaturated diazo compounds obtained through this strategy are routinely lower than their aryl-substituted counterparts. Nonetheless, the efficiencies are based on the starting aldehyde or ketone and correspond to four distinct steps, the last of which is an esterification reaction that conveniently serves to titrate the organic stock solution prior to the diazoalkane's purposeful use or storage. Although a number of methods for titration are known in the literature, few offer a simple procedure that can be executed quickly and with small quantities of the diazo compound. Those based on protonation and the collection of evolved nitrogen gas require large quantities of the nucleophile and elaborate experimental setups [161]. Spectrophotometric methods require preparation of calibration standards and calculation of extinction coefficients for compounds that can easily decompose at room temperature or by light-induced pathways [162]. Esterification with excess benzoic acid and back titration of the unreacted acid is time-consuming, requiring preparation and calibration of stock base solutions in order to obtain accurate results [119]. Esterification and calculation of concentration on the basis of the unpurified yield of benzoate ester is also

[9] Iodobenzene bis(trifluoroacetate) can substitute for lead(IV) acetate with little or no impact on the yield of diazoalkane, but the nonpolar extract becomes contaminated by residual iodobenzene.

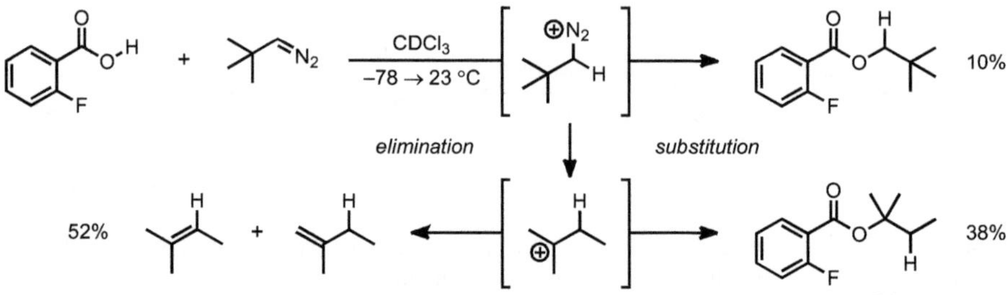


**Scheme 25** Limitation to the scope of a $^{19}$F NMR-based titration method with 2-fluorobenzoic acid

possible, but at times will provide titers of questionable accuracy due to common diazoalkane impurities [139–141].

Arguably, a preferred method would involve quenching a known diazoalkane aliquot with benzoic acid and actually isolating the pure benzoate ester by flash chromatography. The isolated yield could then be used to determine the amount of active nucleophile in the aliquot. Unfortunately, this approach consumes valued experimental time as well, and it can be subject to mechanical losses during the workup, purification, and transfer steps. To confront these and other issues, the Kingsbury group recently developed a new titration method using commercially available 2-fluorobenzoic acid and quantitative $^{19}$F NMR spectroscopy [163]. In a typical run, a carefully weighed quantity of excess reagent is dissolved in just enough $CDCl_3$ to prepare a single NMR sample and chilled. A diazoalkane aliquot is then added rapidly, and nitrogen gas evolution occurs upon warming (<5 min). Conversion, and ultimately concentration, is determined by integrating the only two observable $^{19}$F NMR signals. The assay requires just 10 min of experimental time, can be performed with micromolar amounts of the diazoalkane, and obviates the problem of overlapping resonances occasionally observed for $^{1}$H NMR-based approaches.

The protocol was shown to have a very broad scope, delivering reproducible titers that were far more accurate than those based on a gravimetric benzoylation method and in close agreement with a triplicate $^{1}$H NMR analysis using 1,3,5-trimethoxybenzene as an internal standard [163]. Only one limitation was identified for diazonium cations that can undergo spontaneous rearrangement or elimination after initial protonation [164]. When 1-diazo-2,2-dimethylpropane was exposed to 2-fluorobenzoic acid, rearrangement to the tertiary carbocation occurred, affording predominantly ester **61** and two alkene byproducts (Scheme 25). The anticipated ester resulting from direct substitution only accounted for 10% of the distribution. The titer for the starting *tert*-alkyl diazomethane could still be calculated from the combined $^{1}$H and $^{19}$F NMR data, but not with the same level of precision obtained for other diazoalkanes that cleanly transformed to a single ester product.

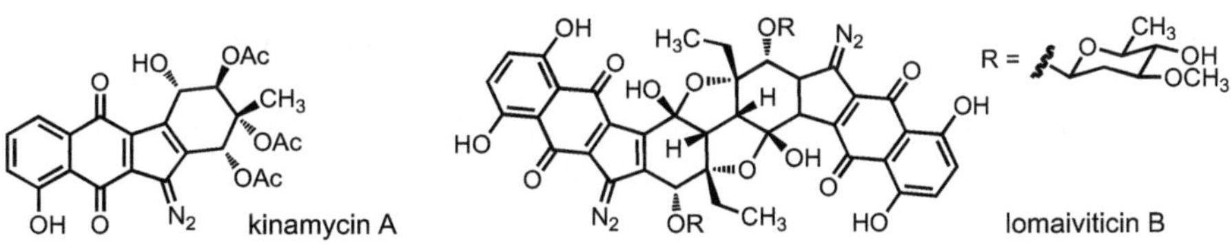


**Fig. 5** Some biologically active natural products containing diazoalkyl carbons

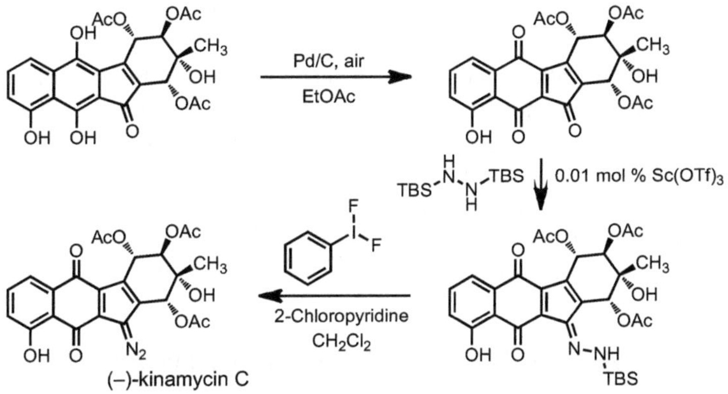


**Scheme 26** Porco's synthesis of kinamycin C featuring the Myers methodology

### 3.4.7 Examples in Total Synthesis

To a large extent, the preceding discussion serves to illustrate the tantalizing nature of diazalkanes – extremely synthetically valuable and atom-economical in theory, but occasionally difficult to justify the risks of instability and/or toxicity in practice. Furthermore, though diazo compounds are usually thought of as reagents or reactive intermediates, they are sometimes the components of complex natural products [165]. The kinamycins and lomaiviticins represent two such medicinally relevant structures whose total syntheses were pursued by several research groups in the last decade (Fig. 5). Despite stabilization of the diazoalkyl moiety by vinylogy with adjacent carbonyl groups in both target molecules, it is noteworthy that hydrazone oxidation methods came to the forefront in successful routes to the kinamycins.

Porco's 2006 approach to (–)-kinamycin C utilized the Myers's procedures in the final stages of the synthesis, both for *N*-TBS-hydrazone installation as well as oxidative unveiling of the stabilized diazoalkane [166] (Scheme 26). An alternate strategy reported by the Nicoloau group involved merging advanced intermediate **62** with tosylhydrazine under acidic conditions [167]. Subsequent oxidation by CAN furnished kinamycin C after desilylation (7). Further manipulation of the acetoxy functions allowed for flexible access to kinamycins F and J by this route. In more recent work, Herzon and coworkers accomplished diazotization within the same family of compounds by diazo transfer to the activated methylene unit [168, 169]. The Yale strategy was repeated en route to a synthesis of the antibiotic (–)-lomaiviticin aglycon [169]. Elaboration of the diazo function at a much earlier stage sets this synthesis apart from those of the kinamycins.

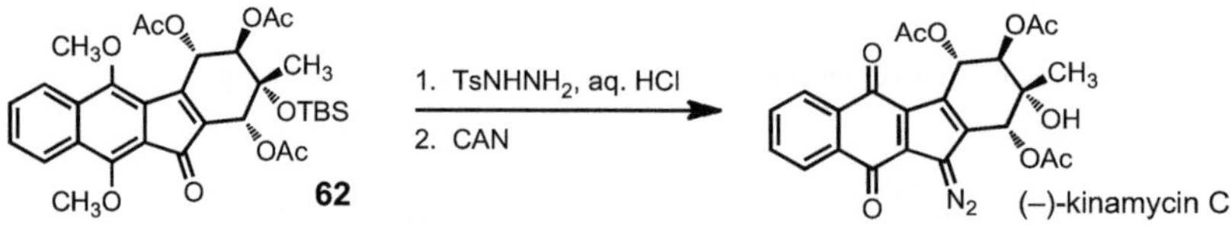


(7)

While no unified synthetic approach to diazoalkanes can furnish all manner of substituted diazo compounds, the various methods and reagents surveyed here go far in enabling access to a rich inventory of these fascinating molecules. Further improvements will serve not only to popularize known applications of the neutral carbon nucleophiles, but also to spur interest in the development of related new reactions. Our focus in the next section is modernized, catalytic carbon insertion methodologies involving both stabilized and nonstabilized diazoalkane reagents.

# 4 Modern Catalytic Diazoalkyl Carbon Insertion Methodology

As stated in the introduction, diazo compounds were reacted with aldehydes not long after their discovery by Buchner and Curtius [170], Staudinger [171], and Schlotterbeck [59, 60]. While the pioneers of diazoalkane chemistry benefited from the high reactivity of unhindered aldehyde acceptors, unspoken hydrogen bonding effects were often in place to govern the reaction course, and Lewis acids were quickly identified as advantageous promoters owing to enhanced reaction rates.

## *4.1 Diazoacetate Homologation of Aldehydes (Roskamp Reactions)*

Although the merger of α-diazo esters and carboxaldehydes had been known, the direct synthesis of β-keto esters was not viewed as a general synthetic method due to the formation of dioxolane byproducts derived from a second equivalent of the carbonyl compound becoming involved in the transformation. Not until 1989, when Roskamp and Holmquist published the Sn(II)-catalyzed addition of EDA to aldehydes, was the direct synthesis of β-keto esters accomplished in a satisfactory manner [5, 6] (Scheme 27). Not long thereafter, the sluggish reactivity with hindered or deactivated substrates was addressed by Nomura and coworkers, who achieved higher yields for neopentylic electrophiles with a methoxycarbonyl-substituted diazoester **63**, albeit by turning to stoichiometric amounts of $ZrCl_4 \cdot (THF)_2$ [172].

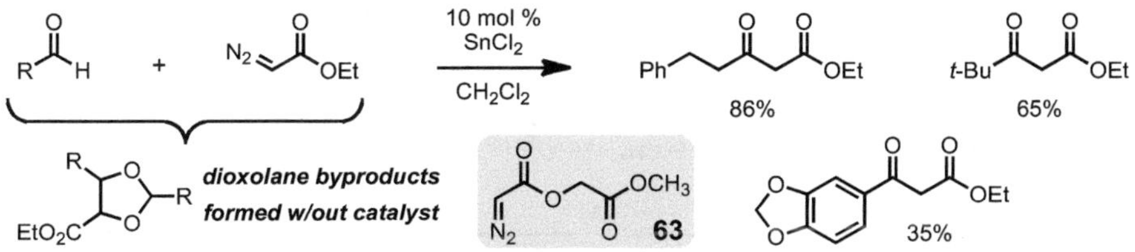

**Scheme 27** Roskamp's tin-catalyzed β-keto ester synthesis, in which Nomura's methoxycarbonyl diazoester **63** afforded slightly improved outcomes

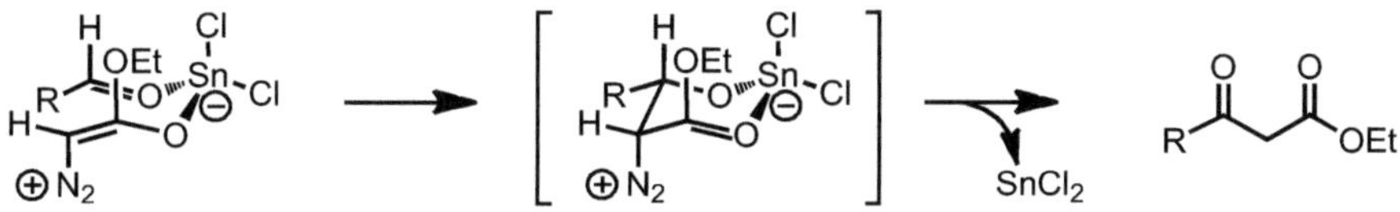

**Scheme 28** A closed transition state, with R group equatorial, has been advanced as a basis for preferential α C–H migration in the Roskamp β-keto ester synthesis

In the reactions above, the homologated products derive from C–H migration within the diazonium tin (or zirconium) alkoxide intermediates. Interestingly, the literature holds cases in which C–C migration has been observed as well. Hossain and coworkers reported the use of iron Lewis acid **64** to effect the coupling of aryl aldehydes and EDA, yielding enol esters in addition to β-keto esters (8) [173]. With this catalyst, typical Roskamp reactivity predominates, but Kanemasa and coworkers later published a system that resulted in an actual turnover of the regiochemical outcome. The combination of TMSCl/ZnCl or TMSOTf allowed for selective formation of enol esters among a sampling of aromatic electrophiles tested [174]. The authors postulate that use of a non-chelating Lewis acid was the basis for 1,2-migration of the aryl substituents in preference to the α hydrogen atom, as is illustrated in Scheme 28 for a prototypical Roskamp homologation.

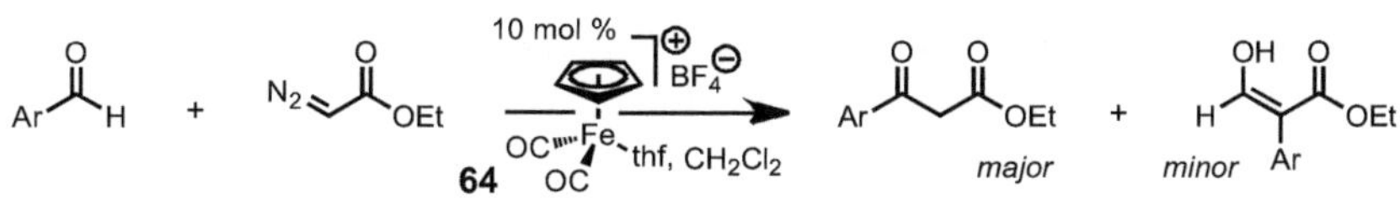 (8)

## 4.2 *Stereoselective Additions of α-Aryl Diazoacetates to Aldehydes*

Maruoka and his research team were the first to capitalize on all the above observations, in particular the potential for stereospecific C–C bond shifts, noting that an application with α-*substituted* diazo esters could furnish chiral quaternary carbons. In a simple but highly effective strategy, internal diazo esters containing an alcohol auxiliary were utilized in stereoselective syntheses of chiral aldoesters

**Table 6** Maruoka's entry to α-quaternary carboxaldehydes by rare non-Roskamp breakdown of protonated hydroxydiazonium triflates

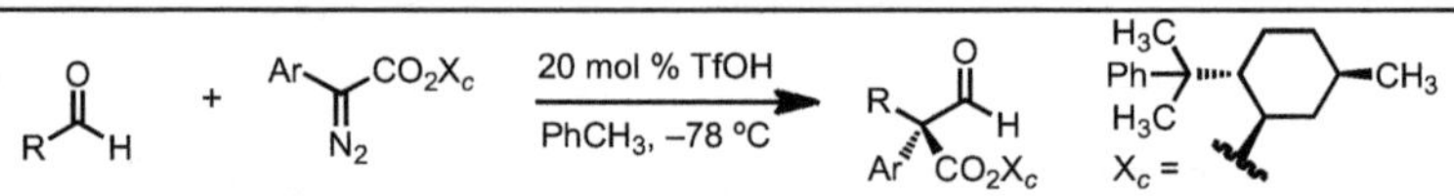


| Entry | R | Ar | Yield (%) | de (%) |
|---|---|---|---|---|
| 1 | 3-Tolyl | Ph | 67 | >95 |
| 2 | $4\text{-MeOC}_6H_4$ | Ph | 42 | 89 |
| 3 | 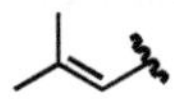 | Ph | 89 | 87 |
| 4 | Ph | $3\text{-MeOC}_6H_4$ | 72 | >95 |
| 5 | Ph | $4\text{-ClC}_6H_4$ | 73 | >95 |

**Table 7** An alternative chiral diazoamide controller in a $BF_3$-catalyzed synthesis of stereochemically labile α-alkyl-β-keto imides

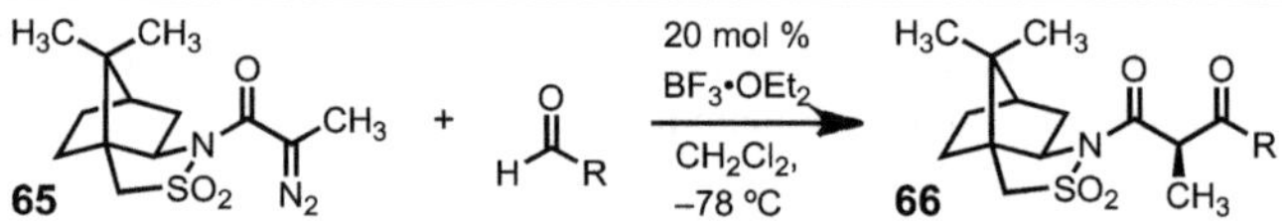


| Entry | R | Yield (%) | | dr |
|---|---|---|---|---|
| 1 | 2-Tolyl | 78 | **66a** | >20:1 |
| 2 | 2-Np | 75 | **66b** | >20:1 |
| 3 | $4\text{-ClC}_6H_4$ | 75 | **66c** | >20:1 |
| 4 | Cy | 74 | **66d** | >20:1 |
| 5[a] | 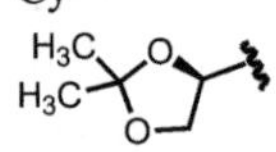 | 70 | **66e** | >20:1 |

[a]Performed with 1.0 equiv. catalyst

[175]. After the identification of triflic acid as a Brønsted acid catalyst and (–)-phenyl-menthyl aryldiazoacetates as optimal nucleophilic reagents, a series of reactions that proceed in good yield and high diastereomeric excess for aromatic aldehydes were identified (Table 6). This chiral controller system is also characterized by high versatility. In a striking case of reversal of chemoselectivity based on catalyst choice, Maruoka's group found that initial 1,4-addition with α-substituted acrolein substrates would give way to stereodefined 1,1,2,2-tetrasubstituted cyclopropanes [176].

As further testimony to the significance of the catalyst in controlling the regiochemical course of diazoalkyl carbon insertion events, Maruoka and coworkers also devised a strategy toward the synthesis of α-alkyl-β-keto imides under not proton but $BF_3 \cdot Et_2O$ catalysis [177]. Beginning with (–)-camphorsultam-derived diazo compound **65**, imides **66a–e** were obtained in good yields and with excellent diastereomer ratios (Table 7). Further elaboration was possible through a diastereoselective allylation at the β carbonyl and either a basic or reductive cleavage of the chiral auxiliary.

**Table 8** Feng's catalytic asymmetric Roskamp synthesis of α-alkyl-β-keto esters

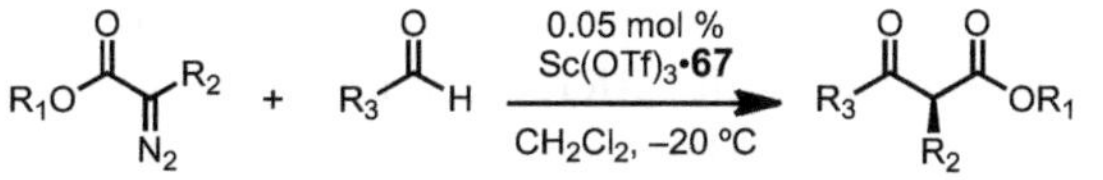


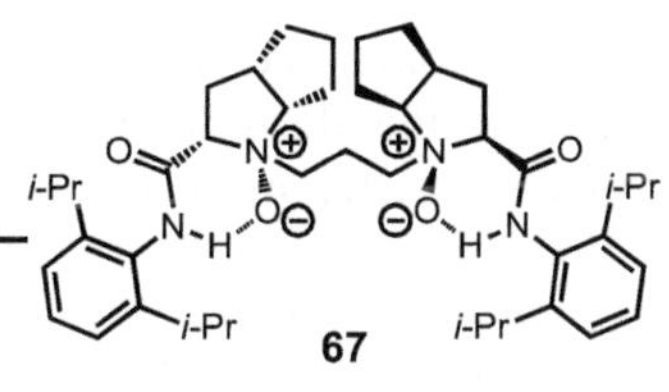

| Entry | $R_1$, $R_2$ | $R_3$ | *ee* (%)[a] | Yield (%)[b] |
|---|---|---|---|---|
| 1 | Et, Bn | Ph | 95 (77) | 97 |
| 2 | Bn, Bn | Ph | 96 (96) | 99 |
| 3 | $CH_3$, Bn | 2-Np | 98 (84) | 90 |
| 4 | Bn, allyl | Ph | 94 (40) | 96 |

[a]Values in parentheses represent eroded optical purity after purification
[b]Isolated yields after chromatography

Despite being catalytic, the previous examples relied on chiral auxiliaries for asymmetric induction. In a more economical approach, the Feng group has developed Sc-catalyzed [105], enantioselective Roskamp reactions that rely on $C_2$-symmetric anilide-substituted amine *N*-oxide ligands [178]. As shown in Table 8, **67** was identified as the premiere ligand. Intriguing is the fact that its structure is neutral overall, but tight coordination to the $Sc(OTf)_3$ catalyst was expected given the ionic character of the trication and the very delocalized nature of the triflate counterions. The compiled data (Table 8) reveals that modifications in the electrophile were well tolerated, but the scope for "$R_2$" was rather limited in this initial report. Moreover, erosion of enantiomeric purity was a complication during purification because of the high acidity of the α proton in the β keto ester. These issues aside, this study represents groundbreaking work in the development of broadly applicable, catalytic asymmetric Roskamp reactions.

## 4.3 *Diazocarbonyl Homologation of Ketones*

We have seen that the traditional Roskamp process and its stereoselective variants allow one to substitute a synthetically valuable unsubstituted or substituted alkoxy ethanoyl function in place of the formyl C–H bond. The analogous homologation of *ketone* acceptors with stabilized diazo compounds promises to become a highly advantageous tool for the synthetic chemist. Unfortunately, since ketones are far less reactive than aldehydes, comparably fewer reports in this area are known that still meet the modern criteria of catalysis, high efficiency, and stereocontrol.

Preliminary work by Gutsche and Kharasch showed that under conditions in which a metallocarbene was likely to be formed (e.g., EDA, $Cu^0$), reaction commenced with electrophilic addition of the carbonyl oxygen atom, giving complex product mixtures dominated by enol ethers [179, 180]. In other words, the main adducts were products of formal O–H insertion via the enol tautomer of the cycloalkanone (no ring expansion). Tai and Warnhoff later reported the $BF_3$-promoted addition of diazoacetic esters to ketones [91]. This disclosure displays the characteristic attribute of the use of such a strong Lewis acid: reaction rates were very rapid and convenient, but efficiency still suffered overall due to decomposition of the diazo compound. Mock and Hartman soon achieved notable improvements, reporting on a novel use of triethyloxonium tetrafluoroborate as an effective promoter of carbon insertion with diazoacetic esters [181]. These authors then extended their method, demonstrating a beautifully successful intramolecular reaction of α-diazoketone **68** (9). Continued expansion of the substrate scope in a later publication revealed the method to be general for non-hindered alkanones and cycloalkanones [182]. The mechanism and regioselectivity of EDA insertion reactions were also explored by Mock and Hartman [183]. Empirical kinetic data showed the rate-determining step to be ketone activation to give ethoxycarbenium tetrafluoroborate ion pairs; subsequent rapid trapping by EDA, 1,2-rearrangement, and dealkylation by diethyl ether furnished the α-alkyl-β-keto ester products. It should be noted that similar reactions of EDA with cyclobutanone, cyclohexanone, and caged dione electrophiles have been published with the workhorse $BF_3 \cdot Et_2O$ as promoter [184–186]. These historically important accounts served to foster a greater awareness of the merits of such "off-the-shelf" ring enlargements, and the typical steric biases mentioned in Sect. 2 as controlling factors in regioselective carbon insertion with diazomethane were further validated. As in Greene's [81] memorable investigation with α-chloro bicyclobutanones (see Scheme 8), Dave and Warnhoff also showed that halogen substitution could be employed as a means of directing ring expansion with EDA [187].

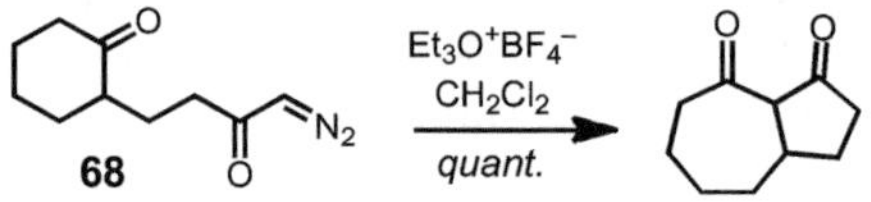


(9)

More recent reports from the Maruoka group show that the stereoselective addition of stabilized diazo nucleophiles to cycloalkanones continues to advance. A logical starting point is the enlargement of six- to seven-membered rings [188], owing to a prevalence of seven-membered ring frameworks in biologically active compounds and the myriad paths for six-membered ring construction available to synthetic chemists. Maruoka began by exploring the stereochemical outcome of the reaction of 4-substituted cyclohexanones and diazoacetates under $BF_3 \cdot Et_2O$ catalysis (Fig. 6) [189]. Similar to what was shown by the prior work of Mock and Hartmann [182, 183], minimization of non-bonding interactions and charge separation coupled with ring conformation analysis were highlighted as factors contributing to stereoselectivity. Initial equatorial attack gives rotamers **69** and **70**. A preferred C–C bond shift via **69** results in the observed stereochemical outcome.

To probe further the relevance of the above Newman projections, silyl ether **71** was examined under identical reaction conditions. The well-established [190, 191] preference for 4-siloxy and alkoxy groups to occupy the axial position in chair

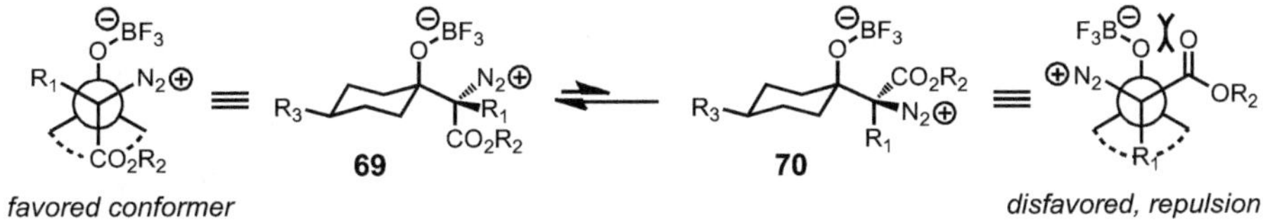


**Fig. 6** Equatorial addition modes for diazoester addition to a prochiral ketone

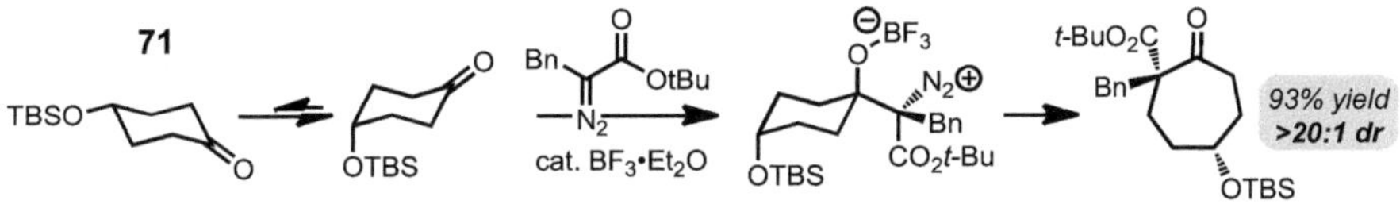


**Scheme 29** With an axially-disposed siloxy substituent, preference for equatorial addition is preserved, furnishing a *trans*-2,4-substituted cycloheptanone in high diastereomeric excess

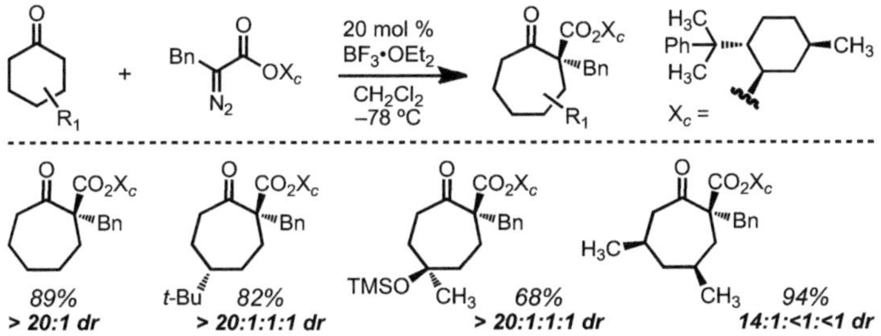


**Scheme 30** Maruoka's versatile auxiliary applied to complex cycloheptyl adducts

cyclohexanones gave the anticipated stereochemistry and bolstered the conclusion made in the opening analysis (Scheme 29).

Having invented the chiral (–)-phenyl-menthyl diazoacetate auxiliary that had served so well in their homologation of aldehydes, Maruoka and coworkers were in a prime position to immediately translate their observations into medium ring syntheses in which relative and absolute configurations could be controlled. As shown in Scheme 30 for a representative α-benzylated reagent, high levels of diastereoselection were transferred to an array of final products, all of which were derived from readily accessible achiral or prochiral cyclohexanones [189]. The power with which these single-step transformations can install remote stereogenic centers, many of which are quaternary, is impressive.

Maruoka and coworkers have further introduced the binaphthol derivative **72** as a chiral ligand for Al-catalyzed desymmetrization reactions with 4-substituted cyclohexanones (Scheme 31) [192]. Though relatively high catalyst loadings were required for proper efficiency, their findings are exceedingly novel in light of the fact that, up until this point, turnover with an Al-based Lewis acid had never been observed. The active species was readily prepared by mixing a twofold excess

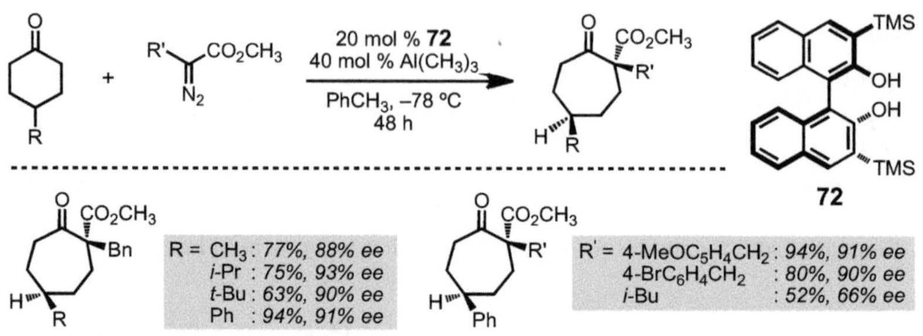


**Scheme 31** Maruoka's Al-catalyzed ring expansion with a promising early scope

of a commercial trimethylaluminum solution with **72**. The illustrated collection of products confirms that good enantioselectivities and yields were obtained for a number of ring substituents and internal diazoester reagents. An additional twist on reaction utility was evident when a terminal diazoacetate insertion was teamed with subsequent ester hydrolysis and decarboxylation. The authors deemed this a "traceless" desymmetrization and obtained 4-phenyl cycloheptanone in 97% *ee* and 54% overall yield.

## 4.4 *Ketone α-Arylation by Ring Expansion with Aryldiazomethanes*

The α-functionalization of "naked" cyclic ketones with control over absolute stereochemistry remains a formidable challenge in synthetic chemistry. Extensive work by Maruoka has gone far to present practical procedures for constructing α-quaternary carbons in substituted cycloalkanones with high levels of both relative and absolute stereochemical control. Still, a universal reliance on α-diazocarbonyl reagents carries unyielding restrictions on the types of products one can prepare. What if a deceptively simple α-monosubstituted cycloheptanone is desired, and how does one counteract the propensity for facile racemization in such structures? Furthermore, how many synthetic tools must be developed before other ring sizes can be accessed in stereochemically pure form? Even now, as multiple research groups have achieved groundbreaking advances in carbonyl homologation, we must stop and remind ourselves of how far we have to go. It is difficult to forecast for how long, but stark deficiencies in scope and generality continue to persist.

As discussed in Sect. 2.3, Kingsbury and Moebius were the first to invoke inexpensive scandium(III) salts as potent catalysts for carbonyl homologation with nonstabilized diazoalkanes. An enduring goal of that program is the development of general methods for elaborating alkyl-, vinyl-, and aryl-bearing stereogenic centers adjacent to carbonyl functions. As a first step in this direction, the authors' lab chose to study α-arylation processes on the basis of three crucial considerations: (1) aryl diazoalkanes are considerably more stable than their alkyl or vinyl

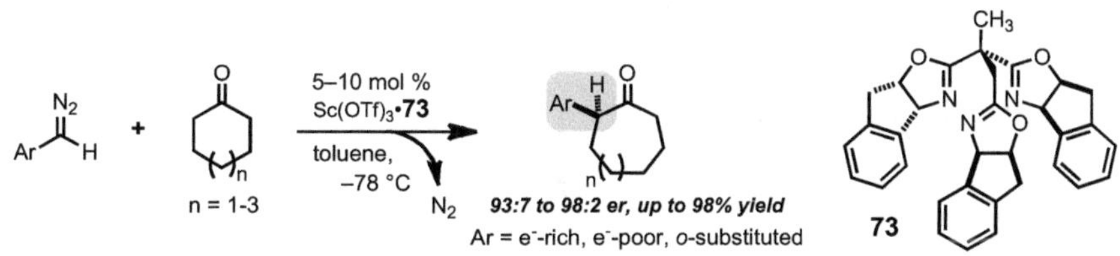


**Scheme 32** Enantioselective medium ring α-arylation with an Sc-"trisox" complex

counterparts, (2) monoaryl diazomethanes offer the best-case scenario for steric differentiation and facial discrimination at the prochiral diazoalkyl carbon, and (3) no synthetic methods exist for enantioselective tertiary α-arylation, as cross-coupling strategies call for metal enolates that promote facile racemization.

A screen of bis(oxazoline) ligands in the Sc-catalyzed union of cyclohexanone and phenyldiazomethane revealed promising but just shy of satisfactory results. Ultimately, the authors deduced that trace amounts of moisture were a detriment to high enantioselectivity. This was not surprising given the tendency of scandium to experience polyvalency, not to mention the fact that only hydrated forms of the trication have been successfully crystallized for X-ray diffraction [193–196]. Extensive optimization of the parent reaction, along with a decision to turn to $C_3$-symmetric tris(oxazolines) [197, 198] to discourage adventitious hydration, has allowed a wealth of enantioselective formal ketone α-arylations to be uncovered [110, 118]. As shown in Scheme 32, the optimal "trisox" ligand **73** derived from the Merck aminoindanol delivered homologated products bearing a diversity of aryl substituents with high selectivities and yields. Reactions with cyclobutanone and cyclopentanone were less selective, so the catalyst system was most effective for medium seven-, eight-, and nine-membered rings. It should be noted that high enantiomeric purity was seen even for powerfully electron-withdrawing *para* substituents such as $NO_2$ and $CF_3$. Resubjection of isolated products to the Sc-ligand complex or excess diazoalkane returned products without any erosion of enantiopurity.

Before proposing a stereochemical model, information was gathered about the trajectory of the diazoalkane reagent. The authors designed a diastereoselective homologation reaction similar to that first conducted by Yamamoto in 1994 (see Scheme 12). Treatment of 4-*tert*-butylcyclohexanone with *p*-tolyldiazomethane in the presence of 10 mol % $Sc(OTf)_3$ led to the formation of *trans* insertion product **74** in 96:5:3.5 dr (by achiral GC analysis; Scheme 33). Crystallization of the major diastereomer allowed confirmation of its relative configuration. Consistent with data reported in the literature [99, 100], the stereoselectivity with stoichiometric trimethylaluminum was lower (82:18 dr). With 10 mol % $Sc(OTf)_3$ and ligand **73**, enantio- and diastereoselective reaction furnished **74** in 93:7 dr with 92.5:7.5 enantiomer ratio (er) for the major enantiomer [118]. This outcome can be rationalized by invoking axial approach of the nucleophile in an orientation that places the proton over the six-membered ring to minimize penalizing steric interactions

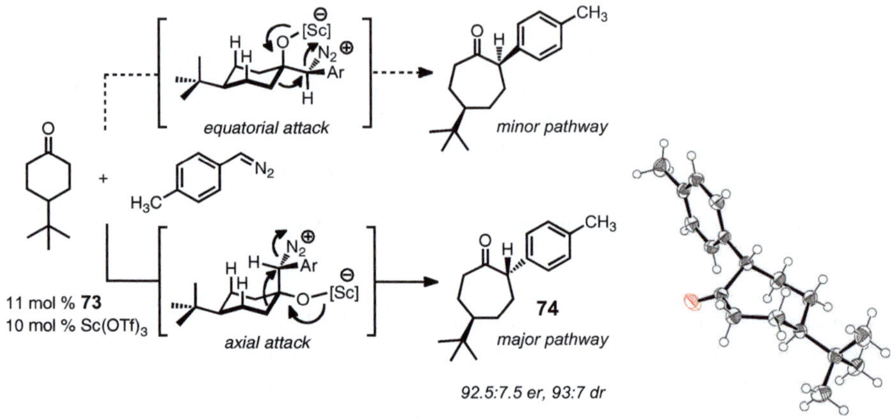


**Scheme 33** Diastereo- and enantioselective Sc-catalyzed ring expansion reaction

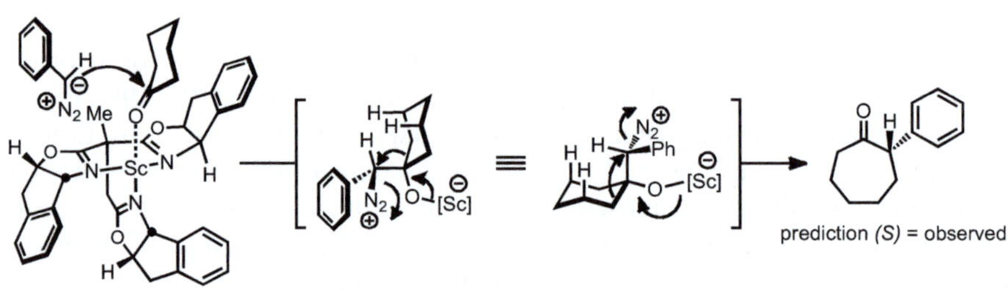


**Scheme 34** Stereochemical model for asymmetric induction

between the aryl group and the ring. The principle of least motion states that "those elementary reactions will be favored that involve the least change in atomic position and electronic configuration" [199]. The major diastereomer is correctly predicted by assuming the betaine intermediate undergoes least motion collapse directly from the drawn conformation and without C–C bond rotation (Scheme 33).

A predictive stereochemical rationale incorporating these features was then advanced (Scheme 34). The enantioselectivity of Sc-catalyzed α-arylation derives from control over the orientation with which aryldiazoalkanes add to symmetric cycloalkyl ketones. The triflate counterions (omitted for clarity) and ligand **73** establish a chiral pocket that forces the nucleophile to approach over the open side of the ligand (from the left), in an orientation such that the large group is directed away from the metal and the proton is centered over the cycloalkanone ring. The newly formed C–C bond initially occupies an axial position, and then concerted collapse with expulsion of nitrogen delivers the *S* product. This prediction was in agreement with the observed selectivity.

# 5 Applications of Single Carbon Insertion in Total Synthesis

Single-carbon ring expansion is a powerful synthetic transform and has been successfully implemented in a number of complex molecule syntheses. As discussed in Sect. 2, diazoalkane-based ring expansions have been subject to incremental, but nonetheless significant, advances over the years. More recent methodologies, based on the findings of Shioiri [92] and Yamamoto [99, 100], have made their way into the total syntheses of naturally-occurring and synthetic bioactive complex molecules. Chemists will often construct or purchase the lower homologue of a ring system, utilize known methods to build up complexity, and then implement a key ring expansion event to access the target ring size.

Polycyclic ether marine natural products, especially those belonging to the brevotoxin family, have been linked to cases of neurotoxic shellfish poisoning [200]. The discovery of these molecules and their corresponding biological effects led to the development of new synthetic strategies to access the *trans*-fused six- to nine-membered polycyclic ether frameworks common to these natural products [201]. In 1997, Mori and coworkers published a strategy based on iterative ring expansion of the corresponding six-membered lower homologue to prepare seven-membered oxepane rings [202–204]. Equation (10) shows the best result after a Lewis acid screen on model substrate **75**. The highest yields and regioselectivities were observed with the Shioiri [92] conditions at −78°C. As expected, preferential migration of the less substituted bond, followed by 1,3-Brook rearrangement [93–96], yielded **76**, which was deprotected with PPTS to afford the target oxepane **77** in 76% yield over two steps.

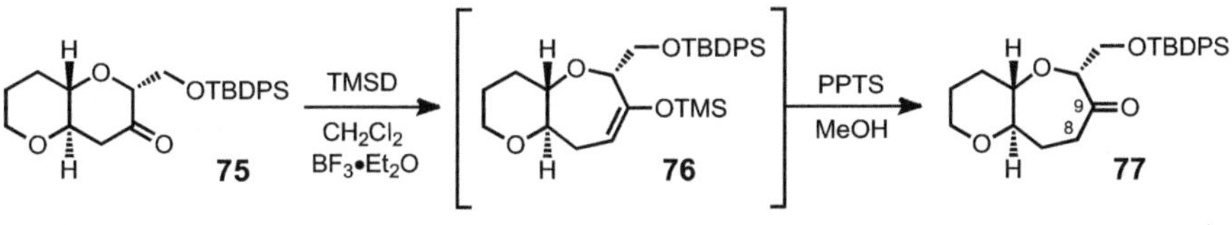


(10)

Satisfied with the model studies, Mori utilized this ring expansion strategy in a formal synthesis of hemibrevetoxin B (Scheme 35). Lewis acid-mediated ring closure of **78** and deoxygenation afforded cyclohexanone homologation substrate **79**. Single carbon ring expansion under highly optimized conditions yielded the first seven-membered ether **80** in a 67% yield. After a series of manipulations, **81** was obtained and subsequently homologated to introduce the second seven-membered ring. Mori was able to execute successfully two regioselective single-carbon ring expansion events and secure intermediate **82**, which could then be elaborated to hemibrevetoxin B.

In Pazos' 2009 synthesis of isolaurepan, a similar homologation strategy was employed to produce the oxepane ring system found in the desired target

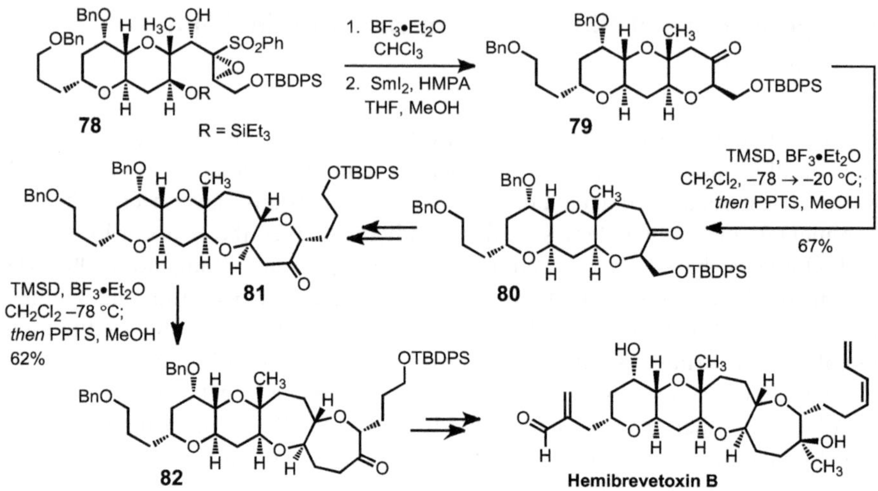


**Scheme 35** Mori's formal synthesis of hemibrevitoxin B featuring iterative ring expansions

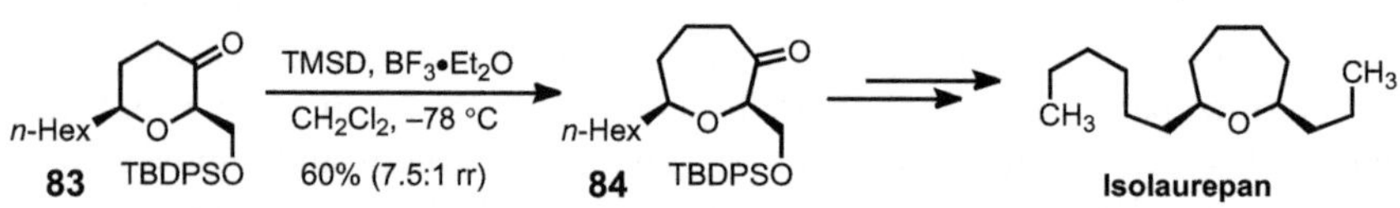


**Scheme 36** Pazos' total synthesis of isolaurepan

(Scheme 36) [205]. Treatment of α-tertiary substituted cyclohexanone **83** with $BF_3 \cdot Et_2O$ and TMSD afforded the seven-membered ether **84** in a respectable 60% isolated yield. Again, preferential migration of the less substituted carbon atom was observed to deliver a 7.5:1 mixture of regioisomers. The late stage homologation product **84** was then advanced to the target isolaurepan with four additional steps.

In Seto's synthesis of 6a-carbabrassinolide, a regioselective ring expansion facilitated concise access to the target steroid derivative (Scheme 37) [206]. Global acetate protection (**85**→**86**) prevented formation of methyl ethers by O–H insertion. Seto proposed that the diazoalkane addition places the bulky TMS group away from the ring fusion and the proton pointed inside of the ring system (**87**). This simple model, based on minimization of steric interactions, correctly predicts the regiochemical outcome in the previous two examples as well. Seto obtained the desired heptanone with 11:1 regioselectivity and an excellent 80% yield. Base-mediated global acetate deprotection delivered 6a-carbabrassinolide.

In Smalley's approach to the novel antiviral compound TAK-779 (Scheme 38), a decagram scale, highly regioselective single carbon ring expansion was employed to form the crucial benzofused seven-membered carbocycle [207]. Starting from inexpensive and commercially available 7-methoxy-1-tetralone, biaryl tetralone **88**

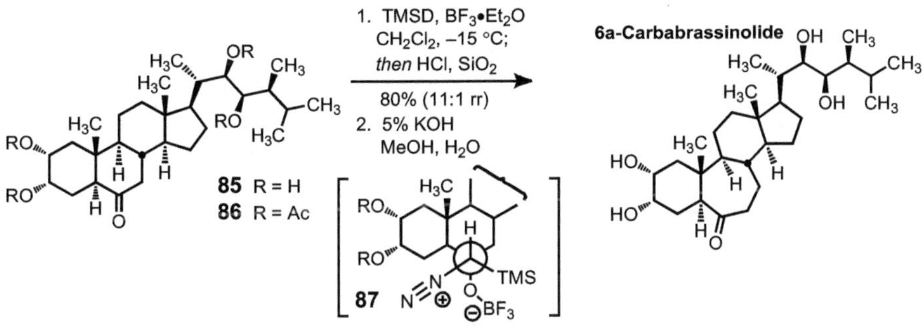


**Scheme 37** Seto's synthesis of 6a-carbabrassinolide

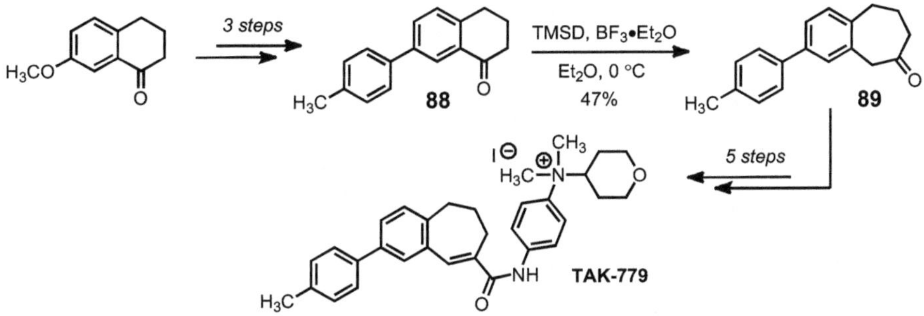


**Scheme 38** Smalley's entry to TAK-779 with highly selective ring enlargement

was quickly accessed through a three-step sequence. Ring expansion with $BF_3 \cdot Et_2O$ and TMSD afforded the desired suberone **89** in multi-gram quantities as a single regioisomer on the basis of $^1$H NMR spectroscopy. The high preference for migration of the aryl bond can be rationalized by an electronic orbital overlap argument. Diazoalkane insertion reactions with aldehydes typically afford ketone products, formed by preferential migration of the carbonyl C–H bond [116, 117]. The spherical, non-directional nature of the hydrogen *s* orbital allows for facile migration. In Smalley's example, the migrating carbon center is $sp^2$ hybridized, resulting in a less directional orbital that can overlap more readily with the C–N σ* orbital. This migration preference is also consistent with a previous report by House that showed a strong preference for phenyl vs alkyl migration with diazomethane and $BF_3 \cdot Et_2O$ [89]. The synthesis was completed in five additional steps, providing scalable access to TAK-779.

The reaction of diazomethane with α,β-unsaturated carbonyl compounds under classical protic conditions has been shown to produce pyrazoline products arising from 1,3-dipolar cycloaddition [3, 4]. Limited examples of α,β-unsaturated carbonyl substrates undergoing ring expansion in the presence of Lewis acid promoters have been reported. It was not until the introduction of Lewis acids for diazoalkane ring expansion that these types of substrates were even reactive. In Drège's

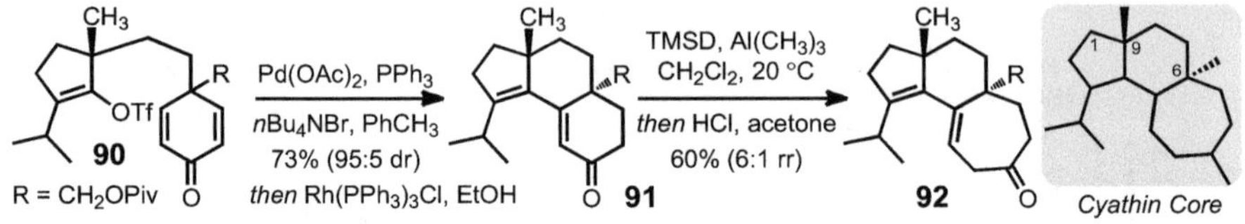


**Scheme 39** Drège's expansion approach to the cyathin terpenoid carbon skeleton

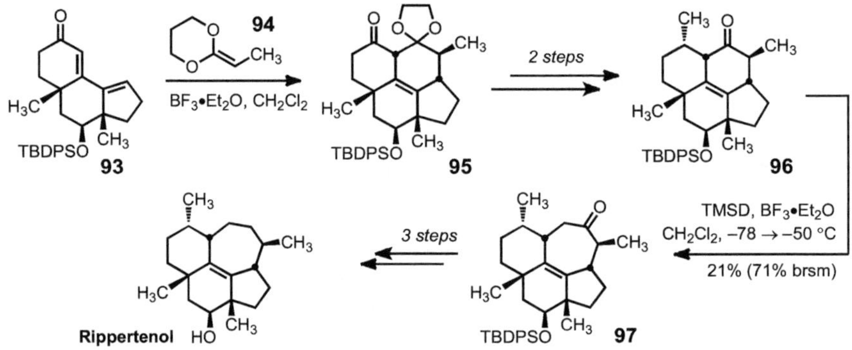


**Scheme 40** Snyder's remarkably concise and convergent synthesis of rippertenol

synthesis of the cyathin terpenoid framework, an intramolecular Heck reaction (**90**→**91**; Scheme 39) set the stage for a rare α,β-unsaturated cyclohexenone ring expansion [208, 209]. Under the Yamamoto [99, 100] conditions, cyclohexenone **91** was smoothly converted to the desired β,γ-unsaturated cycloheptanone **92** in a 60% isolated yield with 6:1 regioselectivity.

In arguably one of the most challenging single carbon homologations to date, the Snyder group attempted to homologate an exceptionally crowded α,α′-disubstituted cyclohexanone during their synthesis of rippertenol (Scheme 40) [210]. A Lewis acid-promoted inverse demand Diels–Alder reaction between electron deficient diene **93** and ketene acetal **94** furnished the carbon skeleton in cyclohexanone ketal **95**. Two additional steps, a Lombardo–Takai olefination with acid workup followed by hydrogenation, unmasked the ketone for ring expansion (**95**→**96**). Extensive screening led to modified Shioiri conditions as the optimum means to obtain cycloheptanone **97**, though it was only recovered in 21% yield under fully optimized conditions. The regiochemical outcome was not anticipated. However, it was of little consequence as the ketone was removed in subsequent steps. To avoid epimerization of the adjacent methyl stereocenter, a two-step reduction/radical deoxygenation strategy preceded silyl group removal to deliver the targeted natural product.

## 6 Conclusions

Snyder's daring approach to rippertenol, and other complex ring expansions just presented, go far to underscore the strategic value of single carbon homologation. In addition, in spite of the fact that a number of the routes showcase successful and high yielding methylene insertions, none of the applications are catalytic! This is no longer due to lack of available promoter or catalyst systems that can be tested under substoichiometric conditions, but rather due to the highly sensitive nature of the reactive diazoalkyl function itself. It is fundamentally a Lewis base that can interact unfavorably with all manner of Lewis acids, especially when a carbonyl acceptor is not present or prohibitively hindered. Our own recent investigations with TMSD show that in the presence of $Sc(OTf)_3$ and the absence of a carbonyl acceptor, the reagent is immediately destroyed upon the introduction of moisture. Therein lies a pressing need for even milder and more selective catalysts for formal methylene insertion.

Continued engagement in diazoalkane research is vital to the field of synthetic chemistry. Chemists have developed safer and more efficient pathways to access all varieties of diazo compounds, and the pace with which novel applications are appearing in the literature has quickened. While stabilized diazocarbonyl reagents already enjoy a strong foothold as precursors to electrophilic metal carbenes and as 1,3-dipolarophiles, only recently have chemists taken a deliberate interest in nonstabilized species, which necessarily confer much greater structural diversity. Future work, particularly in enantioselective catalysis, will undoubtedly add to the ever-growing synthetic chemistry toolbox through the discovery of new ways to harness the reactivity and selectivity of these reagents.

## References


1. Curtius T (1883) Ber Dtsch Chem Ges 16:2230
2. Pechmann HV (1891) Ber Dtsch Chem Ges 27:1888
3. Gutsche CD (1954) Org Reactions 8:364
4. Gutsche CD, Redmore D (1968) In: Hart H, Karabatsos GJ (eds) Carbocyclic ring expansion reactions. Academic, New York, pp 81–98
5. Holmquist CR, Roskamp EJ (1989) J Org Chem 54:3258
6. Holmquist CR, Roskamp EJ (1992) Tetrahedron Lett 33:1131
7. Bug T, Hartnagel M, Schlierf C, Mayr H (2003) Chem Eur J 9:4068
8. Mayr H, Bug T, Gotta MF, Hering N, Irrgang B, Janker B, Kempf B, Loos R, Ofial AR, Remennikov G, Schimmel H (2001) J Am Chem Soc 123:9500
9. Regitz M, Maas G (1986) Diazo compounds–properties and synthesis. Academic, Orlando
10. Doyle MP, McKervey MA, Ye T (1998) Modern catalytic methods for organic synthesis with diazo compounds. Wiley, New York
11. Ye T, McKervey MA (1994) Chem Rev 94:1091
12. De Boer TJ, Backer HJ (1963) Organic Syntheses Collective, vol 4. Wiley, New York, p 250
13. Hudlicky M (1980) J Org Chem 45:5377

14. Clark J, Shah A, Peterson J, Patelis L, Kersten R, Heemskerk A, Grogan M, Camden S (2002) Thermochim Acta 386:65
15. Clark J, Shah A, Peterson J, Patelis L, Kersten R, Heemskerk A (2002) Thermochim Acta 386:73
16. Proctor LD, Warr AJ (2002) Org Process Res Dev 6:884
17. Furrow ME, Myers AG (2004) J Am Chem Soc 126:12222
18. Kühnel E, Laffan DDP, Lloyd-Jones GC, Martínez Del Campo T, Shepperson IR, Slaughter JL (2007) Angew Chem Int Ed 46:7075
19. Roberts JD, Watanabe W (1950) J Am Chem Soc 72:4869
20. Roberts JD, Watanabe W, McMahon RE (1951) J Am Chem Soc 73:760
21. Roberts JD, Watanabe W, McMahon RE (1951) J Am Chem Soc 73:2521
22. Roberts JD, Regan CM (1952) J Am Chem Soc 74:3695
23. Müller E, Bauer M, Rundel W (1959) Z Naturforsch 14b:209
24. Müller E, Rundel W (1958) Angew Chem 70:105
25. Müller E, Meischkeil R, Bauer M (1964) Annalen 677:55
26. Müller E, Rundel W, Huber-Emden H (1957) Angew Chem 69:614
27. Müller E, Huber-Emden H (1961) Annalen 649:70
28. Müller E, Huber-Emden H, Rundel W (1959) Annalen 623:34
29. Neeman M, Caserio MC, Roberts JD, Johnson WS (1959) Tetrahedron 6:36
30. Neeman M, Caserio MC, Roberts JD, Johnson WS (1958) J Am Chem Soc 80:2584
31. Pechmann HV (1895) Ber Dtsch Chem Ges 28:855
32. Smith LI (1938) Chem Rev 23:193
33. Huisgen R, Grashey R, Sauer J (1964) In: Patai S (ed) The chemistry of alkenes. Interscience, London, p 739
34. Huisgen R (1968) J Org Chem 33:2291
35. Atherton JH, Fields R (1968) J Chem Soc C 1507
36. Barlow MG, Hazseldine RN, Morton WD (1969) Chem Commun 931
37. Manecke G, Schenck HG (1968) Tetrahedron Lett 9:2061
38. Tabushi I, Takagi K, Okano M, Oda R (1967) Tetrahedron 23:2621
39. D'Yakonov IA, Repinskaya IB, Golodnikov GV (1966) Zh Org Khim 2:2256
40. Kadaba PK (1966) Tetrahedron 22:2453
41. Ledwith A, Parry D (1966) J Chem Soc B 1408
42. Doyle MP (1986) Chem Rev 86:919
43. Doyle MP, Forbes DC (1998) Chem Rev 98:911
44. Aggarwal VK, Ferrara M, O'Brien C, Thompson A, Jones R, Fieldhouse R (2001) J Chem Soc Perk T 1 1635
45. Aggarwal VK, Alonso E, Fang G, Ferrara M, Hynd G, Porcelloni M (2001) Angew Chem Int Ed 40:1433
46. Aggarwal VK, Alonso E, Bae I, Hynd G, Lydon KM, Palmer MJ, Patel M, Porcelloni M, Richardson J, Stenson RA, Studley JR, Vasse J, Winn CL (2003) J Am Chem Soc 125:10926
47. Acevedo O, Ross B, Andrews RS, Springer R, Cook PD (1995) Apparatus and processes for the large scale generation and transfer of diazomethane (Isis Pharmaceuticals). US Patent 549243
48. Archibald TG, Huang DS, Pratton MH, Barnard JC (1998) Large scale batch process for diazomethane (Aerojet General Corporation). US Patent 5,817,778
49. Archibald TG, Barnard JC, Harlan RF (1998) Continuous process for diazomethane from an N-methyl-N-nitrosoamine and from methylurea through *N*-methyl-*N*-nitrosourea (Aerojet General Corporation). US Patent 5,854,405
50. Maas G (2009) Angew Chem Int Ed 48:8186
51. Myers EL, Raines RT (2009) Angew Chem Int Ed 48:2359
52. Holton TL, Shechter H (1995) J Org Chem 60:4725
53. McGuiness M, Shechter H (2002) Tetrahedron Lett 43:8425
54. Javed MI, Brewer M (2007) Org Lett 9:1789

55. Wender PA, Handy ST, Wright DL (1997) Chem Ind 765
56. Buchner E, Curtius T (1885) Ber Dtsch Chem Ges 18:2377
57. Pechmann HV, Frobenius L (1895) Ber Dtsch Chem Ges 28:170
58. Meyer H (1905) Monatsh Chem 26:1295
59. Schlotterbeck F (1907) Ber Dtsch Chem Ges 40:479
60. Schlotterbeck F (1909) Ber Dtsch Chem Ges 42:2559
61. Eistert B (1948) Newer methods of preparative organic chemistry. Interscience, New York, p 521, English edition
62. Meerwein H, Burneleit W (1928) Ber Dtsch Chem Ges 61:1840
63. Meerwein H (1933) Verfahren zur Umsetzung Organischer Verbindungen mit Diazomethan. German Patent 579,309
64. Smith PAS, Baer DR (1960) Org Reactions 11:157
65. Woodward RB, Gosteli J, Ernest I, Friary RJ, Nestler G, Raman H, Sitrin R, Suter C, Whitesell JK (1973) J Am Chem Soc 95:6853
66. Mosettig E, Burger A (1930) J Am Chem Soc 52:3456
67. Heller G (1919) Ber Dtsch Chem Ges 52:741
68. Heller G (1926) Ber Dtsch Chem Ges 59:704
69. Giraitis AP, Bullock JL (1937) J Am Chem Soc 59:951
70. Kohler EP, Tishler M, Potter H, Thompson HT (1939) J Am Chem Soc 61:1057
71. Adamson DW, Kenner J (1939) J Chem Soc 181
72. Gutsche CD (1949) J Am Chem Soc 71:3513
73. Gutsche CD, Strohmayer HF, Chang JM (1958) J Org Chem 23:1
74. Hammett LP (1937) J Am Chem Soc 59:96
75. Goodman RM, Kishi Y (1998) J Am Chem Soc 120:9392
76. Nakamura K, Osamura Y (1993) J Am Chem Soc 115:9112
77. Berson JA, Suzuki S (1959) J Am Chem Soc 81:4088
78. Gutsche CD, Johnson HE (1955) J Am Chem Soc 77:109
79. Gutsche CD, Johnson HE (1955) Org Synth 35:91
80. Gutsche CD, Jason EF (1956) J Am Chem Soc 78:1184
81. Greene AE, Depres JP (1979) J Am Chem Soc 101:4003
82. Jaz J, Davreux JP (1965) Bull Chim Soc Belg 74:370
83. Roberts JD, Gorham WF (1952) J Am Chem Soc 74:2278
84. Reeder LM, Hegedus LS (1999) J Org Chem 64:3306
85. Black TH (1983) Aldrichim Acta 16:3
86. Sammakia T (1995) In: Paquette LA (ed) Encyclopedia of reagents for organic synthesis. Wiley, Chichester, UK, p 1512
87. Bamford WR, Stevens TS (1952) J Chem Soc 4735
88. Müller E, Bauer M, Rundel W (1960) Tetrahedron Lett 1:30
89. House HO, Grubbs EJ, Gannon WF (1960) J Am Chem Soc 82:4099
90. Goubeau J, Rohwedder KH (1957) Liebigs Ann Chem 604:168
91. Tai WT, Warnhoff EW (1964) Can J Chem 42:1333
92. Hashimoto N, Aoyama T, Shioiri T (1980) Tetrahedron Lett 21:4619
93. Brook AG (1974) Acc Chem Res 7:77
94. Moser WH (2001) Tetrahedron 57:2065
95. Brook AG, Limburg WW, MacRae DM, Fieldhouse SA (1967) J Am Chem Soc 89:704
96. Corey EJ, Rücker C (1984) Tetrahedron Lett 25:4345
97. Shioiri T, Aoyama T, Mori S (1990) Org Synth 68:1
98. Kemsley JN (2011) Chem Eng News 89:15
99. Maruoka K, Concepcion AB, Yamamoto H (1994) Synthesis 1283
100. Maruoka K, Concepcion AB, Yamamoto H (1994) J Org Chem 59:4725
101. Müller E, Bauer M (1962) Liebigs Ann Chem 654:92
102. Hoberg H (1966) Angew Chem Int Ed 5:513
103. Youn J-H, Lee J, Cha JK (2001) Org Lett 3:2935

104. Johnson WS, Neeman M, Birkeland SP, Fedoruk NA (1962) J Am Chem Soc 84:989
105. Moebius DC, Kingsbury JS (2009) J Am Chem Soc 131:878
106. Moebius DC (2011) Ph.D. Dissertation, Boston College, Chestnut Hill
107. Evans DA, Fandrick KR, Song H-J, Scheidt KA, Xu R (2007) J Am Chem Soc 129:10029
108. Wu J (2005) Ph.D. Dissertation, Harvard University, Cambridge
109. Shankar BKR, Shechter H (1982) Tetrahedron Lett 23:2277
110. Rendina VL, Kaplan HZ, Kingsbury JS (2012) Synthesis 686
111. Dabrowski JA, Moebius DC, Wommack AJ, Kornahrens AF, Kingsbury JS (2010) Org Lett 12:3598
112. Seyferth D, Dow AW, Menzel H, Flood TC (1968) J Am Chem Soc 90:1080
113. Hashimoto N, Aoyama T, Shioiri T (1982) Chem Pharm Bull 30:119
114. Shioiri T, Aoyama T, Mori S (1990) Org Synth 68:1
115. Fleming I, Sanderson PEJ (1987) Tetrahedron Lett 28:4229
116. Wommack AJ, Moebius DC, Travis AL, Kingsbury JS (2009) Org Lett 11:3202
117. Wommack AJ, Kingsbury JS (2013) J Org Chem 78:10573–10587
118. Rendina VL, Moebius DC, Kingsbury JS (2011) Org Lett 13:2004
119. Arndt F (1935) Org Synth 15:3
120. Harman WW, Phillips R (1933) Org Synth 13:84
121. de Boer TJ, Backer HJ (1956) Org Synth 36:16
122. Lieser T, Beck G (1950) Berichte 83:137
123. Hart H, Brewbaker JL (1969) J Am Chem Soc 91:706
124. Wiseman JR, Chan H (1970) J Am Chem Soc 92:4749
125. Miyashita M, Yoshikoshi A (1974) J Am Chem Soc 96:1917
126. Thomas RC, Fritzen EL (1988) J Antibiot 41:1445
127. Fulton J, Aggarwal VK, de Vincente J (2005) Eur J Org Chem 1479
128. Aggarwal VK, Ford JG, Thompson A, Jones RVH, Standen MCH (1996) J Am Chem Soc 118:7004
129. Aggarwal VK (1998) Synlett 329
130. Aggarwal VK, Vasse J (2003) Org Lett 5:3987
131. Adams LA, Aggarwal VK, Bonnert RV, Bressel B, Cox RJ, Shepherd J, de Vincente J, Walter M, Whittingham WG, Winn CL (2003) J Org Chem 68:9433
132. Aggarwal VK, de Vincente J, Bonnert RV (2001) Org Lett 3:2785
133. Angle SR, Neitzel ML (2000) J Org Chem 65:6458
134. Aggarwal VK, de Vincente J, Pelotier B, Holmes IP, Bonnert RV (2000) Tetrahedron Lett 41:10327
135. Trost BM, Higuchi RI (1996) J Am Chem Soc 118:10094
136. Zhang J, Chan P, Che C (2003) Tetrahedron Lett 44:8733
137. Hu W, Timmons DJ, Doyle MP (2002) Org Lett 4:901
138. Pross A, Sternhill S (1970) Aust J Chem 23:989
139. Overberger CG, Anselme J (1964) J Org Chem 29:1188
140. Abelt CJ, Pleier JM (1989) J Am Chem Soc 111:1795
141. Smith LI, Howard KL (1944) Org Synth 24:53
142. Kotali A (2002) Curr Org Chem 6:965
143. Gale D, Middleton WJ, Krespan C (1966) J Am Chem Soc 88:3617
144. Ciganek E (1970) J Org Chem 35:862
145. Bernard RE (1967) Ph.D. Dissertation, The Ohio State University, Columbus
146. McBee ET, Sienkowski KJ (1973) J Org Chem 38:1340
147. Ciganek E (1965) J Am Chem Soc 87:652
148. Creary X (1986) Org Synth 64:207
149. McGuiness M, Shechter H (1990) Tetrahedron Lett 31:4987
150. Barton DHR, O'Brien RE, Sternhell S (1962) J Chem Soc 470
151. Adamson J, Bywood R, Eastlick DT, Gallagher G, Walker D, Wilson EM (1975) J Chem Soc Perk T 1:2030

152. Gallagher G (1978) US Patent 4,083,837
153. Eastlick DT (1978) US Patent 4,092,306
154. ter Wiel MKJ, Vicario J, Davey SG, Meetsma A, Feringa BL (2005) Org Biomol Chem 3:28
155. Nicolaou KC, Mathison CJN, Montagnon T (2004) J Am Chem Soc 126:5192
156. Weiss R, Seubert J (1994) Angew Chem Int Ed 33:891
157. Lapatsanis L, Milias G, Paraskewas S (1985) Synthesis 513
158. Furrow ME, Myers AG (2004) J Am Chem Soc 126:5436
159. Brewer M (2006) Tetrahedron Lett 47:7731
160. Justo de Pomar JC, Soderquist JA (2000) Tetrahedron Lett 41:3285
161. Javed MI, Brewer M (2008) Org Synth 85:189
162. Gassman PG, Greenlee WJ (1973) Org Synth 53:38
163. Rendina VL, Kingsbury JS (2012) J Org Chem 77:1181
164. Curtin DY, Gerber SM (1952) J Am Chem Soc 74:4052
165. Köpke T, Zaleski JM (2008) Anticancer Agents Med Chem 8:292
166. Lei X, Porco JA (2006) J Am Chem Soc 128:14790
167. Nicolaou KC, Li H, Nold AL, Pappo D, Lenzen A (2007) J Am Chem Soc 129:10356
168. Woo CM, Lu L, Gholap SL, Smith DR, Herzon SB (2010) J Am Chem Soc 132:2540
169. Herzon SB, Lu L, Woo CM, Gholap SL (2011) J Am Chem Soc 133:7260
170. Buchner E, Curtius T (1885) Ber Deut Bot Ges 18:2371
171. Staudinger H, Gaule A (1916) Ber Deut Bot Ges 49:1951
172. Nomura K, Iida T, Hori K, Yoshi E (1994) J Org Chem 59:488
173. Mahmood S, Hossain M (1998) J Org Chem 59:488
174. Kanemasa S, Kanai T, Araki T, Wada E (1999) Tetrahedron Lett 40:5055
175. Hashimoto T, Naganawa Y, Kano T, Maruoka K (2008) J Am Chem Soc 130:2434
176. Hashimoto T, Naganawa Y, Kano T, Maruoka K (2007) Chem Commun 5143
177. Hashimoto T, Miyamoto H, Naganawa Y, Maruoka K (2009) J Am Chem Soc 131:11280
178. Li W, Wang J, Hu X, Shen K, Wang W, Chu Y, Lin L, Liu X, Feng X (2010) J Am Chem Soc 132:8532
179. Gutsche CD, Hillman M (1954) J Am Chem Soc 76:2236
180. Kharasch MS, Rudy T, Nudenberg W, Büchi G (1953) J Org Chem 18:1030
181. Mock WL, Hartman ME (1970) J Am Chem Soc 92:5767
182. Mock WL, Hartman ME (1977) J Org Chem 42:459
183. Mock WL, Hartman ME (1977) J Org Chem 42:466
184. Liu HJ, Ogina T (1973) Tetrahedron Lett 14:4937
185. Liu HJ, Majumdar SP (1975) Synthetic Commun 5:125
186. Marchand AP, Annapurna P, Reddy SP, Watson WH, Nagl A (1989) J Org Chem 54:187
187. Dave V, Warnhoff EW (1983) J Org Chem 48:2590
188. Kantorowski EJ, Kurth MJ (2000) Tetrahedron 56:4317
189. Hashimoto T, Naganawa Y, Maruoka K (2009) J Am Chem Soc 131:6614
190. Nagao Y, Goto M, Ochiai M, Shiro M (1990) Chem Lett 19:1503
191. Baghdasarian G, Woerpel KA (2006) J Org Chem 71:6851
192. Hashimoto T, Naganawa Y, Maruoka K (2011) J Am Chem Soc 133:8834
193. Evans DA, Sweeney ZK, Rovis T, Tedrow JS (2001) J Am Chem Soc 123:12095
194. Evans DA, Scheidt KA, Fandrick KR, Lam HW, Wu J (2003) J Am Chem Soc 125:10780
195. Liu Y, Shang D, Zhou X, Liu X, Feng X (2009) Chem Eur J 15:2055
196. Ishikawa S, Hamada T, Manabe K, Kobayashi S (2004) J Am Chem Soc 126:12236
197. Bellemin-Laponnaz S, Gade LH (2002) Chem Commun 1286
198. Gade LH, Bellemin-Laponnaz S (2008) Chem Eur J 14:4142
199. Tee OS (1969) J Am Chem Soc 91:7144
200. Watkins SM, Reich A, Fleming LE, Hammond R (2008) Mar Drugs 6:431
201. Nicolaou KC, Yang Z, Shi G, Gunzner JL, Agrios KA, Gärtner P (1998) Nature 392:264
202. Mori Y, Yaegashi K, Furukawa H (1997) Tetrahedron 53:12917
203. Mori Y, Yaegashi K, Furukawa H (1997) J Am Chem Soc 119:4557

204. Mori Y, Nogami K, Hayashi H, Noyori R (2003) J Org Chem 68:9050
205. Pazos G, Pérez M, Gándara Z, Gómez G, Fall Y (2009) Tetrahedron Lett 50:5285
206. Seto H, Fujioka S, Koshino H, Hayasaka H, Shimizu T, Yoshida S, Watanabe T (1999) Tetrahedron Lett 40:2359
207. Smalley TL (2004) Synthetic Commun 34:1973
208. Drège E, Morgant G, Desmaële D (2005) Tetrahedron Lett 46:7263
209. Drège E, Tominiaux C, Morgant G, Desmaële D (2006) Eur J Org Chem 4825
210. Snyder SA, Wespe DA, von Hof JM (2011) J Am Chem Soc 133:8850

Top Curr Chem (2014) 346: 163–194
DOI: 10.1007/128_2013_505
© Springer-Verlag Berlin Heidelberg 2014
Published online: 15 February 2014

# Asymmetric Transformations via C–C Bond Cleavage

**Laetitia Souillart, Evelyne Parker, and Nicolai Cramer**

**Abstract** Catalytic asymmetric transformations operating by carbon–carbon (C–C) bonds cleavages have emerged as intriguing strategies to access transient organometallic species from different reaction pathways. The reactions and the applicable substrate range have expanded considerably over the last decade. This overview covers the main developments in this field. A major focus is placed on $\beta$-carbon eliminations of strained *tert*-alcohols and related processes which have been shown to be particularly versatile in a broad range of transformations. Furthermore, exciting developments of asymmetric processes based on direct oxidative C–C bond insertion reactions, for instance into the acyl C–C bond of ketones or the C–CN bond of nitriles, are discussed.

**Keywords** $\beta$-Carbon elimination · Asymmetric catalysis · C–C cleavage · Ligands · Oxidative insertion · Palladium · Rhodium · Strain release

## Contents



---

L. Souillart, E. Parker and N. Cramer (✉)
Laboratory of Asymmetric Catalysis and Synthesis, Institute of Chemical Sciences and Engineering, School of Basic Sciences, Ecole Polytechnic Fédérale de Lausanne, 1015 Lausanne, Switzerland
e-mail: nicolai.cramer@epfl.ch

## Abbreviations

| | |
|---|---|
| acac | Acetylacetonate |
| Ac | Acetyl |
| Alk | Alkyl |
| Ar | Aryl |
| BARF | Tetrakis(3,5-bis(trifluoromethyl)phenyl)borate |
| Binap | 2,2′-Bis(diphenylphosphino)-1,1′-binaphthyl |
| Bn | BENZYL |
| cod | Cyclooctadiene |
| Cp | Cyclopentadienyl |
| d | Day(s) |
| dba | Tris(dibenzylideneacetone) |
| dppb | 1,4-Bis(diphenylphosphino)butane |
| dr | Diastereomer ratio |
| DCE | Dichloroethane |
| DME | Dimethoxy ethane |
| DMF | Dimethyl formamide |
| DMPU | 1,3-Dimethyl-3,4,5,6-tetrahydro-2(1*H*)-pyrimidinone |
| DTBM | 3,5-Di-*tert*-butyl-4-methoxyphenyl |
| *ee* | Enantiomeric excess |
| equiv | Equivalent(s) |
| Et | Ethyl |
| h | Hour(s) |
| *i*Pr | Isopropyl |
| L | Ligand |
| LA | Lewis acid |
| mol | Mole(s) |
| M | Metal |
| Me | Methyl |
| Mes | Mesityl, 2,4,6-trimethylphenyl |
| MS | Molecular sieve |
| Np | Naphthyl |
| Ph | Phenyl |
| PMB | 4-Methoxyphenyl |
| quant | Quantitative |
| *t*Bu | *tert*-butyl |
| *tert* | Tertiary |
| TBS | *tert*-Butyldiphenylsilyl |
| Tol | 4-Methylphenyl |
| Ts | Tosyl, 4-toluenesulfonyl |

# 1 Introduction

Over the past decade, the catalytic enantioselective activation and cleavage of carbon–carbon (C–C) bonds has gained increasing interest. Such reactions have the potential to unlock a complementary substrate portfolio and reaction pathways. They offer different methods to difficult-to-access transient organometallic species. In this respect, C–C bonds can be regarded as latent C-[M] equivalents [1–4]. In many cases, strain release plays a major role as the driving force of carbon–carbon bond cleavages. For instance, three- and four-membered rings largely contribute to the success of such cleavages with a freed strain between 18 and 28 kcal $mol^{-1}$ [5]. However, reactions with less strained molecules are known as well. The generation of the strong carbonyl C=O bond as driving force is also a viable and well exploited strategy. Furthermore, the generation of intermediates having strong carbon metal bonds is an additional factor. Besides the low intrinsic propensity of C–C single bonds towards cleavage, catalytic enantioselective processes add another layer of complexity. The catalyst needs not only to promote the reaction but also to imprint efficiently the chiral environment of the catalyst onto the formed product. Often, elevated reaction temperatures are required for the reactivity of such transformations –usually working against high enantioselectivity – making these processes challenging. Nevertheless, significant advances have been made during the decade. Generally, one can distinguish between two different forms of asymmetric transformation via C–C bond cleavages. In the first, the C–C bond cleavage step is the selectivity determining event (Scheme 1). The substrate can be achiral having two enantiotopic C–C bonds of which one is then selectively addressed. By the same strategy, *meso*-compounds are desymmetrized. With a racemic chiral starting material, kinetic resolutions can be realized by a faster C–C bond cleavage of the preferred enantiomer. For the second type of reactions, the selectivity determining step is located after the C–C bond cleavage event. This usually involves the addition of the intermediate across a carbon–carbon or carbon–hetero atom multiple bond. Eventually, both reaction types can be combined into a single process. Such double selection can result in enhanced enantioselectivities, sometimes at the expense of the diastereoselectivity.

Formally, reactions involving the rearrangement and cleavage of acyl–carbon bonds such as Baeyer–Villiger oxidations or Wolff rearrangements are also C–C cleavage processes. Though a number of excellent developments have rendered these reactions enantioselective, they are beyond the scope of this chapter. For the interested reader, several reviews cover this topic [6–10]. Herein, the focus is placed exclusively on transition metal-catalyzed reactions involving discrete carbon–metal intermediates. Several previous overviews also cover the asymmetric aspect of such transformations [11–15]. The asymmetric processes described herein are discussed in detail in individual sections sharing a similar mechanism. At first, reactions proceeding by direct cleavages of the acyl–carbon bond of cyclobutanone substrates are described. This is followed by a larger section mechanistically based on $\beta$-carbon eliminations. Then, transformations occurring by 1,2-shifts are featured, followed by a section on ring-opening reactions of vinyl cyclopropanes and

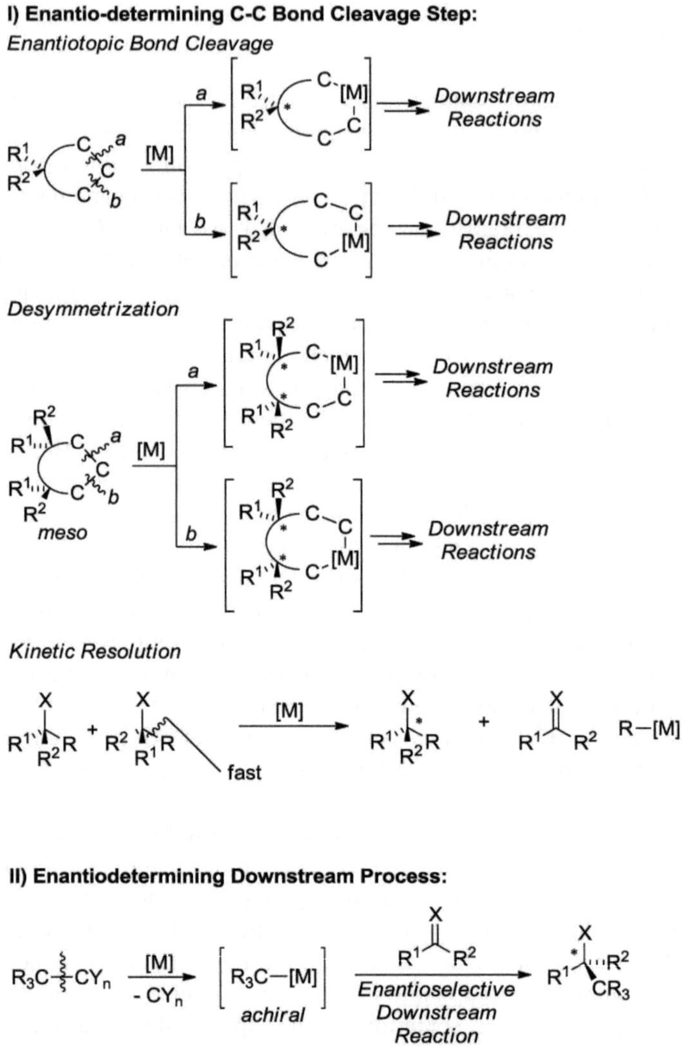


**Scheme 1** Different selectivity modes of enantioselective transformations via C–C cleavages

vinylidene cyclopropanes. Finally, processes initiated by the cleavage of C–CN bonds are covered.

## 2 Cyclobutanones in C–C Cleavage Processes

The direct approach to cleave a C–C $\sigma$-bond consists of the oxidative addition of this C–C bond to a transition metal. This is the uncommon reverse pathway of the ubiquitous reductive elimination step and is quite difficult to achieve, even with

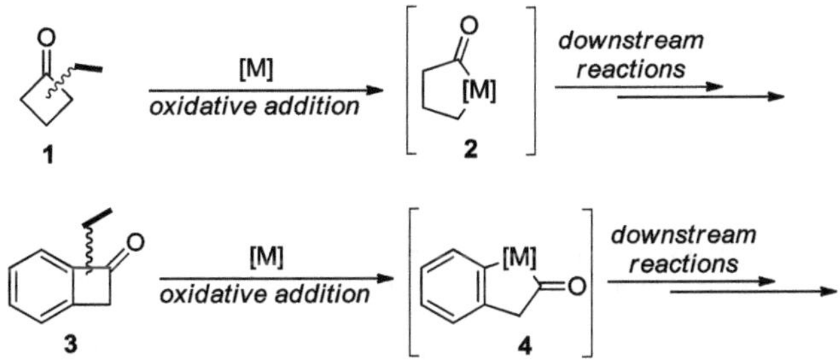


**Scheme 2** Oxidative addition of C–C bonds as direct path for C–C bond cleavage

strong directing groups. The C–C bond in the $\alpha$-position of a carbonyl moiety has evolved as a promising target for direct C–C bond cleavages. The arising acyl–metal bond is stronger than a corresponding alkyl–metal bond. However, in most cases, this is not a sufficient driving force and, by and large, the reverse reductive elimination pathway is observed. To drive the reaction towards the C–C cleavage direction, the energy of ring strain release is highly important. Not surprisingly, cyclobutanone **1** and benzocyclobutenone **3** exhibit, as the smallest stable saturated cyclic ketones, the best reactivity toward undirected C–C bond insertions (Scheme 2). The arising metallocyclopentanones **2** and **4** are relatively stable, and, importantly, less prone toward decarbonylation than other ring sizes. These intermediates were exploited as a platform for an array of downstream reactions.

The general ability of transition metal promoted C–C bond cleavages of cyclobutenones to give transient metallacyclopentenones was studied by Liebeskind [16–19]. This reactivity was exploited for nickel-catalyzed alkyne insertion into the acyl–carbon bond. Later, Kondo and Mitsudo expanded the process to olefin insertions with rhodium and ruthenium catalysts [20–23]. Recently, Dong explored the rhodium(I)-catalyzed intramolecular carboacylation between benzocyclobutenones and olefins by cleavage of the usually less reactive C1–C2 bond [24]. They further reported a highly enantioselective version of this process (Scheme 3) [25]. The critical C–C bond cleavage event occurs as the initial step of the transformation. Selective oxidative insertion into the aryl acyl C–C bond of benzocyclobutenones **5** is likely to be directed by the pendant olefin coordination and formation of the more stable C(sp$^2$)–Rh bond and generates the rhodacyclopentenone **6**. Subsequent migratory insertion of the appended olefin leads to rhodacycloheptanone **7**, which in turn undergoes reductive elimination to afford cyclohexanone **8**. With DTBM-Segphos (**L1**), tricyclic products **8** having a quaternary stereocenter in the $\beta$-position of the carbonyl group were accessed in moderate to excellent yields and excellent enantioselectivities of up to 99% *ee*.

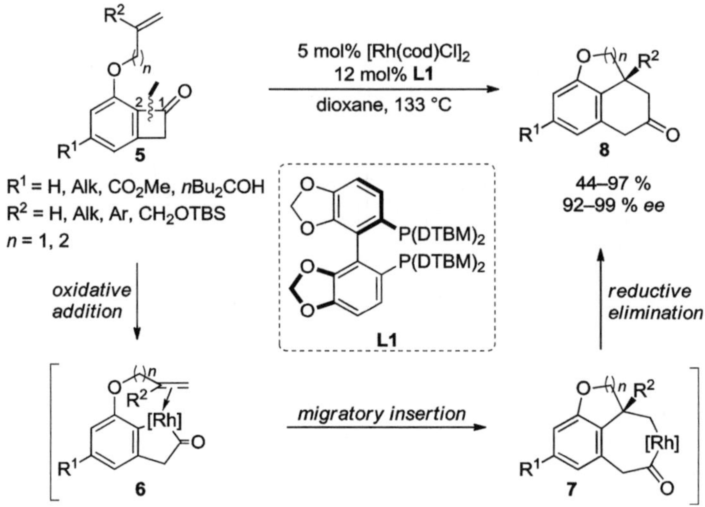


**Scheme 3** Olefin insertion through Rh(I)-catalyzed selective C–C cleavage of benzocyclobutenones

## 3 C–C Bond Cleavages by *β*-Carbon Eliminations of *tert*-Alcoholates

Whereas the oxidative addition is the direct way for C–C bond cleavage, *β*-carbon elimination emerged as a viable alternative pathway. Generally, *β*-carbon elimination is the carbon analogue of the ubiquitous *β*-hydride elimination [26]. In analogy, after formation of the appropriate alkyl-, aza- or alkoxy-metal species **9** (Scheme 4), the C–C bond between the *β*- and *γ*-carbon atom is cleaved, providing a C=X portion and an alkyl-metal species **10** which is predisposed for a host of downstream reactions. Asymmetric versions of such processes are covered in this section.

In particular, *tert*-cyclobutanolates **12** (Scheme 5) have proven to be a versatile substrate class for *β*-carbon eliminations. The critical *β*-carbon elimination step that delivers alkyl-metal species **14** is favored due to ring strain release of the four-membered ring and the formation of a strong C=O bond. Generally, two methods have been described to access metal-cyclobutanolate **13**: (1) addition of an organometallic reagent across a cyclobutanone precursor **11** and (2) starting directly from the corresponding *tert*-cyclobutanol **12** via deprotonation and ligand exchange with a transition metal alkoxide [3, 12, 27].

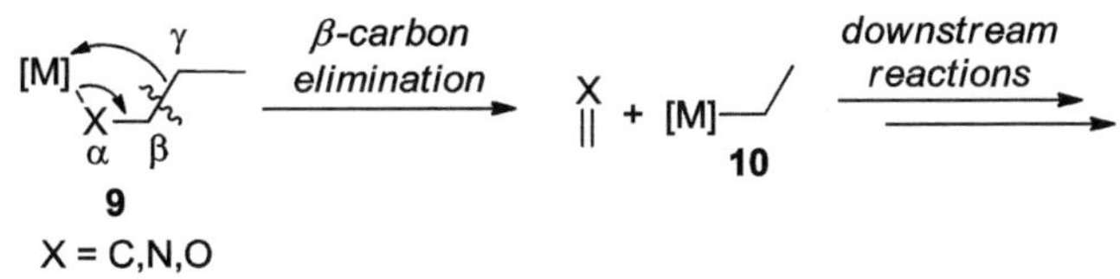


**Scheme 4** C–C bond cleavages by *β*-carbon elimination

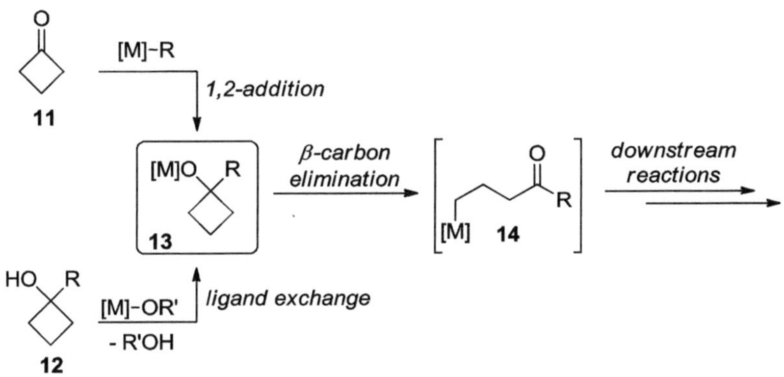


**Scheme 5** *β*-Carbon eliminations from *tert*-cyclobutanolates

## 3.1 *Metal* tert-*Cyclobutanolates by 1,2-Additions of Cyclobutanones*

In a pioneering study, the feasibility of rhodium-catalyzed C–C single bond cleavages of cyclobutanones was reported by Ito in 1994 [28]. Later, the synthetic potential of the transient rhodacycles for olefin insertion of a tethered styryl group was unlocked [29, 30]. Besides rhodium(I)-catalysts, nickel(0)-complexes also proved to be viable for C–C cleavages of four-membered ketones and various intriguing achiral or racemic insertion reactions were developed [31–35]. In contrast to alkynes, olefins are less reactive and require the benefits of an intramolecular transformation [36]. On the other hand, they offer the possibility of an asymmetric reaction. In this respect, Murakami achieved an enantioselective synthesis of benzobicyclo[2.2.2]octenones **19** from styryl-substituted cyclobutanones **15** (Scheme 6) [37]. Nickel-catalyzed C–C bond cleavages are mechanistically different from their rhodium-catalyzed counterparts. For instance, the reaction starts with an intramolecular oxidative cyclization event between the appended olefin and the carbonyl group of the cyclobutanone on the Ni(0)-complex to give oxanickelacyclopentane **17**. The stereochemistry of this step determines the enantioselectivity of the whole process. It is governed by Feringa-type phosphoramidite **L2** favoring a coordination of one of the two enantiotopic faces of the C–C double bond to nickel. Due to geometric constraints, intermediate **17** undergoes a highly diastereoselective *β*-carbon elimination, leading to the

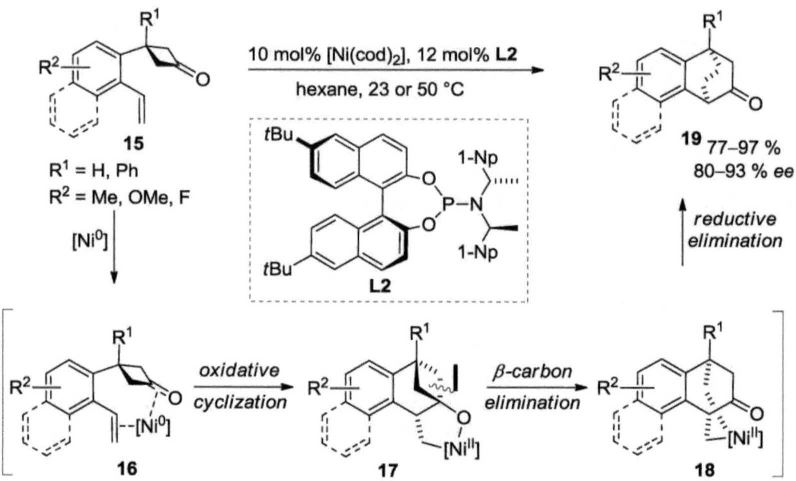


**Scheme 6** Nickel-catalyzed intramolecular insertions of olefins into a cyclobutanone C–C bond

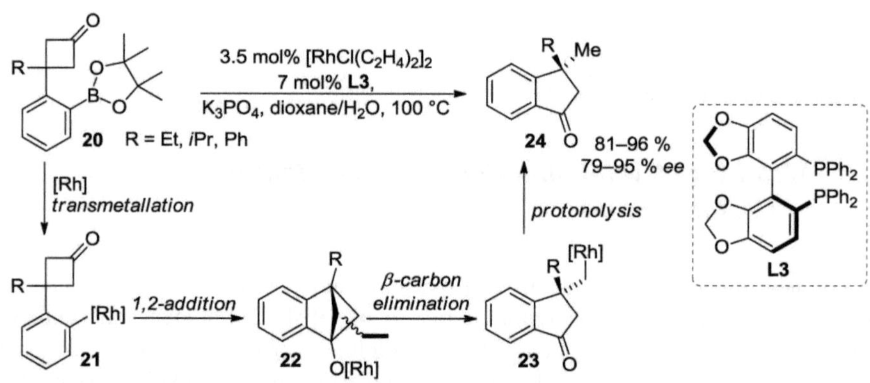


**Scheme 7** Rhodium-catalyzed intramolecular 1,2-addition followed by enantioselective β-carbon elimination

nickelacycle **18**. Reductive elimination closes the catalytic cycle delivering benzobicyclo[2.2.2]octenones **19** in good yields and high enantioselectivities of up to 93% *ee*.

In 2006, Murakami and co-workers described enantioselective rhodium(I)-catalyzed C–C bond cleavages by the 1,2-addition pathway (Scheme 7) [38]. Starting from aryl boronic ester **20**, intramolecular 1,2-addition across the cyclobutanone moiety gives strained rhodium-alkoxide **22**. In turn, β-carbon elimination leads to alkyl-rhodium species **23**. Subsequent protonation of the neopentylic alkyl-rhodium yields indanones **24**. Differentiation of the two enantiotopic C–C bonds of **22** during the β-carbon elimination step enables a highly enantioselective process. Segphos (**L3**) proved to be the best ligand providing indanones **24** in high yields and selectivities of up to 95% *ee*.

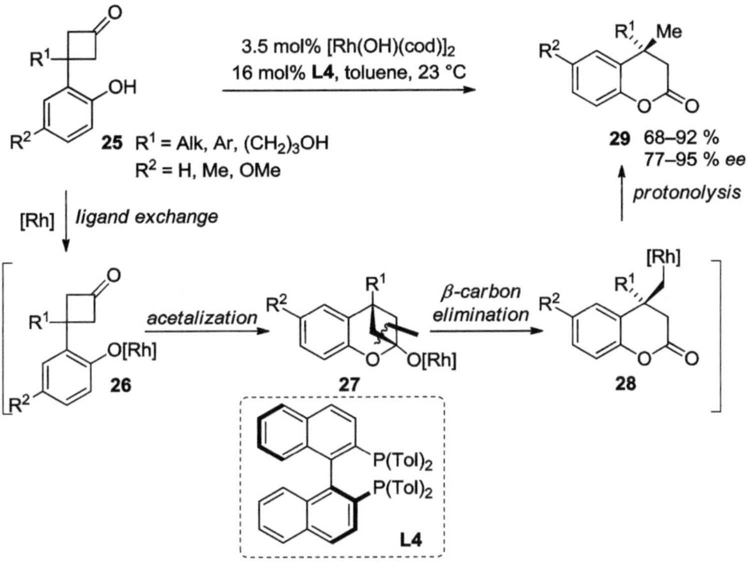


**Scheme 8** Aldolization promoted by rhodium-phenolate formation and subsequent enantioselective *β*-carbon elimination

This concept was extended to intramolecular 1,2-additions of rhodium-phenolates (Scheme 8) [39]. After the initial ligand exchange giving rhodium species **26** from phenols **25**, 1,2-addition delivers hemiacetalate **27**. The following *β*-carbon elimination from **27** is conducted in a highly enantioselective manner by Tol-Binap (**L4**) as chiral ligand. Protodemetalation of the formed alkyl-rhodium species **28** provides dihydrocoumarins **29** in good yields and enantiomeric excesses ranging from 77 to 95% *ee*.

## *3.2 Metal* tert-*Cyclobutanolates by Ligand Exchange*

Uemura reported a first example of palladium-catalyzed *β*-carbon elimination from *tert*-cyclobutanols **30**[40] and demonstrated the trapping of the alkylpalladium **32** formed by aryl bromides [41]. Subsequently, they extended this process to an enantioselective version (Scheme 9) [42, 43]. The bulky bidentate *P,N*-ferrocenyl ligand **L5** was critical for obtaining high enantioselectivities. For instance, *γ*-arylated ketones **33** bearing a tertiary stereogenic center were obtained in excellent enantioselectivities of up to 95% *ee*. However, corresponding 3,3-disubstituted cyclobutanols gave significantly lower enantioselectivities. In addition to aryl halides, vinyl or propargyl halides furnished the corresponding *γ*-vinylated and allenylated ketones.

In 2008, Cramer described a rhodium(I)-catalyzed enantioselective rearrangement of allenyl cyclobutanols (Scheme 10) [44]. The formation of rhodium(I)-alkoxide **35** triggered a *β*-carbon elimination, providing the corresponding alkyl–rhodium

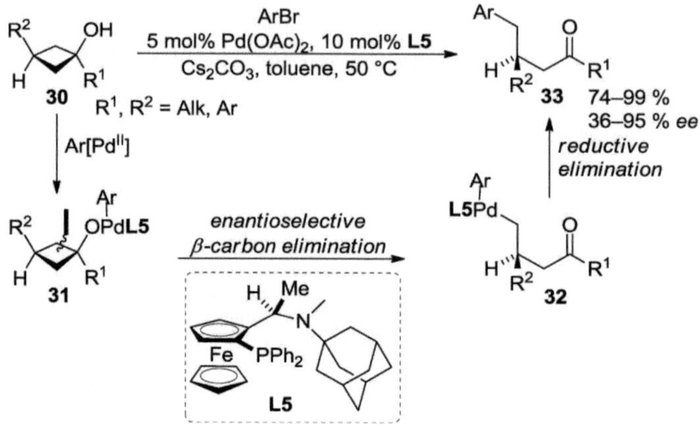


**Scheme 9** Palladium-catalyzed $\beta$-carbon elimination from *tert*-cyclobutanols and subsequent arylative trapping

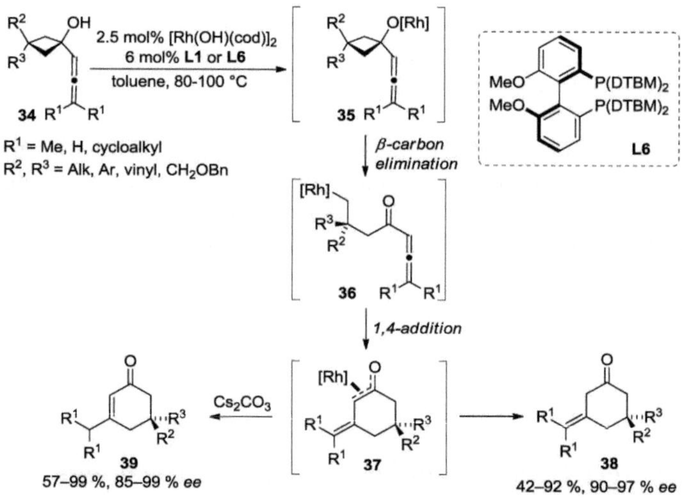


**Scheme 10** Enantioselective rhodium(I)-catalyzed rearrangement of allenyl cyclobutanols

intermediate **36** bearing a quaternary stereogenic center. Bulky biarylphosphine ligands such as DTBM-MeOBiphep (**L6**) or DTBM-Segphos (**L1**) enable excellent enantioselectivities of up to 99% *ee* for the $\beta$-carbon elimination step. Subsequent cyclization of intermediate **36** by 1,4-addition leads to cyclohexanones **38** with an exocyclic double bond. In the presence of cesium carbonate, double bond migration into conjugation occurs, furnishing cyclohexenones **39** in excellent yields (57–99%) and enantioselectivities (85–99% *ee*).

The scope of this reaction was extended by one pot domino processes (Scheme 11). For instance, in situ hydrosilylation affords directly protected cyclohexanols **41**. Saturated cyclohexanones **42** were obtained by a copper catalyzed 1,4-reduction

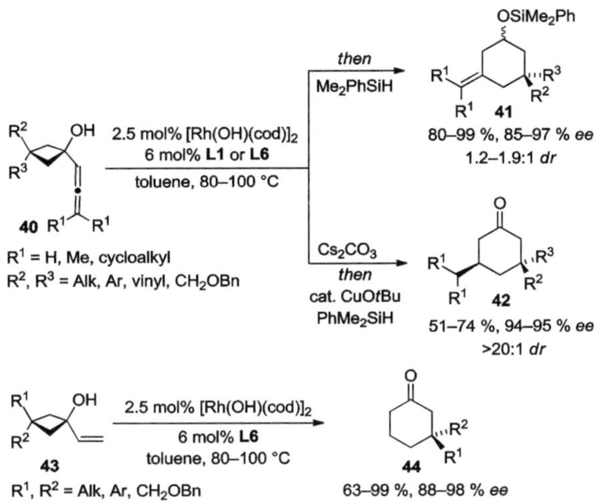


**Scheme 11** Extension of the rhodium(I)-catalyzed $\beta$-carbon elimination to vinyl cyclobutanols and to domino processes

reaction. Besides allenyl cyclobutanols, vinyl *tert*-cyclobutanols **43** are suited as well for the cyclization, giving cyclohexanones **44** [45].

Independently, Cramer and Murakami reported that *tert*-cyclobutanols bearing an aryl substituent at the 3-position can undergo a 1,4-rhodium shift leading to indanols **49** (Scheme 12) [46, 47]. At first, $\beta$-carbon elimination occurs, giving rise to alkyl-rhodium species **46**. This key intermediate then undergoes the 1,4-rhodium shift leading to aryl-rhodium species **47**. The driving force for this shift is the generation of a more stable aryl-rhodium species **47**. Mechanistically, it is believed to proceed via a C-H activation/reductive elimination pathway. A subsequent intramolecular 1,2-addition gives rise to indanols **49** in high diastereo- and enantioselectivities. Furthermore, substrates with an electron-rich $R^1$ substituent such as 1-thienyl are prone to a second, unstrained, $\beta$-carbon elimination from intermediate rhodium indanolate **48** yielding indanones **50** [48].

Due to the transposition of the aromatic hydrogen atom to the methylene group, in this transformation the product scope is limited to 3-methyl substituted indanols **49**. This shortcoming was addressed by a related rhodium(I)-catalyzed 1,4-silicon shift of trialkyl silanes (Scheme 13) [49]. The typical $\beta$-carbon elimination generates the alkyl-rhodium species **52** which, in turn, undergoes an oxidative addition/reductive elimination sequence with the appended silyl group, leading to aryl-rhodium species **54** with a transposed silyl group. Similar to the aforementioned reaction, subsequent 1,2-addition of the formed aryl-rhodium species **54** across the generated ketone moiety affords silyl indanols **55**. Complete selectivity for the C–Si activation over any competing C–H activation was observed for a variety of trialkylsilyl groups. Conducting the reaction with difluorphos **L7** yields indanols **55** in high diastereo- and enantioselectivities of up to 99% *ee*. Sterically more demanding silyl groups are completely transferred with diene ligand **L8**. Along the same lines, diene ligands provide the opposite indanol diastereoisomer **56**, allowing for a second C–Si bond activation yielding silacycle **57**.

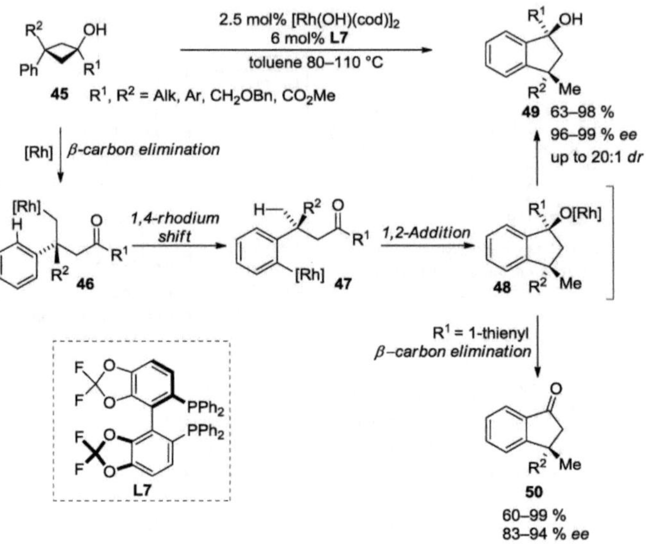


**Scheme 12** Formation of indanols and indanones via $\beta$-carbon elimination induced C–H activation/1,2-addition reaction cascade

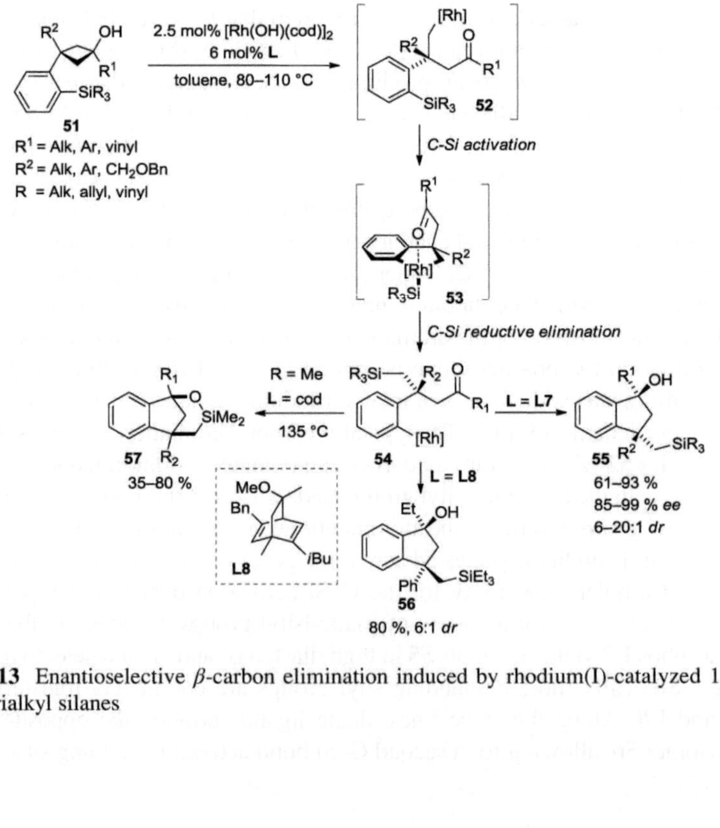


**Scheme 13** Enantioselective $\beta$-carbon elimination induced by rhodium(I)-catalyzed 1,4-silicon shift of trialkyl silanes

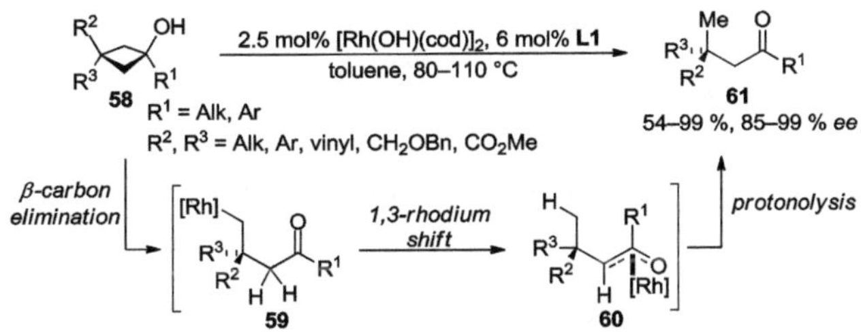


**Scheme 14** $\beta$-Carbon elimination and intramolecular 1,3-rhodium shift leading to acyclic ketones

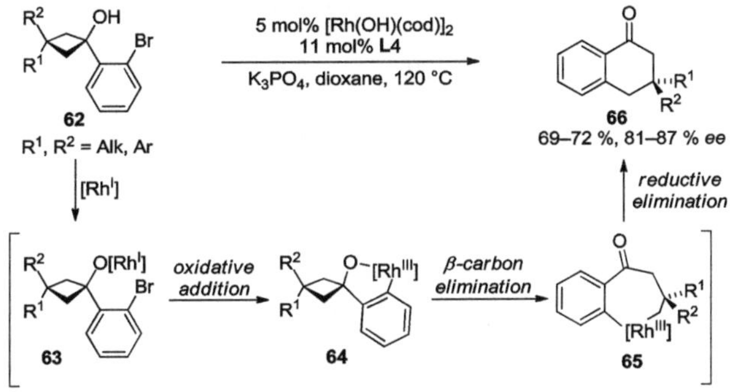


**Scheme 15** Incorporation of an Rh(I)–Rh(III) redox process in the $\beta$-carbon elimination pathway to obtain $\alpha$-tetralones

In the case where neither C($sp^2$)–H nor C($sp^2$)–$SiR_3$ groups are available for the alkyl-rhodium species **59** to get stabilized by a 1,4-shift, an alternative 1,3-rhodium shift is induced, leading to the oxa-$\pi$-allylrhodium species **60** (Scheme 14) [50]. Subsequent protonation affords acyclic methyl substituted quaternary stereogenic centers **61** in excellent selectivities ranging between 85 and 99% *ee*.

Extending the range of follow-up pathways, Murakami reported a Rh(I)-Rh(III) redox process from bromoaryl cyclobutanols to access $\alpha$-tetralones **66** (Scheme 15) [51]. Initial ligand exchange with bromoaryl *tert*-cyclobutanols **62** generates the usual rhodium(I)-alkoxide precursor **63**. Presumed oxidative addition gives rise to the five-membered rhoda(III)cycle **64**. In turn, $\beta$-carbon elimination occurs, providing the seven-membered rhoda(III)cycle **65**. Reductive elimination affords **66** in good yields. The use of Tol-Binap (**L4**) enables enantioselective $\beta$-carbon elimination and furnishes tetralones **66** in good selectivities of up to 87% *ee*.

## 3.3 *Retro-Allylations of Tertiary Homoallyl Alcohols*

The metal-catalyzed allylation of carbonyl groups is a classical reaction in organic chemistry. However, the reverse reaction pathway, the retro-allylation, is

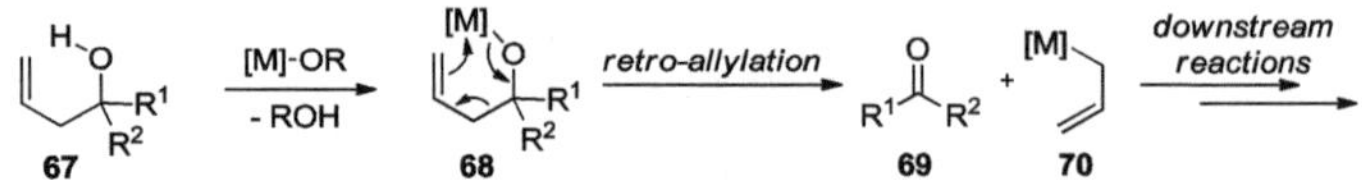


**Scheme 16** Metal-promoted C–C bond cleavage by retro-allylations

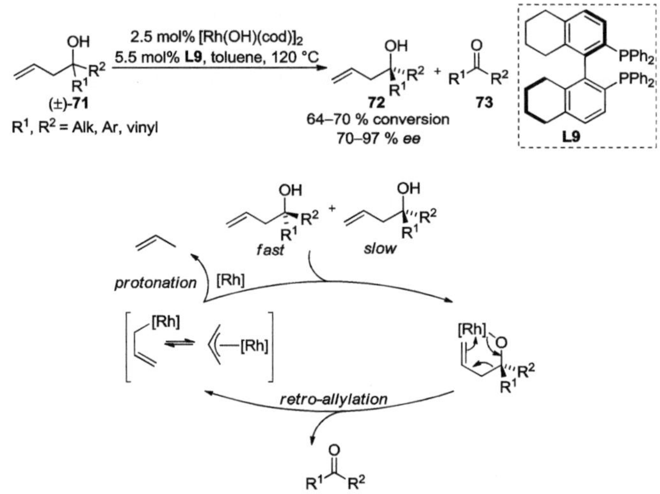


**Scheme 17** Rhodium-catalyzed kinetic resolution of racemic tertiary homoallylic alcohols

sometimes observed with transition metals and consists of a specific case of *β*-carbon elimination (Scheme 16).

In general, metal-catalyzed retro-allylation of homoallylic *tert*-alcohols **67** proceed by a six-membered transition state **68**, and have become a relatively unorthodox method to generate allylmetal species **70** and ketones **69** [52].

Hayashi and co-workers exploited this transformation and reported a kinetic resolution of racemic tertiary homoallylic alcohols **71** with a chiral rhodium catalyst (Scheme 17) [53]. For instance, with (*R*)-H8-Binap (**L9**), selectivity factors of up to 12 were achieved, and the remaining enantioenriched tertiary alcohols **72** were obtained in moderate to excellent enantioselectivities of up to 97% *ee*.

Cramer and co-workers reported an enantioselective palladium(0)-catalyzed arylative retro-allylation of norbornene-derived tertiary alcohol **74** (Scheme 18) [54]. The *meso*-starting material is desymmetrized by enantiotopic retro-allylation leading to intermediate palladium(II) species **76**. In turn, this intermediate is trapped with aryl halides in a highly regio- and diastereoselective fashion, leading to tetrasubstituted cyclohexenes. The Taddol-based phosphoramidite ligand **L10** gave product **77** in 89% yield and 64% *ee* which was subsequently increased to 92% *ee* by trituration.

Intriguingly, rhodium catalysts display with the same substrates a complementary reactivity profile (Scheme 19) [55]. Moreover, the pathway after the retro-allylation step is heavily dependent on the conditions employed and substrate substitution. For instance, enantioselective retro-allylative C–C bond cleavage

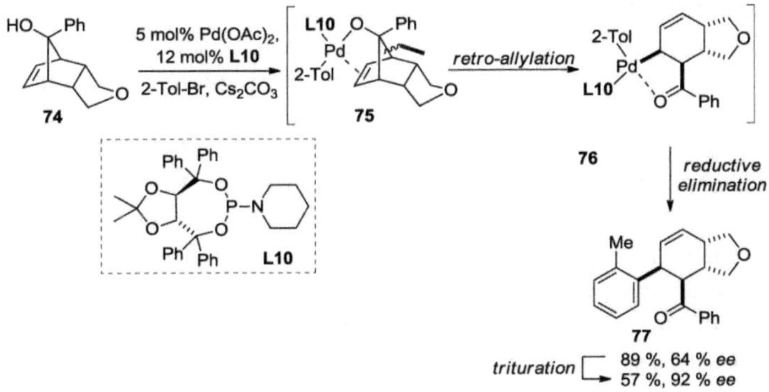


**Scheme 18** Desymmetrization of a *meso-tert*-norbornenol by palladium-catalyzed retro-allylation

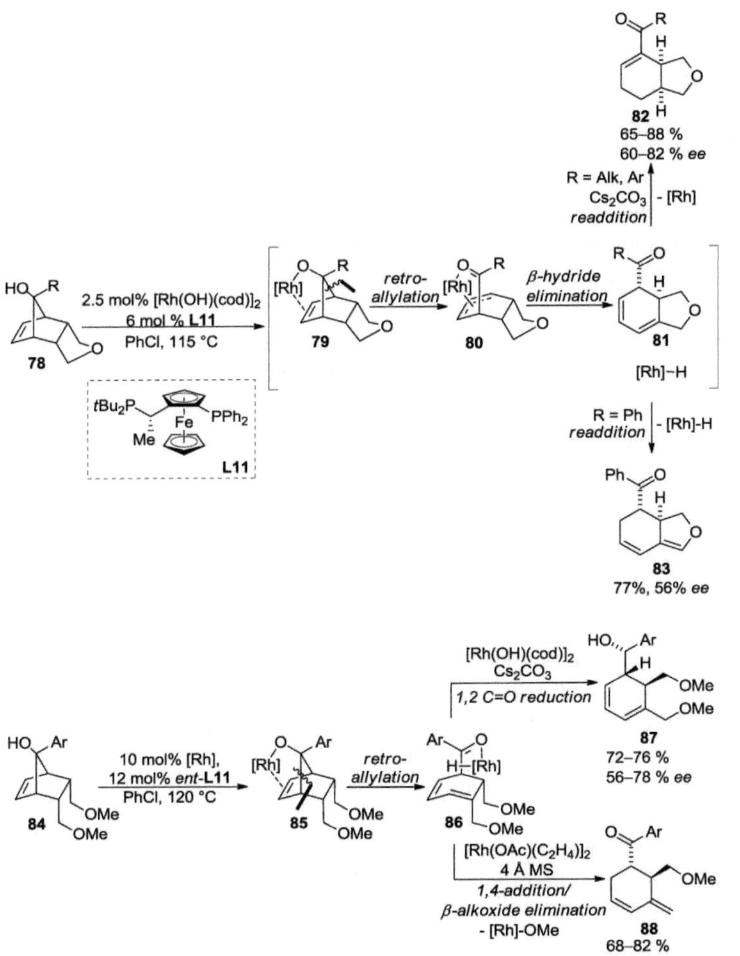


**Scheme 19** Desymmetrization pathways of *meso-tert*-norbornenols induced by rhodium(I)-catalyzed retro-allylation

forms intermediate **80**. Subsequent $\beta$-hydride elimination of this primary intermediate presumably gives diene **81** and a rhodium hydride species. Re-addition of the rhodium hydride generates an oxa-$\pi$-allylrhodium intermediate which in turn undergoes isomerization in the presence of cesium carbonate to enone **82**. In the absence of a base, 1,4-addition of the rhodium hydride and elimination to the vinyl ether ultimately delivers diene **83**. Modification of the bicyclic framework had a large impact on the reaction path. For instance, *tert*-norbornenols **84** resulted in the presence of cesium carbonate in a selective rhodium hydride mediated reduction of the carbonyl group, providing secondary alcohols **87**. Switching to a different rhodium precursor and using molecular sieves gave diene **88** via 1,4-hydrorhodation and subsequent $\beta$-alkoxide elimination.

## *3.4 β-Carbon Eliminations from Unstrained* tert-*Alcohols*

Although the release of ring strain is a good and well exploited driving force for $\beta$-carbon elimination, it is not always necessary to trigger $\beta$-carbon eliminations. For instance, metal-*tert*-alkoxides (**89** and **92**) that can eliminate an [M]-fragment bearing a C(sp)- or C($sp^2$)-moiety (**90** and **93**) do not require this strain-release boost as the generated C($sp^2$)–[M] and C(sp)–[M] bonds are significantly stronger than the corresponding C($sp^3$)–[M] bond (Scheme 20).

In this respect, Miura and co-workers extensively studied the palladium-catalyzed arylation of $\alpha,\alpha$-disubstituted aryl methanols with aryl bromides and chlorides [56]. Furthermore, Hartwig and co-workers showed the propensity of triarylcarbinols to undergo $\beta$-carbon eliminations in the presence of a rhodium complex [57]. Hayashi exploited this reactivity for rhodium-catalyzed enantioselective conjugate arylations of $\alpha,\beta$-unsaturated ketones, generating the required aryl-rhodium species by $\beta$-carbon elimination from trisubstituted aryl carbinol precursors **95** (Scheme 21) [58]. The rhodium-alkoxide **96** formed undergoes $\beta$-carbon elimination, releasing acridinone **97** and aryl-rhodium species **98**. Subsequent standard 1,4-addition across the enone provides the oxa-$\pi$-allylrhodium intermediate **99**, which after protonolysis furnish $\beta$-aryl ketones **100**. High selectivities of up to 94% *ee* were obtained using diene ligand **L12**.

This concept was extended to an asymmetric conjugate alkynylation of enones [59]. In this case, $\beta$-alkynyl elimination from the rhodium-alkoxide complex **101** (Scheme 22), delivers not only alkynyl-rhodium species **103** but simultaneously also reveals the reacting substrate, $\alpha,\beta$-unsaturated ketones **102**. Subsequent conjugate addition gives rise to $\beta$-alkynylketones **105** in high yields and enantioselectivities. This "on-demand" release and generation of the alkynyl-rhodium and the enone substrate bypasses the common dimerization issues associated with the formation of alkynyl-rhodium species from terminal alkynes.

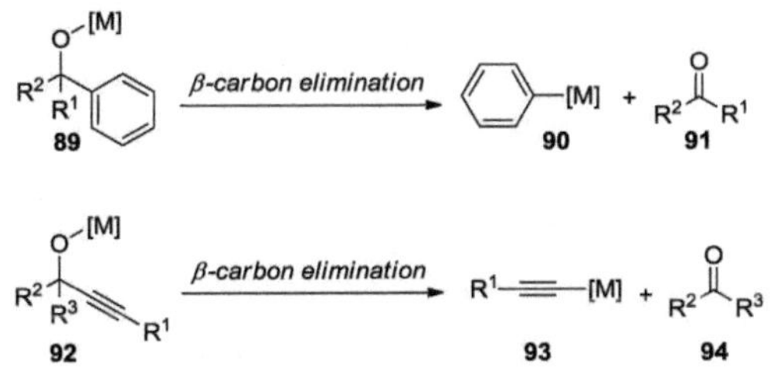


**Scheme 20** Stronger C(sp$^2$)–[M] and C(sp)–[M] bonds allow for *β*-carbon elimination from unstrained *tert*-alcoholate substrates

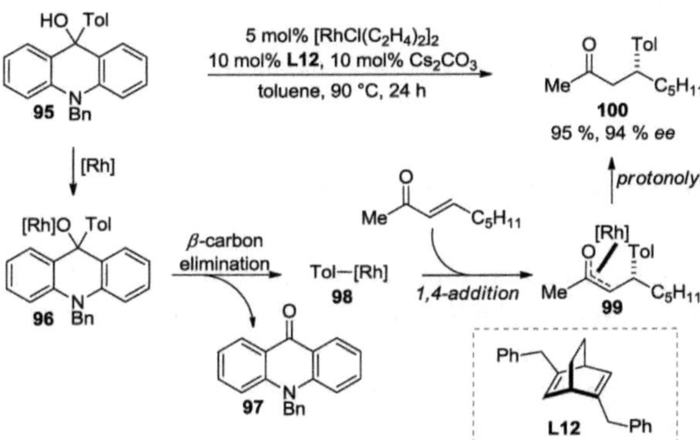


**Scheme 21** Conjugate arylation of *α*,*β*-unsaturated ketones by rhodium-catalyzed *β*-carbon elimination from unstrained *tert*-alcohols

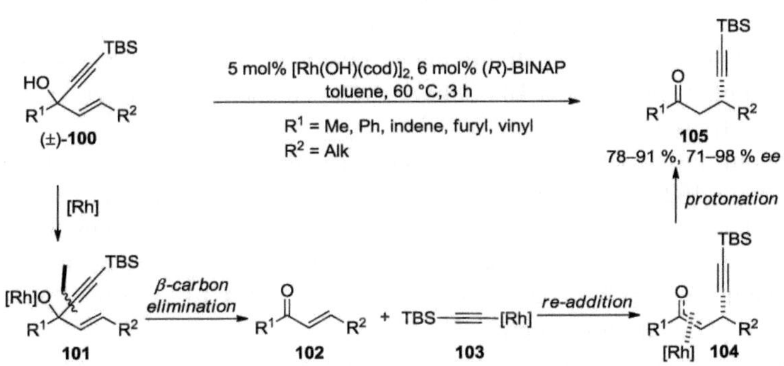


**Scheme 22** Conjugate alkynylation of *α*,*β*-unsaturated ketones by rhodium-catalyzed *β*-carbon elimination from unstrained *tert*-alcohols

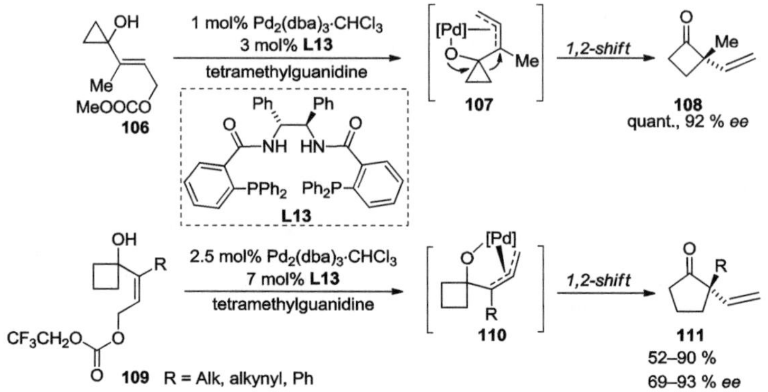


**Scheme 23** Enantioselective palladium-catalyzed ring expansion of *tert*-cyclopropanols and *tert*-cyclobutanols

# 4 Ring Expansions via 1,2-Carbon Shift Reactions

Besides the discussed $\beta$-carbon elimination pathway, strained *tert*-alcohols like *tert*-cyclopropanol or *tert*-cyclobutanol also engage in mechanistically different rearrangement reactions. Such ring expansions, consisting formally of a Wagner–Meerwein type 1,2-shift, are triggered by several transition metals via their $\pi$-allyl metal complexes. Nemoto proved the viability of this approach [60] and, subsequently, Trost developed an asymmetric ring expansion of *tert*-cyclopropanols **106** bearing an allylic carbonate leaving group (Scheme 23) [61]. In the presence of a palladium(0)-complex and tetramethylguanidine, $\pi$-allyl intermediate **107** is formed from allylic carbonates **106**. The Trost ligand (**L13**) dictates a highly selective migration of one of the two enantiotopic C–C bonds, giving rise to cyclobutanone **108** in quantitative yield and up to 92% *ee*. Similarly, corresponding cyclobutanols **109** deliver cyclopentenones **111** in good yields and enantioselectivities.

Later on, the reactivity was extended to electron-rich alkoxyallenyl *tert*-cyclobutanols [62, 63]. With this substrate class, hydropalladation of the allene generates the required $\pi$-allylpalladium intermediate **114** to trigger the 1,2-shift (Scheme 24). Again, ring expanded cyclopentanones **115** were obtained in high yields, enantioselectivity, and diastereoselectivity with **L13**. A critical point of the initial hydropalladation is careful pH control.

Toste reported an enantioselective gold-catalyzed ring expansion from 1-allenylcyclopropanols **116** (Scheme 25) [64]. The mechanistic picture for this reaction is complementary to the previously described palladium-catalyzed process. The chiral cationic gold(I)-complex generated by anion metathesis of xylyl-MeOBiphep (**L14**) with NaBARF activates the allene moiety by $\pi$-coordination, forming complex **117**. Ring expansion subsequently delivers vinyl-gold species **118**. Protodemetalation yields cyclobutanones **119** in excellent yields and enantioselectivities of up to 94% *ee*.

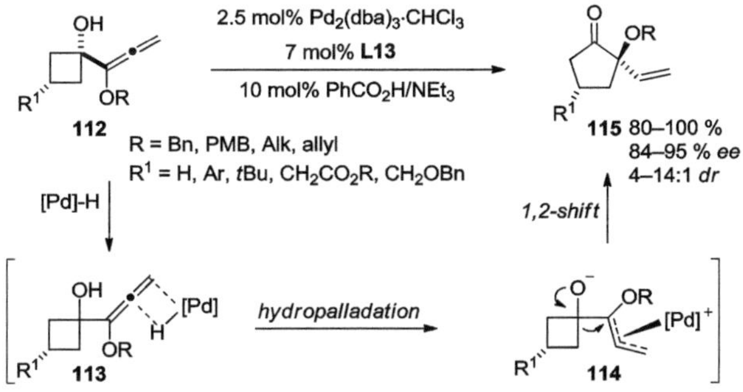


**Scheme 24** Enantioselective palladium-catalyzed ring expansion of allenyl *tert*-cyclobutanols

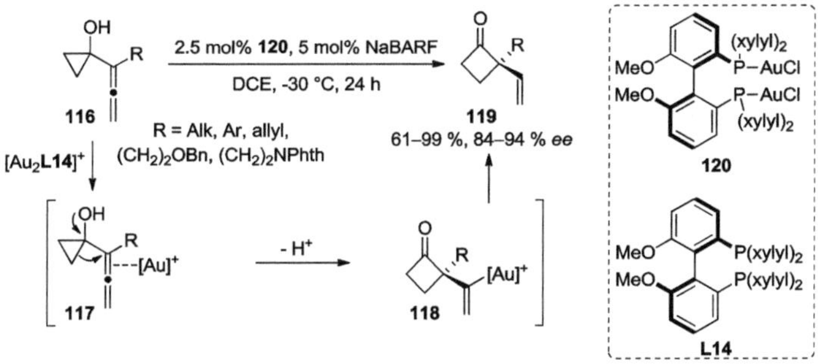


**Scheme 25** Enantioselective gold-catalyzed ring expansion of allenyl *tert*-cyclopropanols

# 5 Vinylcyclopropanes in C–C Bond Cleavage Processes

Vinylcyclopropane (VCP, **121**) is a well appreciated substrate in metal catalysis. For instance, it enables access to cyclopropylmethyl metal species **122** which in turn undergoes facile strain-release accelerated $\beta$-carbon elimination to the corresponding ring-opened form **123** (Scheme 26).

Wender extensively exploited this behavior for higher order cycloaddition reactions. They pioneered rhodium(I)-catalyzed [5+2] cycloadditions of vinylcyclopropanes and $\pi$-systems [65, 66]. As a steering phosphine ligand diminishes the reactivity for the [5+2]-cycloaddition process, only few asymmetric examples of this versatile process have been reported so far. For instance, Wender reported examples of enantioselective intramolecular [5+2] cycloaddition (Scheme 27) [67]. The cationic [Rh(Binap)$SbF_6$] complex was identified as an efficient catalyst for the transformation. Formation of rhodacyclopentane intermediate **125** in an enantioselective manner promotes strain-driven $\beta$-carbon elimination to give rhodacyclooctene **126**. Reductive elimination yields cycloheptenes **127** in excellent yields and selectivities from 52 to 99% *ee*. However, the reaction times are very long and switching from alkenes to alkynes is detrimental for the selectivity.

Later, Hayashi extended the scope of rhodium-catalyzed asymmetric intramolecular [5+2] cycloaddition to alkynyl-vinylcyclopropanes (Scheme 28) [68].

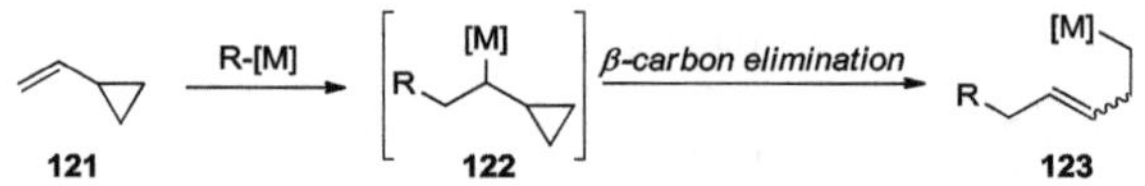


**Scheme 26** Metal-promoted C–C bond cleavage from vinylcyclopropanes **121**

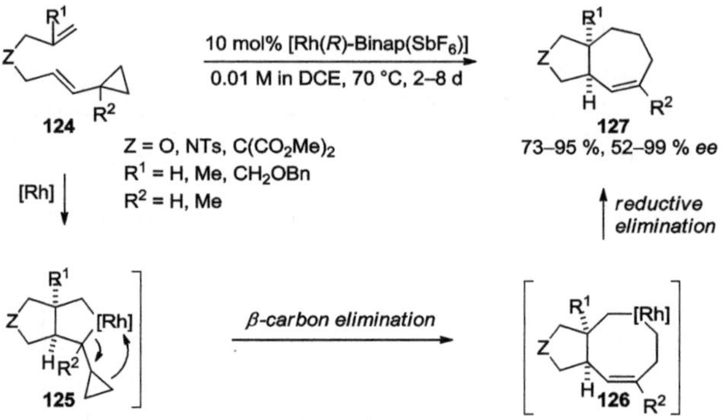


**Scheme 27** Enantioselective rhodium catalyzed [5+2] cycloaddition of alkenyl-tethered vinylcyclopropanes

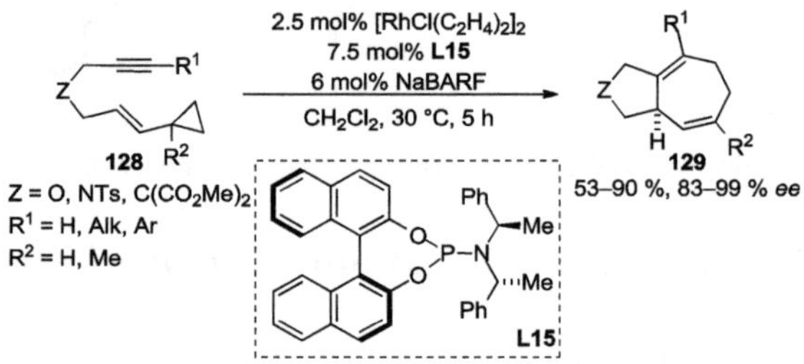


**Scheme 28** Enantioselective rhodium-catalyzed [5+2] cycloaddition of alkynyl-tethered vinylcyclopropanes

The monodentate Feringa ligand **L15** enables the synthesis of cycloheptadienes **129** in high yields and enantioselectivities of up to 99% *ee*. The use of the non-coordinating BARF-counteranion is critical to ensure sufficiently high reactivity.

# 6 Alkylidenecyclopropanes in C–C Bond Cleavage Processes

The highly strained alkylidenecyclopropanes (ACPs) were extensively used for synthetic transformations as they exhibit a unique reactivity [69–71]. These particularly versatile substrates are of strong interest for metal-catalyzed reactions and a few examples of enantioselective transformations have been described. The general

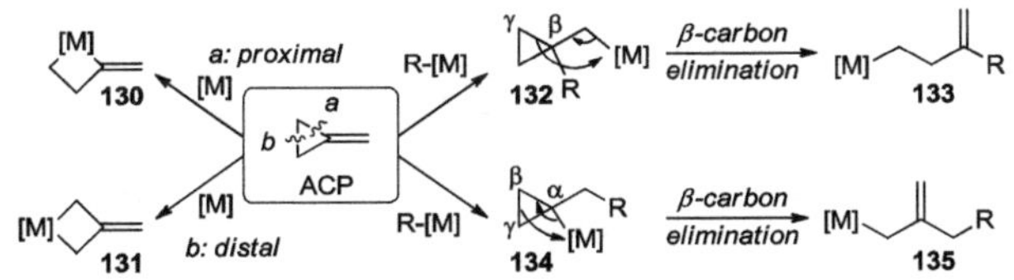


**Scheme 29** Different pathways of metal-catalyzed C–C bond cleavages of alkylidenecyclopropanes

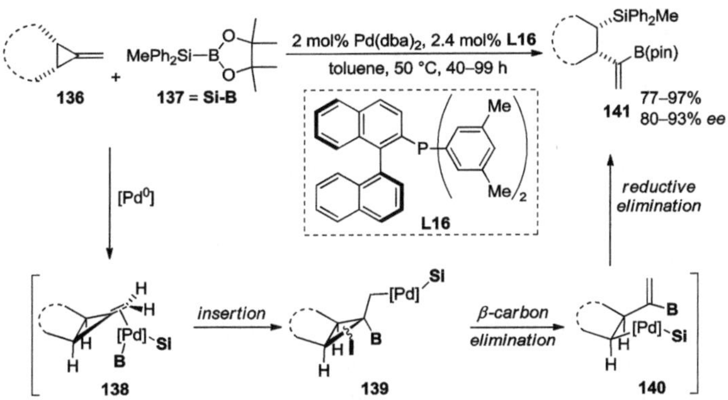


**Scheme 30** Desymmetrization of *meso*-methylenecyclopropanes via enantioselective β-carbon elimination

reactivity of ACPs with transition metal is characterized by two main processes (Scheme 29). The first pathway arises from the insertion of the transition metal into the cyclopropane proximal (*a*) or distal bond (*b*), leading to metallacyclobutanes **130** and **131**, respectively. The second pathway involves the carbometallation of the *exo*-methylene double bond giving either **132** or **134** which subsequently ring-open to an allyl-metal species **135** or homoallyl-metal species **133**.

Being part of their investigations of silaborations of olefins, Suginome and Ito reported on regioselective palladium-catalyzed silaborations of methylene-cyclopropanes (MCPs) occurring under C–C bond cleavage [72]. An asymmetric version of this process was devised for *meso*-MCPs **136** using the monodentate ligand **L16** (Scheme 30) [73]. The process is suggested to be initiated by the oxidative addition of silylborane **137** to the palladium(0)-complex providing a (boryl)(silyl)Pd(II) species. Coordination of the double bond of *meso*-MCP substrate to the palladium-complex (**138**) and subsequent regioselective migratory insertion placing the palladium at the favorable terminal carbon atom leads to (cyclopropylmethyl)palladium species **139**. An excellent differentiation between the two enantiotopic cyclopropane C–C bonds for the following β-carbon elimination step gives rise to intermediate **140** in a highly enantioselective fashion. Reductive elimination affords 2-boryl-4-silyl-1-butene derivatives **141** in good yields and high enantioselectivities between 81 and 91% *ee*. Later, the same group developed a polymer-based chiral ligand (PQXphos) which improved palladium-catalyzed asymmetric silaboration [74]. This new catalyst system

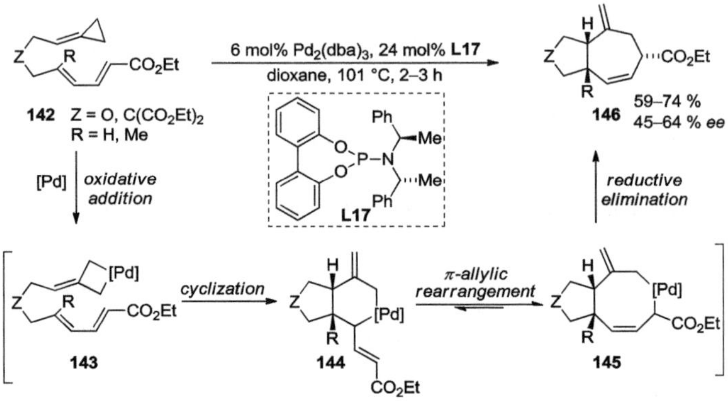


**Scheme 31** Enantioselective palladium-catalyzed [4+3] cycloaddition involving ACPs

enhanced significantly the reaction rates to reduce the previously required overly long reaction times and improved the enantioselectivity up to 96% *ee*.

Mascareñas and co-workers exploited the C–C insertion potential of tethered ACPs for novel higher order cycloadditions. In this respect, they reported a highly diastereoselective intramolecular palladium-catalyzed [4+3] cycloaddition of ACPs and tethered electron-poor dienes generating bicyclic 5,7-ring systems in a stereo-defined manner (Scheme 31) [75]. The anticipated mechanism of this process involves as the initial step the oxidative insertion of the palladium(0)-catalyst into the distal C–C bond of the alkylidenecyclopropane to give palladacyclobutane **143**. Subsequent cyclization gives rise primarily to palladacyclohexane **144**. Though it is possible to form a five-membered ring by a direct reductive elimination, a π-allylic rearrangement proceeds, forming palladacyclooctane **145** in which the palladium is stabilized by the ester moiety. Favorable reductive elimination yields *cis*-fused bicyclic products **146** in good yields. The potential for a catalytic asymmetric process was demonstrated with phosphoramidite **L17**. With this ligand, cycloadducts **146** were formed as single diastereoisomers in moderate enantioselectivities of up to 64% *ee*.

Evans and co-workers disclosed a highly stereoselective rhodium-catalyzed [3+2+1] carbocyclization of ACPs with carbon monoxide providing *cis*-fused bicyclohexenones (Scheme 32) [76]. Mechanistically, the reaction shares similarities with the previous palladium-catalyzed reaction. The in situ formed Rh-CO complex undergoes oxidative addition of the distal C–C bond of the vinylidenecyclopropane **147** affording the rhodacyclobutane **148**. Subsequent enantioselective carbometallation of the tethered olefin selectively leads to *cis*-fused rhodacyclohexane **149**. Migratory insertion of carbon monoxide into the rhodium-carbon bond gives acyl-rhodium intermediate **150**. Subsequent reductive elimination furnishes the cyclohexanone derivative **151** with the rhodium complex presumably still coordinated to the exocyclic olefin. Base- or metal-catalyzed migration of the double bond into conjugation delivers stable cyclohexenones **152**. With Foxap derivative **L18**, the reaction occurs in high enantioselectivity, delivering the cycloadduct **152** in 75% yield and 89% *ee*.

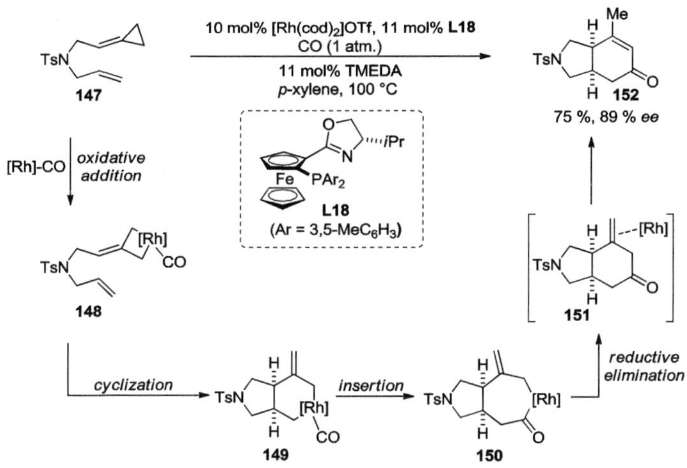


**Scheme 32** Enantioselective rhodium-catalyzed [3+2+1]-carbocyclization of alkylidencyclopropanes

# 7 Metal-Catalyzed Cleavages of C–CN Bonds

A particular case of C–C bond cleavage which is not dependent on strain assistance is the oxidative addition of low-valent transition metal complexes to a C–CN bond. This general reactivity is exploited in the nickel(0)-catalyzed DuPont adiponitrile process [77]. It was found that a Lewis-acid additive is beneficial for the reactivity of the nickel-based catalyst system [78]. Generally, the carbocyanation via C–CN bond cleavage proceeds by the following catalytic cycle (Scheme 33). Coordination of the Lewis-acid enables oxidative addition of the nickel(0)-complex leading to **155**. Coordination and subsequent migratory insertion of an unsaturated acceptor (alkene, alkyne, or allene) gives rise to intermediate **157**. The catalytic cycle is closed by a reductive C–CN bond formation, liberating the catalyst and product **158**. An efficient process has been devised by Nakao and Hiyama, exploiting the coordination of a boron- or aluminum-based Lewis acid which enhances the reactivity by facilitating the rate-limiting oxidative addition of the C–CN bond [79–83].

Whereas the carbocyanation of alkyne acceptors proceeds efficiently intermolecularly, less reactive olefin acceptors rely on intramolecular processes. Independently, Hiyama and co-workers and Jacobsen and co-workers developed an asymmetric intramolecular carbocyanation of olefins (Scheme 34) [84–86]. Depending upon the substrate structure and the reaction conditions used, three ligands were shown to be highly efficient for this process, providing the product in high yields and enantioselectivities of over 90% *ee*. For instance, an in situ reduction of a nickel(II) precursor by elemental zinc in combination with the Tangphos ligand (**L21**) and triphenyl boron is an efficient match [84]. Alternatively, [Ni(cod)$_2$] could be used directly as a nickel(0) source and provided with *i*Pr-Foxap (**L19**) or *i*Pr-Phox (**L20**) as chiral ligand and dimethylaluminum chloride as Lewis acid – a combination of comparable efficiency [85].

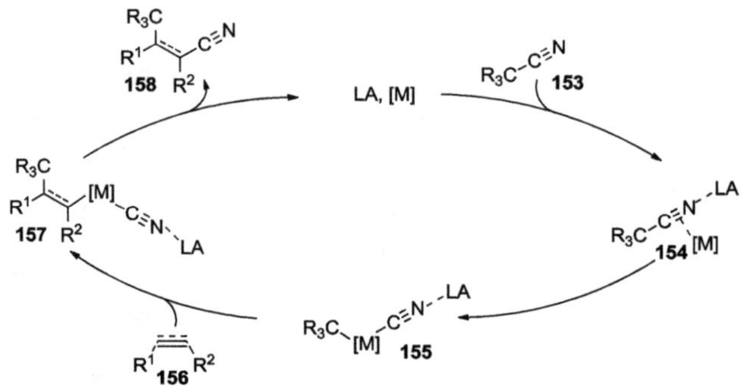


**Scheme 33** Carbocyanations induced by metal- and Lewis-acid-catalyzed C–CN bond cleavage process

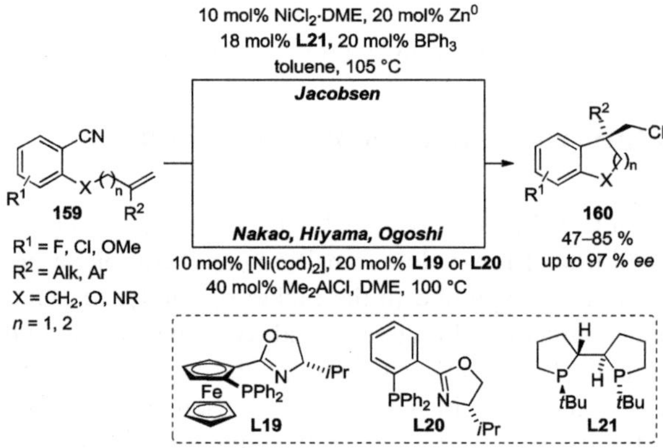


**Scheme 34** Asymmetric intramolecular carbocyanation of olefins

Subsequently, the synthetic use of this asymmetric carbocyanation process was demonstrated by synthesis of (−)-esermethole (**163**) and (−)-eptazocine (**166**) (Scheme 35) [87]. In particular, the method was used to construct the key quaternary stereocenter of both molecules from precursors **161** and **164**.

Related to the nickel-catalyzed process illustrated above, an acyl nitrile bond cleavage was developed by Takemoto and co-workers with palladium catalysts. The oxidative addition of the C(O)–CN bond by low-valent palladium(0) catalysts gives rise to intermediate **168** (Scheme 36) [88]. Using carbamoyl cyanides as starting material, the resulting carbamoyl palladium-complex is more stable than the corresponding acyl-palladium complex arising from acyl nitrile. For the latter, decarbonylation to **170** is a dominant side reaction, limiting so far its synthetic applicability. Both the carbamoyl group and the cyano group are transferred to an acceptor olefin (**171**), resulting in an overall cyanocarbamoylation (**172**).

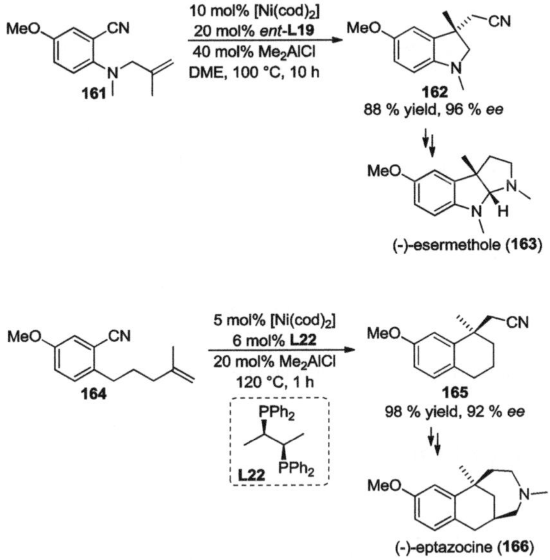


**Scheme 35** Synthetic applications of the asymmetric cyanoarylation technology

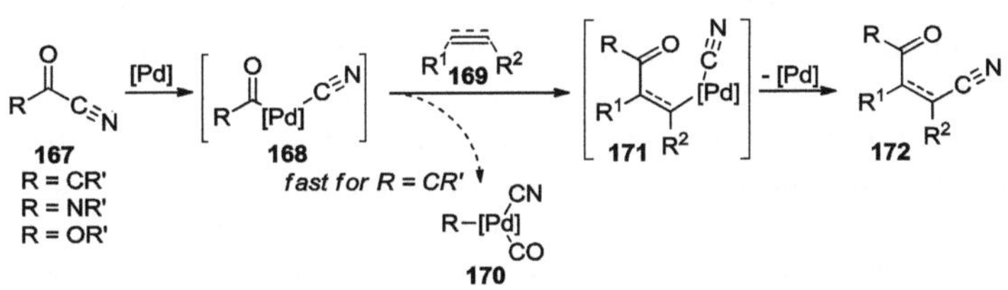


**Scheme 36** Carbocyanations induced by the C–CN bond cleavage process

This reactivity was first studied with more reactive alkyne acceptors to mitigate undesired decarbonylations [88]. Takemoto and co-workers subsequently devised an enantioselective palladium(0)-catalyzed intramolecular cyanoamidation of olefins, yielding synthetically versatile 3,3-disubstituted oxindoles in an asymmetric fashion (Scheme 37) [89, 90]. In line with the general mechanism, the reaction is initiated by an oxidative addition of the acyl cyanide moiety, providing palladium (II)-carbamoyl complex **174**. Carbamoylopalladation of the adjacent double bond occurs, delivering alkyl-palladium species **175**. Without the possibility of $\beta$-hydride elimination, reductive elimination forms the C–CN bond and closes the catalytic cycle by releasing a palladium(0) species. The process proceeds generally in excellent yields with 2 mol.% palladium catalyst. Good enantioselectivities of up to 86% *ee* are obtained with the Feringa ligand **L15**. The reaction works well with a range of substituents $R^2$ and is also tolerant to aryl chlorides. However, without the rigidifying aryl backbone, the reaction largely becomes less efficient.

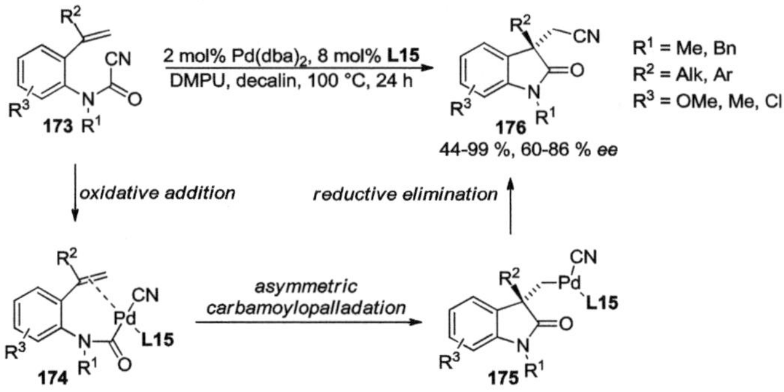


**Scheme 37** Asymmetric Pd-catalyzed intramolecular cyanocarbamoylations

# 8 Conclusions and Outlook

The selective cleavage of carbon–carbon bonds remains a challenging transformation due to weak interactions of the sterically and directionally constrained C–C bond with *d*-orbitals of transition metals. The additional selectivity component for asymmetric reactions requires a fine balance between reactivity and selectivity. High reaction temperatures are often needed for sufficient reactivity, and are considered as the "natural enemy" of enantioselectivity. Nevertheless, a steadily increasing number of highly enantioselective examples have been realized over the past few years. Not surprisingly, most work concentrated on predisposed small ring systems, exploiting strain release as a major driving force. Even with this rather limited substrate set, an impressive range of different transformations leading to valuable structures became accessible. On the other hand, unstrained C–CN bond cleavages opened novel reactivity windows. In the future, asymmetric transformations induced by C–C bond cleavages hold great opportunities for the discovery of novel reactivity patterns and practical implementation of the transformations in strategic synthesis planning.

# References

1. Jun C-H (2004) Transition metal-catalyzed carbon–carbon bond activation. Chem Soc Rev 33 (9):610–618. doi:10.1039/b308864m
2. Muzart J (2005) Palladium-catalysed reactions of alcohols. Part D: rearrangements, carbonylations, carboxylations and miscellaneous reactions. Tetrahedron 61(40):9423–9463. doi:10.1016/j.tet.2005.06.103
3. Murakami M, Makino M, Ashida S, Matsuda T (2006) Construction of carbon frameworks through beta-carbon elimination mediated by transition metals. Bull Chem Soc Jpn 79 (9):1315–1321. doi:10.1246/bcsj.79.1315
4. Park YJ, Park J-W, Jun C-H (2008) Metal–organic cooperative catalysis in C–H and C–C bond activation and its concurrent recovery. Acc Chem Res 41(2):222–234. doi:10.1021/ar700133y

5. Khoury PR, Goddard JD, Tam W (2004) Ring strain energies: substituted rings, norbornanes, norbornenes and norbornadienes. Tetrahedron 60(37):8103–8112. doi:10.1016/j.tet.2004.06.100
6. Russo A, De Fusco C, Lattanzi A (2012) Organocatalytic asymmetric oxidations with hydrogen peroxide and molecular oxygen. ChemCatChem 4(7):901–916. doi:10.1002/cctc.201200139
7. Leisch H, Morley K, Lau PCK (2011) Baeyer–Villiger monooxygenases: more than just green chemistry. Chem Rev 111(7):4165–4222. doi:10.1021/cr1003437
8. Bolm C (2007) Asymmetric Baeyer-Villiger. In: Asymmetric synthesis – the essentials. Wiley-VCH, Weinheim, pp 62–66
9. ten Brink GJ, Arends IWCE, Sheldon RA (2004) The Baeyer–Villiger reaction: new developments toward greener procedures. Chem Rev 104(9):4105–4124. doi:10.1021/cr030011l
10. Kirmse W (2002) 100 years of the Wolff rearrangement. Eur J Org Chem 2002 (14):2193–2256. doi:10.1002/1099-0690(200207)2002:14<2193::aid-ejoc2193>3.0.co;2-d
11. Seiser T, Saget T, Tran DN, Cramer N (2011) Cyclobutanes in catalysis. Angew Chem Int Ed 50(34):7740–7752. doi:10.1002/anie.201101053
12. Seiser T, Cramer N (2009) Enantioselective metal-catalyzed activation of strained rings. Org Biomol Chem 7(14):2835–2840. doi:10.1039/b904405a
13. Winter C, Krause N (2009) Rhodium(I)-catalyzed enantioselective C–C bond activation. Angew Chem Int Ed 48(14):2460–2462. doi:10.1002/anie.200805578
14. Murakami M (2010) Rhodium-catalyzed restructuring of carbon frameworks. Chem Rec 10(5):326–331. doi:10.1002/tcr.201000023
15. Murakami M, Matsuda T (2011) Metal-catalysed cleavage of carbon–carbon bonds. Chem Commun 47(4):1100–1105. doi:10.1039/c0cc02566f
16. Huffman MA, Liebeskind LS, Pennington WT (1990) Synthesis of metallacyclopentenones by insertion of rhodium into cyclobutenones. Organometallics 9(8):2194–2196. doi:10.1021/om00158a009
17. Huffman MA, Liebeskind LS (1990) Insertion of (.eta.5-indeny)cobalt(I) into cyclobutenones: the first synthesis of phenols from isolated vinylketene complexes. J Am Chem Soc 112 (23):8617–8618. doi:10.1021/ja00179a075
18. Huffman MA, Liebeskind LS (1991) Nickel(0)-catalyzed synthesis of substituted phenols from cyclobutenones and alkynes. J Am Chem Soc 113(7):2771–2772. doi:10.1021/ja00007a072
19. Huffman MA, Liebeskind LS, Pennington WT (1992) Reaction of cyclobutenones with low-valent metal reagents to form.eta.4- and.eta.2-vinylketene complexes. Reaction of.eta.4-vinylketene complexes with alkynes to form phenols. Organometallics 11(1):255–266. doi:10.1021/om00037a047
20. Kondo T, Taguchi Y, Kaneko Y, Niimi M, Mitsudo TA (2004) Ru- and Rh-catalyzed C–C bond cleavage of cyclobutenones: reconstructive and selective synthesis of 2-pyranones, cyclopentenes, and cyclohexenones. Angew Chem Int Ed 43(40):5369–5372. doi:10.1002/anie.200461002
21. Kondo T, Mitsudo TA (2005) Ruthenium-catalyzed reconstructive synthesis of functional organic molecules via cleavage of carbon–carbon bonds. Chem Lett 34(11):1462–1467
22. Kondo T, Niimi M, Nomura M, Wada K, Mitsudo TA (2007) Rhodium-catalyzed rapid synthesis of substituted phenols from cyclobutenones and alkynes or alkenes via C–C bond cleavage. Tetrahedron Lett 48(16):2837–2839, http://dx.doi.org/10.1016/j.tetlet.2007.02.091
23. Kondo T (2008) On inventing catalytic reactions via ruthena- or rhodacyclic intermediates for atom economy. Synlett 5:629–644. doi:10.1055/s-2008-1042807
24. Xu T, Dong G (2012) Rhodium-catalyzed regioselective carboacylation of olefins: a C–C bond activation approach for accessing fused-ring systems. Angew Chem Int Ed 51(30):7567–7571. doi:10.1002/anie.201202771

25. Xu T, Ko HM, Savage NA, Dong G (2012) Highly enantioselective Rh-catalyzed carboacylation of olefins: efficient syntheses of chiral poly-fused rings. J Am Chem Soc 134 (49):20005–20008. doi:10.1021/ja309978c
26. Miura M, Satoh T (2005) Catalytic processes involving β-carbon elimination. In: Tsuji J (ed) Palladium in organic synthesis, vol 14, Topics in organometallic chemistry. Springer, Berlin, Heidelberg, pp 1–20. doi:10.1007/b104133
27. Cramer N, Seiser T (2011) Beta-carbon elimination from cyclobutanols: a clean access to alkylrhodium intermediates bearing a quaternary stereogenic center. Synlett 4:449–460. doi:10.1055/s-0030-1259536
28. Murakami M, Amii H, Ito Y (1994) Selective activation of carbon–carbon bonds next to a carbonyl group. Nature 370(6490):540–541. doi:10.1038/370540a0
29. Matsuda T, Fujimoto A, Ishibashi M, Murakami M (2004) Eight-membered ring formation via olefin insertion into a carbon–carbon single bond. Chem Lett 33(7):876–877
30. Murakami M, Itahashi T, Ito Y (2002) Catalyzed intramolecular olefin insertion into a carbon−carbon single bond. J Am Chem Soc 124(47):13976–13977. doi:10.1021/ja021062n
31. Kumar P, Zhang K, Louie J (2012) An expeditious route to eight-membered heterocycles by nickel-catalyzed cycloaddition: low-temperature C $sp^2$-C $sp^3$ bond cleavage. Angew Chem Int Ed 51(34):8602–8606. doi:10.1002/anie.201203521
32. Ishida N, Yuhki T, Murakami M (2012) Synthesis of enantiopure dehydropiperidinones from α-amino acids and alkynes via azetidin-3-ones. Org Lett 14(15):3898–3901. doi:10.1021/ol3016447
33. Ashida S, Murakami M (2008) Nickel-catalyzed 4+2+2 -type annulation reaction of cyclobutanones with diynes and enynes. Bull Chem Soc Jpn 81(7):885–893. doi:10.1246/bcsj.81.885
34. Murakami M, Ashida S, Matsuda T (2006) Eight-membered ring construction by [4+2+2] annulation involving β-carbon elimination. J Am Chem Soc 128(7):2166–2167. doi:10.1021/ja0552895
35. Murakami M, Ashida S, Matsuda T (2005) Nickel-catalyzed intermolecular alkyne insertion into cyclobutanones. J Am Chem Soc 127(19):6932–6933. doi:10.1021/ja050674f
36. Murakami M, Ashida S (2006) Nickel-catalysed intramolecular alkene insertion into cyclobutanones. Chem Commun 44:4599–4601. doi:10.1039/b611522e
37. Liu L, Ishida N, Murakami M (2012) Atom- and step-economical pathway to chiral benzobicyclo[2.2.2]octenones through carbon–carbon bond cleavage. Angew Chem Int Ed 51(10):2485–2488. doi:10.1002/anie.201108446
38. Matsuda T, Shigeno M, Makino M, Murakami M (2006) Enantioselective C–C bond cleavage creating chiral quaternary carbon centers. Org Lett 8(15):3379–3381. doi:10.1021/ol061359g
39. Matsuda T, Shigeno M, Murakami M (2007) Asymmetric synthesis of 3,4-dihydrocoumarins by rhodium-catalyzed reaction of 3-(2-hydroxyphenyl)cyclobutanones. J Am Chem Soc 129 (40):12086–12087. doi:10.1021/ja075141g
40. Nishimura T, Ohe K, Uemura S (1999) Palladium(II)-catalyzed oxidative ring cleavage of tert-cyclobutanols under oxygen atmosphere. J Am Chem Soc 121(11):2645–2646. doi:10.1021/ja984259h
41. Nishimura T, Uemura S (1999) Palladium-catalyzed arylation of tert-cyclobutanols with aryl bromide via C–C bond cleavage: new approach for the γ-arylated ketones. J Am Chem Soc 121 (47):11010–11011. doi:10.1021/ja993023q
42. Nishimura T, Matsumura S, Maeda Y, Uemura S (2002) Palladium-catalysed asymmetric arylation of tert-cyclobutanols via enantioselective C–C bond cleavage. Chem Commun 7 (1):50–51. doi:10.1039/b107736h
43. Matsumura S, Maeda Y, Nishimura T, Uemura S (2003) Palladium-catalyzed asymmetric arylation, vinylation, and allenylation of tert-cyclobutanols via enantioselective C–C bond cleavage. J Am Chem Soc 125(29):8862–8869. doi:10.1021/ja035293l

44. Seiser T, Cramer N (2008) Enantioselective C–C bond activation of allenyl cyclobutanes: access to cyclohexenones with quaternary stereogenic centers. Angew Chem Int Ed 47(48):9294–9297. doi:10.1002/anie.200804281
45. Seiser T, Cramer N (2010) Rhodium(I)-catalyzed enantioselective activation of cyclobutanols: formation of cyclohexane derivatives with quaternary stereogenic centers. Chem-Eur J 16(11):3383–3391. doi:10.1002/chem.200903225
46. Seiser T, Roth OA, Cramer N (2009) Enantioselective synthesis of indanols from tert-cyclobutanols using a rhodium-catalyzed C–C/C–H activation sequence. Angew Chem Int Ed 48(34):6320–6323. doi:10.1002/anie.200903189
47. Shigeno M, Yamamoto T, Murakami M (2009) Stereoselective restructuring of 3-arylcyclobutanols into 1-indanols by sequential breaking and formation of carbon–carbon bonds. Chem-Eur J 15(47):12929–12931. doi:10.1002/chem.200902593
48. Seiser T, Cathomen G, Cramer N (2010) Enantioselective construction of indanones from cyclobutanols using a rhodium-catalyzed C–C/C–H/C–C bond activation process. Synlett 11:1699–1703. doi:10.1055/s-0029-1219959
49. Seiser T, Cramer N (2010) Rhodium(I)-catalyzed 1,4-silicon shift of unactivated silanes from aryl to alkyl: enantioselective synthesis of indanol derivatives. Angew Chem Int Ed 49(52):10163–10167. doi:10.1002/anie.201005399
50. Seiser T, Cramer N (2010) Rhodium-catalyzed C–C bond cleavage: construction of acyclic methyl substituted quaternary stereogenic centers. J Am Chem Soc 132(15):5340–5341. doi:10.1021/ja101469t
51. Ishida N, Sawano S, Murakami M (2012) Synthesis of 3,3-disubstituted α-tetralones by rhodium-catalysed reaction of 1-(2-haloaryl)cyclobutanols. Chem Commun 48(14): 1973–1975. doi:10.1039/c2cc16907j
52. Hayashi S, Hirano K, Yorimitsu H, Oshima K (2006) Palladium-catalyzed stereo- and regiospecific allylation of aryl halides with homoallyl alcohols via retro-allylation: selective generation and use of σ-allylpalladium. J Am Chem Soc 128(7):2210–2211. doi:10.1021/ja058055u
53. Shintani R, Takatsu K, Hayashi T (2008) Rhodium-catalyzed kinetic resolution of tertiary homoallyl alcohols via stereoselective carbon – carbon bond cleavage. Org Lett 10(6):1191–1193. doi:10.1021/ol800120p
54. Waibel M, Cramer N (2010) Palladium-catalyzed arylative ring-opening reactions of norbornenols: entry to highly substituted cyclohexenes, quinolines, and tetrahydroquinolines. Angew Chem Int Ed 49(26):4455–4458. doi:10.1002/anie.201001752
55. Waibel M, Cramer N (2011) Desymmetrizations of meso-tert-norbornenols by rhodium(I)-catalyzed enantioselective retro-allylations. Chem Commun 47(1):346–348. doi:10.1039/c0cc01950j
56. Terao Y, Wakui H, Nomoto M, Satoh T, Miura M, Nomura M (2003) Palladium-catalyzed arylation of α, α-disubstituted arylmethanols via cleavage of a C – C or a C – H bond to give biaryls. J Org Chem 68(13):5236–5243. doi:10.1021/jo0344034
57. Zhao P, Incarvito CD, Hartwig JF (2006) Direct observation of β-aryl eliminations from Rh(I) alkoxides. J Am Chem Soc 128(10):3124–3125. doi:10.1021/ja058550q
58. Nishimura T, Katoh T, Hayashi T (2007) Rhodium-catalyzed aryl transfer from trisubstituted aryl methanols to α, β-unsaturated carbonyl compounds. Angew Chem Int Ed 46(26):4937–4939. doi:10.1002/anie.200700902
59. Nishimura T, Katoh T, Takatsu K, Shintani R, Hayashi T (2007) Rhodium-catalyzed asymmetric rearrangement of alkynyl alkenyl carbinols: synthetic equivalent to asymmetric conjugate alkynylation of enones. J Am Chem Soc 129(46):14158–14159. doi:10.1021/ja076346s
60. Nemoto H, Yoshida M, Fukumoto K (1997) Allenylcyclobutanol, a versatile initiator for the palladium-catalyzed cascade reaction: a novel route to bicyclo[5.3.0] and -[6.3.0] frameworks. J Org Chem 62(19):6450–6451. doi:10.1021/jo9712310
61. Trost BM, Yasukata T (2001) A catalytic asymmetric Wagner – Meerwein shift. J Am Chem Soc 123(29):7162–7163. doi:10.1021/ja010504c

62. Trost BM, Xie J (2006) Palladium-catalyzed asymmetric ring expansion of allenylcyclobutanols: an asymmetric Wagner – Meerwein shift. J Am Chem Soc 128(18):6044–6045. doi:10.1021/ja0602501
63. Trost BM, Xie J (2008) Palladium-catalyzed diastereo- and enantioselective Wagner – Meerwein shift: control of absolute stereochemistry in the C – C bond migration event. J Am Chem Soc 130(19):6231–6242. doi:10.1021/ja7111299
64. Kleinbeck F, Toste FD (2009) Gold(I)-catalyzed enantioselective ring expansion of allenylcyclopropanols. J Am Chem Soc 131(26):9178–9179. doi:10.1021/ja904055z
65. Wender PA, Gamber GG, Williams TJ (2005) Rhodium(I)-catalyzed [5+2], [6+2], and [5+2+1] cycloadditions: new reactions for organic synthesis. In: Modern rhodium-catalyzed organic reactions. Wiley-VCH Verlag GmbH & Co. KGaA, pp 263–299. doi:10.1002/3527604693.ch13
66. Ylijoki KEO, Stryker JM (2012) [5+2] Cycloaddition reactions in organic and natural product synthesis. Chem Rev 113(3):2244–2266. doi:10.1021/cr300087g
67. Wender PA, Haustedt LO, Lim J, Love JA, Williams TJ, Yoon J-Y (2006) Asymmetric catalysis of the [5+2] cycloaddition reaction of vinylcyclopropanes and π-systems. J Am Chem Soc 128(19):6302–6303. doi:10.1021/ja058590u
68. Shintani R, Nakatsu H, Takatsu K, Hayashi T (2009) Rhodium-catalyzed asymmetric [5+2] cycloaddition of alkyne–vinylcyclopropanes. Chem Eur J 15(35):8692–8694. doi:10.1002/chem.200901463
69. Brandi A, Cicchi S, Cordero FM, Goti A (2003) Heterocycles from alkylidenecyclopropanes. Chem Rev 103(4):1213–1270. doi:10.1021/cr010005u
70. Rubin M, Rubina M, Gevorgyan V (2007) Transition metal chemistry of cyclopropenes and cyclopropanes. Chem Rev 107(7):3117–3179. doi:10.1021/cr050988l
71. Masarwa A, Marek I (2010) Selectivity in metal-catalyzed carbon–carbon bond cleavage of alkylidenecyclopropanes. Chem Eur J 16(32):9712–9721. doi:10.1002/chem.201001246
72. Suginome M, Matsuda T, Ito Y (2000) Palladium- and platinum-catalyzed silaboration of methylenecyclopropanes through selective proximal or distal C – C bond cleavage. J Am Chem Soc 122(44):11015–11016. doi:10.1021/ja002885k
73. Ohmura T, Taniguchi H, Kondo Y, Suginome M (2007) Palladium-catalyzed asymmetric silaborative C – C cleavage of meso-methylenecyclopropanes. J Am Chem Soc 129(12):3518–3519. doi:10.1021/ja0703170
74. Akai Y, Yamamoto T, Nagata Y, Ohmura T, Suginome M (2012) Enhanced catalyst activity and enantioselectivity with chirality-switchable polymer ligand PQXphos in Pd-catalyzed asymmetric silaborative cleavage of meso-methylenecyclopropanes. J Am Chem Soc 134(27):11092–11095. doi:10.1021/ja303506k
75. Gulías M, Durán J, López F, Castedo L, Mascareñas JL (2007) Palladium-catalyzed [4+3] intramolecular cycloaddition of alkylidenecyclopropanes and dienes. J Am Chem Soc 129(36):11026–11027. doi:10.1021/ja0756467
76. Mazumder S, Shang D, Negru DE, Baik M-H, Evans PA (2012) Stereoselective rhodium-catalyzed [3+2+1] carbocyclization of alkenylidenecyclopropanes with carbon monoxide: theoretical evidence for a trimethylenemethane metallacycle intermediate. J Am Chem Soc 134(51):20569–20572. doi:10.1021/ja305467x
77. McKinney RJ (1992) The applications and chemistry of catalysis by soluble transition metal complexes. In: Homogeneous catalysis. Wiley, New York, pp 42–50
78. Tolman CA, Seidel WC, Druliner JD, Domaille PJ (1984) Catalytic hydrocyanation of olefins by nickel(0) phosphite complexes – effects of Lewis acids. Organometallics 3(1):33–38. doi:10.1021/om00079a008
79. Nakao Y, Oda S, Hiyama T (2004) Nickel-catalyzed arylcyanation of alkynes. J Am Chem Soc 126(43):13904–13905. doi:10.1021/ja0448723
80. Nakao Y, Yukawa T, Hirata Y, Oda S, Satoh J, Hiyama T (2006) Allylcyanation of alkynes: regio- and stereoselective access to functionalized di- or trisubstituted acrylonitriles. J Am Chem Soc 128(22):7116–7117. doi:10.1021/ja060519g

81. Nakao Y, Yada A, Ebata S, Hiyama T (2007) A dramatic effect of Lewis-acid catalysts on nickel-catalyzed carbocyanation of alkynes. J Am Chem Soc 129(9):2428–2429. doi:10.1021/ja067364x
82. Nakao Y, Hirata Y, Tanaka M, Hiyama T (2008) Nickel/BPh3-catalyzed alkynylcyanation of alkynes and 1,2-dienes: an efficient route to highly functionalized conjugated enynes. Angew Chem Int Ed 47(2):385–387. doi:10.1002/anie.200704095
83. Nakao Y (2012) Nickel/Lewis acid-catalyzed carbocyanation of unsaturated compounds. Bull Chem Soc Jpn 85(7):731–745
84. Watson MP, Jacobsen EN (2008) Asymmetric intramolecular arylcyanation of unactivated olefins via C − CN bond activation. J Am Chem Soc 130(38):12594–12595. doi:10.1021/ja805094j
85. Nakao Y, Ebata S, Yada A, Hiyama T, Ikawa M, Ogoshi S (2008) Intramolecular arylcyanation of alkenes catalyzed by nickel/AlMe2Cl. J Am Chem Soc 130 (39):12874–12875. doi:10.1021/ja805088r
86. Nájera C, Sansano JM (2009) Asymmetric intramolecular carbocyanation of alkenes by C–C bond activation. Angew Chem Int Ed 48(14):2452–2456. doi:10.1002/anie.200805601
87. Hsieh J-C, Ebata S, Nakao Y, Hiyama T (2010) Asymmetric synthesis of indolines bearing a benzylic quaternary stereocenter through intramolecular arylcyanation of alkenes. Synlett 2010(11):1709–1711. doi:10.1055/s-0029-1219964
88. Murahashi S, Naota T, Nakajima N (1986) Palladium-catalyzed decarbonylation of acyl cyanides. J Org Chem 51(6):898–901. doi:10.1021/jo00356a029
89. Yasui Y, Kamisaki H, Takemoto Y (2008) Enantioselective synthesis of 3,3-disubstituted oxindoles through Pd-catalyzed cyanoamidation. Org Lett 10(15):3303–3306. doi:10.1021/ol801168j
90. Yasui Y, Kamisaki H, Ishida T, Takemoto Y (2010) Synthesis of 3,3-disubstituted oxindoles through Pd-catalyzed intramolecular cyanoamidation. Tetrahedron 66(11):1980–1989, http://dx.doi.org/10.1016/j.tet.2010.01.073

Top Curr Chem (2014) 346: 195–232
DOI: 10.1007/128_2014_527
© Springer-Verlag Berlin Heidelberg 2014
Published online: 6 May 2014

# Transition Metal-Catalyzed Cycloadditions of Cyclopropanes for the Synthesis of Carbocycles: C–C Activation in Cyclopropanes

**Yang Gao, Xu-Fei Fu, and Zhi-Xiang Yu**

**Abstract** Transition metal-catalyzed cycloadditions of cyclopropanes have been well developed over the past several decades, leading to numerous new types of cycloadditions which are complementary to the traditional cycloadditions for the synthesis of carbocycles. Cycloadditions of vinylcyclopropanes (VCPs) and methylenecyclopropanes (MCPs) constitute two main aspects of this field. VCPs can act either as five-carbon synthons or three-carbon synthons, depending on whether the vinyl substituent is acting as an additional two-carbon synthon or not. As five-carbon synthons, VCPs are involved in [5+1], [5+2], [5+2+1], and [5+1+2+1] cycloadditions. As three-carbon synthons, VCPs are mainly involved in [3+2] and [3+2+1] cycloadditions. MCPs mostly act as three-carbon synthons and can have [3+2] cycloadditions with different π systems. Other types of cycloadditions involving MCPs are also reviewed, such as [3+1], [3+2+2], and [4+3+2] cycloadditions. Cycloadditions of some other unusual cyclopropane derivatives are also introduced briefly. The cycloadditions of VCPs and MCPs have found applications in total synthesis and some representative molecules are tabulated as selected examples.

**Keywords** C-C activation · Cycloaddition · Cyclopropane · Transition metal

## Contents



Y. Gao, X.-F. Fu, and Z.-X. Yu (✉)
College of Chemistry, Peking University, Beijing 100871, China
e-mail: yuzx@pku.edu.cn

## 1 Introduction

Carbocycles can be accessed by many methods, among which the most important is perhaps the cycloaddition reaction. Due to its high efficiency and atom/step economy, the traditional cycloaddition reaction, which mostly involves (conjugated) unsaturated carbon–carbon bonds, has been widely used to synthesize carbocycles (or polycarbocycles). However, it also suffers from some notorious limitations, such as harsh conditions, poor selectivity, and limited substrate scope. Introducing activating functional groups and/or transition metal catalysis can overcome these disadvantages to a great extent, but there is still room to improve. For example, there are limited approaches to odd-membered carbocycles owing to restrictions of Woodward–Hoffman rules or even-membered synthons used. Therefore, introducing new synthons, especially odd-membered synthons, would provide opportunities to develop new reaction patterns to synthesize more types of carbocycles.

It has long been recognized that cyclopropanes have reactivities similar to alkenes, and cyclopropanes can be viewed as homologues of alkenes. Nevertheless, cyclopropanes, as unique three-membered synthons, were not able to participate in cycloaddition reactions via C–C activation strategy until the development of transition metal chemistry. The past several decades have witnessed the great advance of cycloaddition chemistry of cyclopropanes along with development of the transition metal-catalyzed organic chemistry. As a result, combinations of cyclopropanes and other new or traditional synthons in the presence of transition metals have led to many new types of cycloadditions, and C–C activation of cyclopropanes strategies in cycloadditions have also found many applications in the synthesis of complex molecules. Generally, a vinyl or a methylene substituent was necessary to activate (or direct) the C–C bond activation in cyclopropanes so vinylcyclopropanes (VCPs) and methylenecyclopropanes (MCPs) chemistries constitute the two main aspects of cycloadditions of cyclopropanes via the C–C activation mechanism. This chapter will mainly review the transition metal-catalyzed cycloadditions of VCPs and MCPs (via a metallacarbocycle intermediate) for synthesis of carbocycles, and is organized by the type of cyclopropanes (VCPs, MCPs, and others), including the illustration of selected examples of their applications in total synthesis. Given space limitations, cycloadditions of cyclopropanes with heteroatom-containing unsaturated partners leading to heterocycles, cycloadditions of cyclopropanes via radical or dipolar intermediates, and cycloadditions of cyclopropenes and related compounds will not be included.

# 2 VCP Chemistry in Transition Metal-Catalyzed Cycloadditions

Vinylcyclopropane (VCP) can be viewed as the homologue of 1,3-butadiene, which is supported by theoretical computations and spectroscopy analysis [1]. However, it's the experimental fact that VCP can participate in cycloadditions as does 1,3-butadiene that may make the viewpoint more convincing. The vinyl substituent of VCP is usually essential to activate or direct the cyclopropane opening, but it does not have to participate in cycloadditions. As a result, VCP can act as either a five-carbon synthon or a three-carbon synthon. Combination of many other components with VCP (as a five- or three-carbon synthon) under transition metal catalysis or mediation has led to development of numerous new cycloaddition patterns. The cycloadditions of VCP introduced will be classified according to its role as a five- or three-carbon synthon. Under each category, further elaboration is arranged by different cycloaddition patterns and transition metals.

## *2.1 VCP as a Five-Carbon Synthon in Transition Metal-Catalyzed Cycloadditions*

### 2.1.1 [5+1] Cycloadditions

VCP as a five-carbon synthon participating in transition metal-catalyzed cycloadditions can date back to 1969 when Sarel reported the first example of [5+1] reaction involving VCP in the presence of $Fe(CO)_5$ under thermal conditions [2]. When 1,1-dicyclopropylethylene was heated with $Fe(CO)_5$ at a high temperature for a prolonged time, cyclohexenone-Fe complex was unexpectedly obtained, in addition to the desired diene π-complex (see (1)). The cyclohexenone-Fe complex arose from [5+1] carbonylation of 1,1-dicyclopropylethylene with CO from $Fe(CO)_5$ followed by opening of the second cyclopropane ring. The authors systematically studied the reaction under milder photochemical conditions to gain more mechanistic insights and to achieve cleaner conversion as well [3, 4]. They found that a wide range of vinylcyclopropanes performed well, giving the corresponding cyclohexenones as the major products together with minor quantities of π-complexes (see (2)). Based on their results, they also proposed a plausible mechanism, as described in Scheme 1. Vinylcyclopropane compound **1** undergoes an oxidative cycloaddition of an iron carbonyl bond [Fe=C=O] to form a labile, spectroscopically identifiable metallacyclic species **2**. Then **2** could undergo

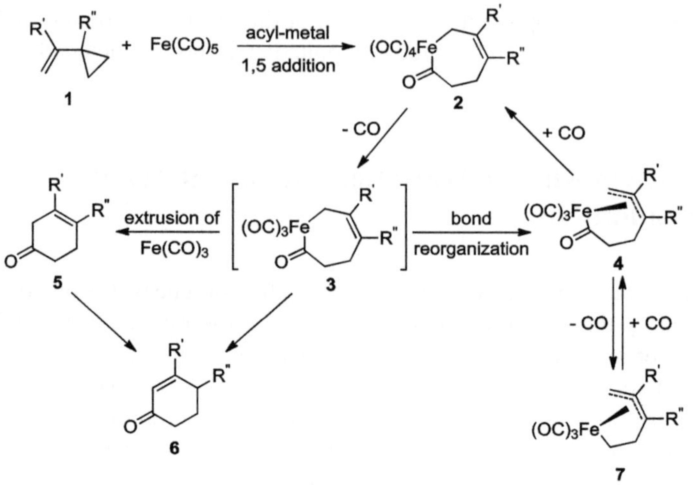


**Scheme 1** The proposed mechanism of the photochemical [5+1] reaction by Sarel

reversible carbon monoxide dissociation to give a non-isolable metallacycloalkenone **3**, which gives cyclohexenone products **5** and/or **6** through reductive elimination or an isolable species **4** through bond reorganization. The σ,π-allyl species **4** could also reversibly lose carbon monoxide to give the isolable, well-characterized σ,π-allyl iron tricarbonyl complex **7**. Later in 1974, Aumann also reported the similar reactivities of vinylcyclopropane with $Fe(CO)_5$ under photochemical conditions [5]. The thermal iron-catalyzed [5+1] cycloaddition of VCP was also reported by Schulze in 1996 [6], and the mechanism under applied conditions was studied by deuterium labeled experiments [7].

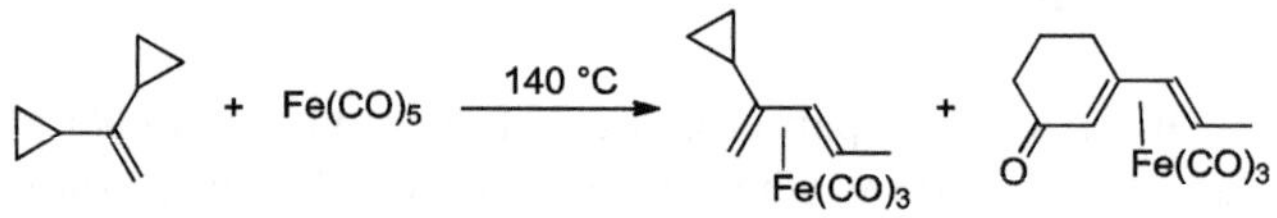


(1)

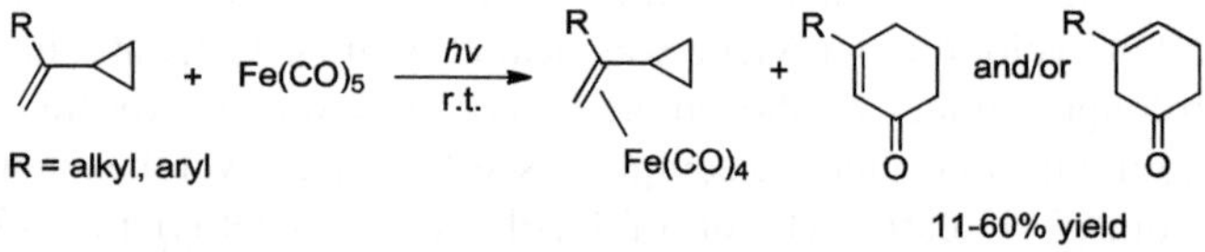


(2)

The group of Taber studied the regioselectivity of opening of the unsymmetrically substituted cyclopropane ring in the iron-catalyzed photochemical [5+1] cycloaddition (Scheme 2) [8, 9]. They found the cleavage of the less hindered C–C bond (bond b in Scheme 2) was favored. Consequently, the stereogenic center of the cyclopropane in the [5+1] reaction would not be destroyed. As a result, 5-alkyl cyclohexenones with high enantiomeric purity can be prepared from enantiomerically pure alkenyl cyclopropanes using the [5+1] cycloaddition method. Since there are many available methods to prepare enantiomerically pure vinylcyclopropanes, this [5+1] cycloaddition may provide a versatile method for

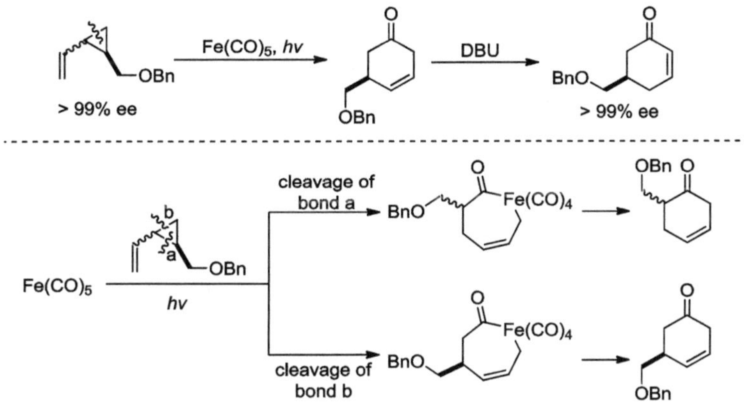

**Scheme 2** Enantiomerically pure cyclohexenones by Fe-catalyzed carbonylation of VCPs

the construction of cyclohexane derivatives. de Meijere's group disclosed another example of [5+1] cycloaddition involving VCP catalyzed by $Co_2(CO)_8$ (see (3)) [10]. They found that α-enyl cyclopropanes, including (cyclopropylmethylene) cyclopropanes and vinylcyclopropanes, can be converted into non-conjugated cyclohexenones in the presence of a stoichiometric quantity of $Co_2(CO)_8$. In some cases, the reaction can also be conducted catalytically under a CO atmosphere.

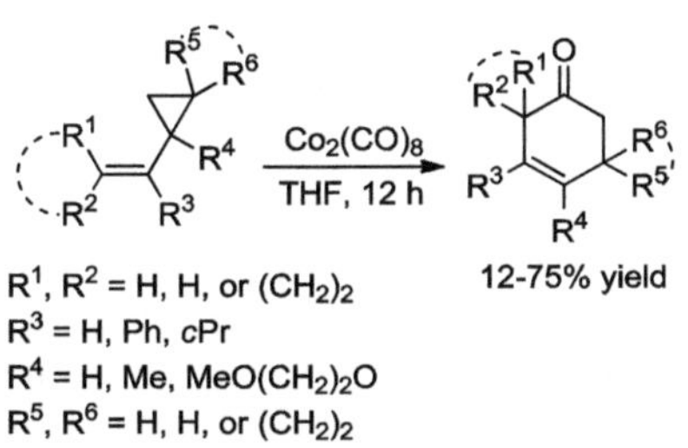

(3)

In 2012, the first general example of Rh(I)-catalyzed [5+1] cycloaddition of vinylcyclopropanes with CO was reported by Yu's group [11]. The reaction was conducted under two different conditions (conditions **A** and **B**, Scheme 3) to provide selectively nonconjugated or conjugated products, respectively. The presence of 4-Å molecular sieves was found to be essential to the reactions. The reaction can accommodate VCP substrates with different functional groups and substitution patterns, as illustrated in Scheme 3.

Related allenylcyclopropanes can also participate in the [5+1] cycloaddition. For example, Iwasawa and co-workers found that allenylcyclopropanols reacted with stoichiometric $Co_2(CO)_8$, giving hydroquinone derivatives in moderate to good yields (see (4)) [12, 13]. Murakami reported an example of iridium-catalyzed [5+1] cycloaddition of multisubstituted allenylcyclopropanes with carbon monoxide for synthesis of methylenecyclohexenones (see (5)) [14]. It was found that the presence of one, or preferably two, substituents at the allenic terminus was necessary.

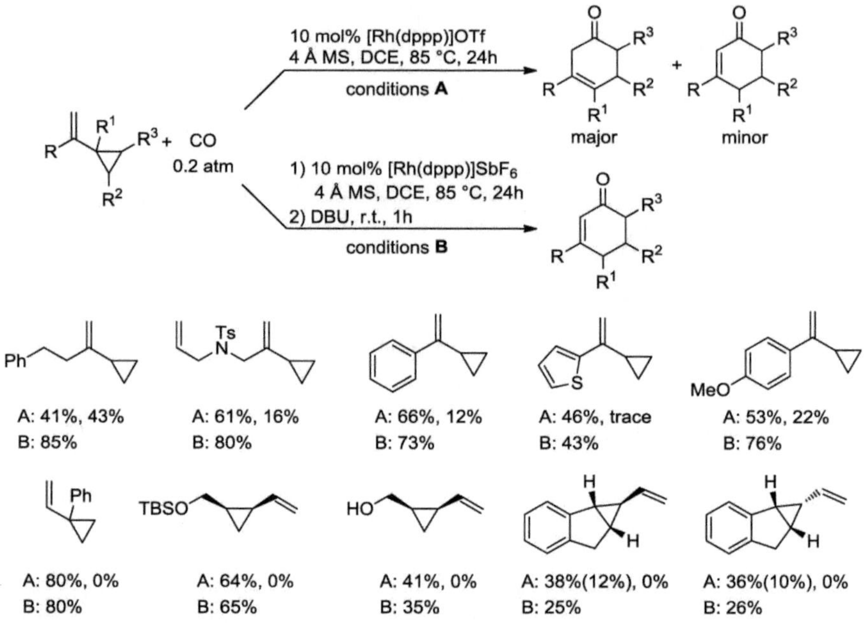


**Scheme 3** Rh(I)-catalyzed [5+1] cycloaddition of VCPs and CO

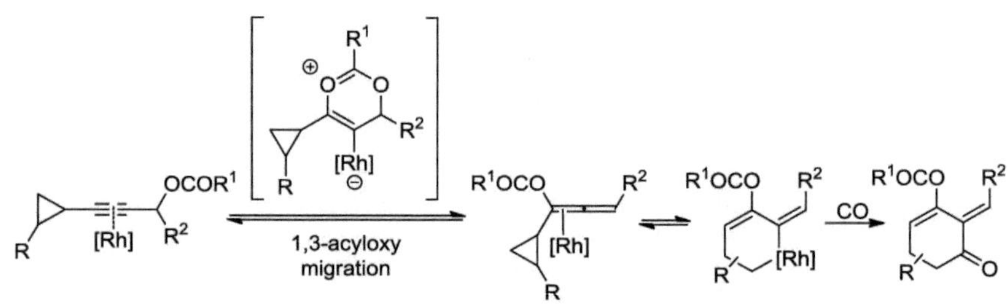


**Scheme 4** Rh(I)-catalyzed [5+1] cycloaddition of allenylcyclopropanes and CO

Tang found that in the presence of $[Rh(CO)_2Cl]_2$, allenylcyclopropanes can be generated from 1,3-acyloxy migration of cyclopropyl propargyl ester in situ, which then underwent the [5+1] cycloaddition to produce highly functionalized cyclohexenones efficiently (see (6)) (Scheme 4) [15, 16]. Mostly, regioselective cleavage of the less substituted cyclopropane C–C σ-bond can be achieved for various substituted substrates and the stereogenic center on the cyclopropane can remain intact. Another type of cyclopropyl substituted propargyl esters with acyloxy placed between the cyclopropane and the alkyne can also undergo similar transformation (see (7)).

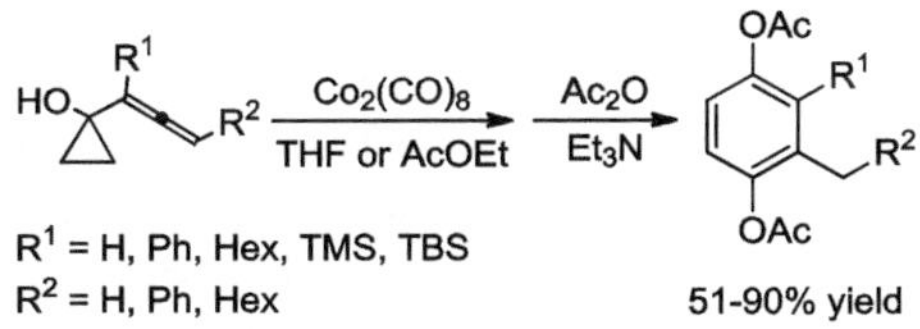


(4)

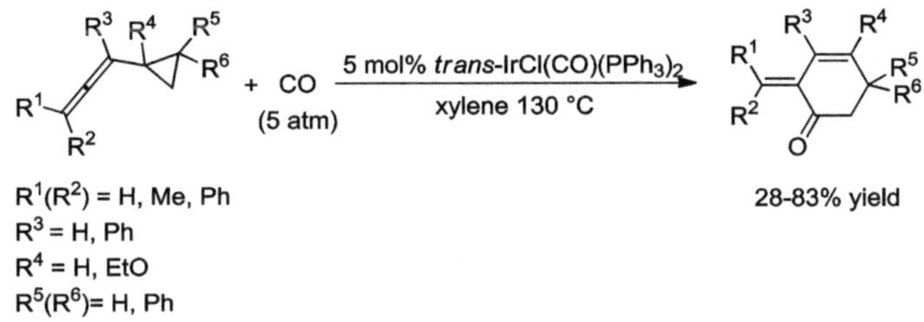


(5)

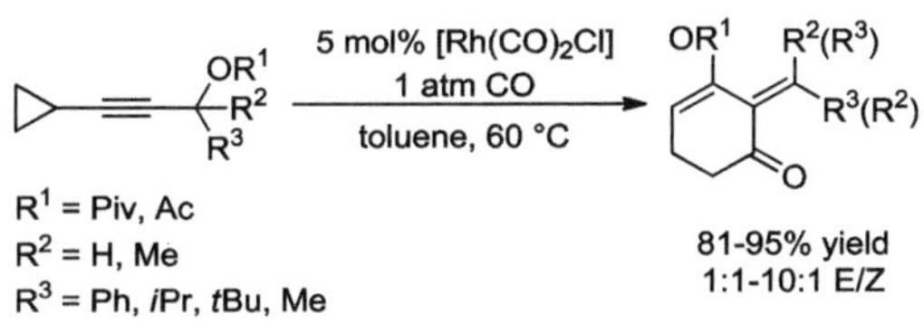


(6)

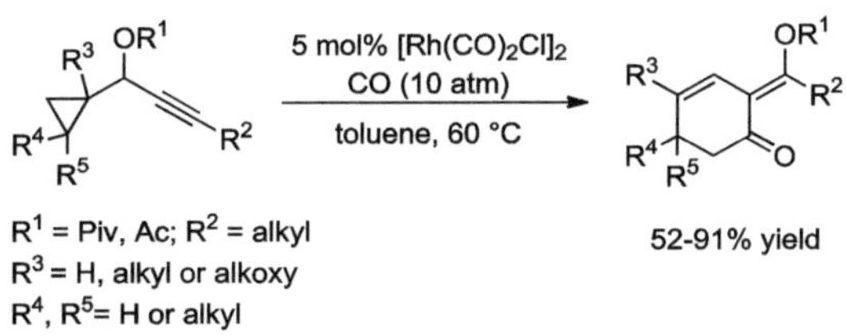


(7)

### 2.1.2 [5+2] Cycloadditions

#### Rhodium-Catalyzed [5+2] Cycloadditions

Developing a formal homologue of Diels–Alder reaction between VCPs and two-carbon π systems for synthesis of seven-membered rings had been a long-term pursuit of many chemists, but it was not until 1995 when Wender's group reported the first efficient and reliable example of rhodium-catalyzed [5+2] cycloaddition reaction of vinylcyclopropanes (VCPs) and alkynes [17]. Seven-membered carbocycles can be obtained from the intramolecular [5+2] reaction in good to excellent yields, which can tolerate different tethers, various substituents on the alkyne or alkene, and both Z and E configuration of the double bond (see (8)) [17]. One limitation was the isomerization of the initially formed double bond from the opening of the cyclopropane ring, which was observed in some cases. The intramolecular [5+2] cycloaddition can also be applied to VCPs with alkenes (see (9)) [18, 19] and VCPs with allenes (see (10)) [20]. In the intramolecular [5+2] reactions of VCPs with alkenes as shown in (9), only the *cis* diastereoisomer was produced with the exception of four-atom tethered substrate which gave the *trans* diastereoisomer exclusively (see (9)). At this time, the authors found that $[Rh(CO)_2Cl]_2$ was a superb catalyst for the [5+2] reactions, benefiting from its high reactivity and suppression of double bond isomerization [21]. In the case of VCPs reacting with allenes, formation of a *cis* ring-fusion was favored, and the level of selectivity can be controlled and amplified by using different catalysts (see (10)) [20]. The reaction displayed complete selectivity for the internal double bond of the allene and it was also proved that the chirality of allene can be transferred to the cycloadduct completely (see (10)).

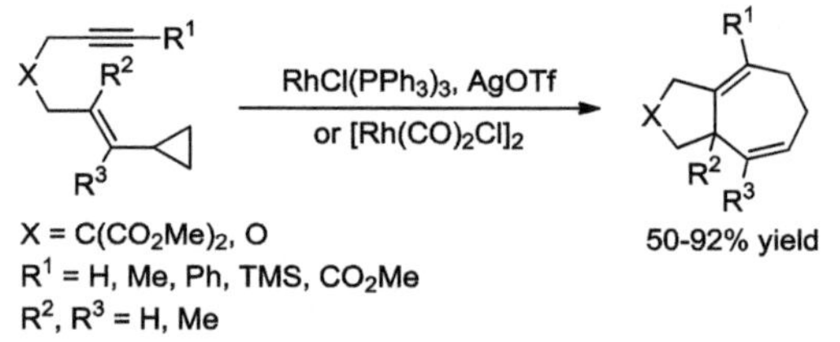


(8)

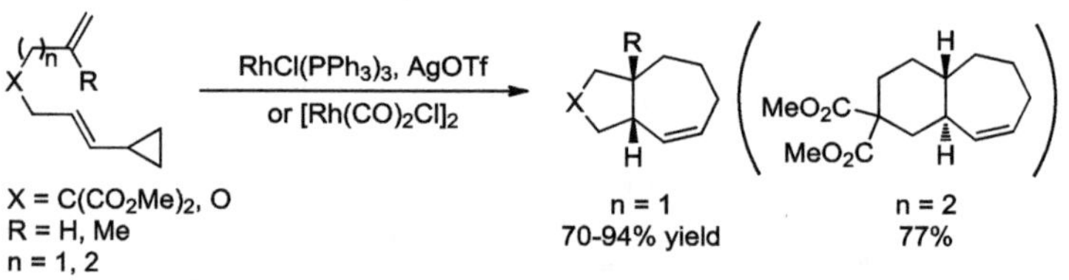


(9)

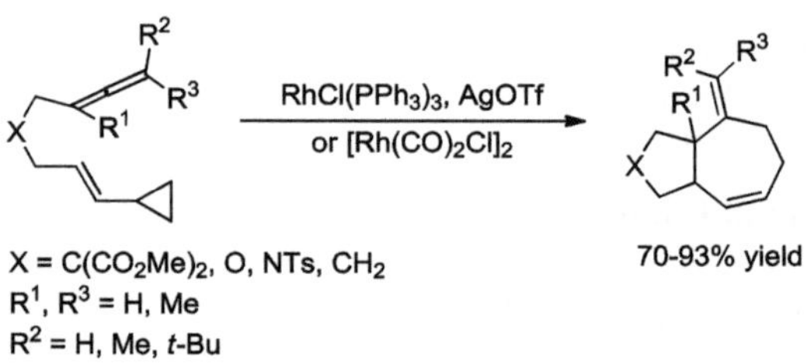


(10)

On the regio- and stereoselectivity of the Rh-catalyzed [5+2] cycloadditions with substituted cyclopropanes, Wender's group had conducted detailed studies [22]. For substrates with 1,2-disubstituted cyclopropanes, the isomer derived from cleavage of the less substituted cyclopropane carbon–carbon σ-bond was the major (if not the only) product for electron-neutral or electron-donating substituents, but for substrates with electron-withdrawing substituents, the regioselectivity was poorer and can even be reversed under different conditions. It was thus concluded that different regioselectivities in cyclopropane bond cleavage can be obtained through judicious selection of substituents and/or catalyst [22]. Additionally, for almost all these substrates, the relative stereochemistry of vicinal cyclopropane substituents is retained through the cycloaddition, translating into a 1,4-stereorelationship in the product (3,4(5)-stereorelationship to 1,4(5)-stereorelationship; see (11)) [23].

**Fig. 1** Examples of rhodium catalysts and ligand

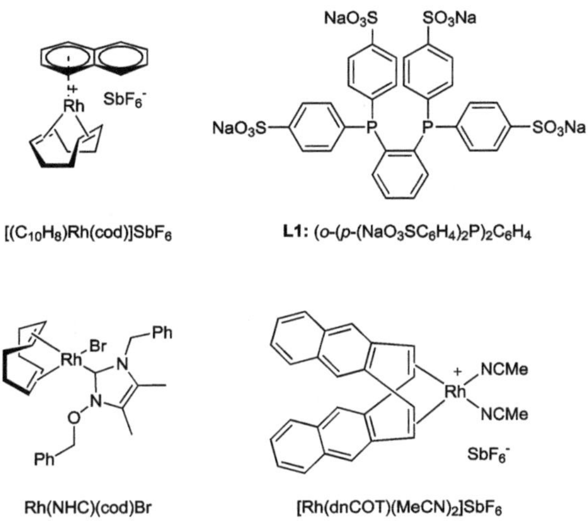


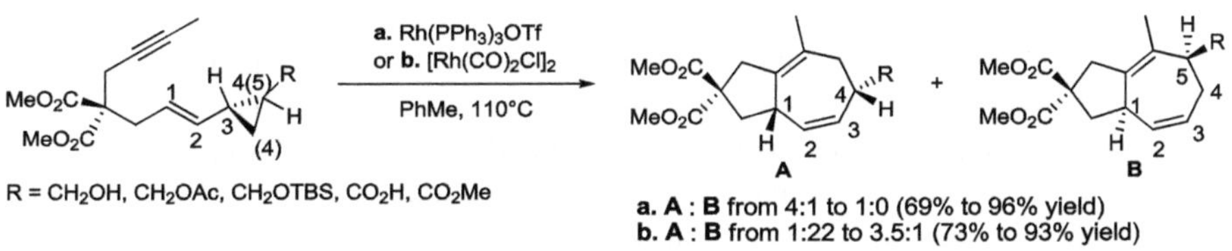


(11)

Since Wender first employed the Wilkinson's catalysts [$RhCl(PPh_3)_3$] and [$RhCl(CO)_2]_2$ to catalyze [5+2] cycloadditions, many other types of rhodium-based catalysts have been developed by Wender and others to effect the [5+2] reactions. These include [$Rh(CH_2Cl_2)_2$(dppe)]$SbF_6$ by Gilbertson [24], [RhCl(dppb)]$_2$ by Zhang [25], [Rh(NHC)(cod)]$SbF_6$ by Chung [26], [Rh(cod)Cl]$_2$ by Saito and Hanzawa [27], [Rh(arene)(cod)]$SbF_6$ [28], water-soluble [Rh(nbd)**L1**]$SbF_6$ [29], NHC complex Rh(NHC)(cod)Br [30], and rhodium dinaphthocyclooctatetraene (dnCOT) complex [Rh(dnCOT)(MeCN)$_2$]$SbF_6$ [31, 32] by Wender (Fig. 1). In particular, Rh(NHC)(cod)Br and [Rh(dnCOT)(MeCN)$_2$]$SbF_6$ proved to be most efficient so far, making the intramolecular or intermolecular [5+2] cycloadditions complete in just minutes at room temperature with high efficiency.

As early as 1998, Wender's group reported their preliminary studies towards an asymmetric version of the [5+2] cycloadditions, but the use of (2*S*,3*S*)-bis(diphenylphosphino)butane as a chiral bisphosphine ligand resulted in only moderate enantioselectivity (up to 63% *ee*) in a single substrate studied [19]. Eight years later they reported the first highly enantioselective [5+2] cycloadditions employing (*R*)-BINAP as the chiral ligand [33]. High enantioselectivity (up to >99% *ee*) was achieved for alkene-VCPs (see (12)) but only moderate enantioselectivity for alkyne-VCPs. In 2009, Hayashi and co-workers reported the Rh-catalyzed asymmetric [5+2] cycloaddition of alkyne-VCPs using a chiral phosphoramidite ligand to deliver the cycloadduct in excellent enantiomeric excess (up to >99.5% *ee*), thus addressing the shortcoming of Wender's protocol (see (13)) [34].

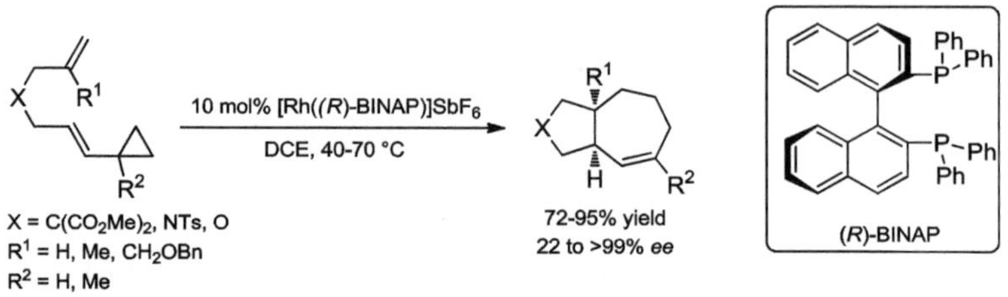


(12)

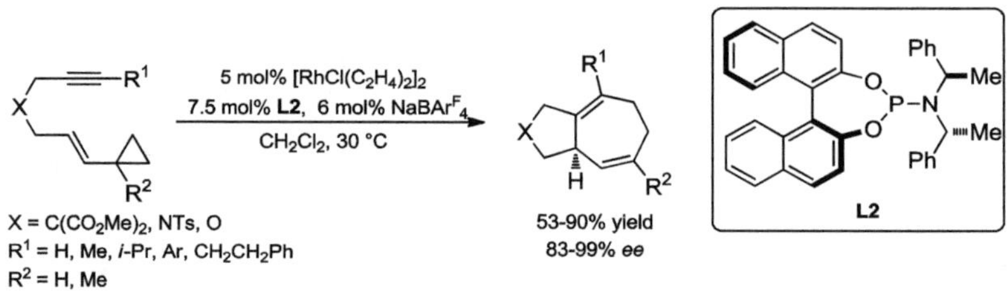


(13)

More effort from others has expanded the types of substrates involved in the [5+2] cycloadditions. For example, Saito and Hanzawa et al. reported the rhodium-catalyzed [5+2] cycloaddition of ester-tethered alkyne-VCPs using fluorinated alcohols as solvents (see (14)) [27]. Mukai's work demonstrated that alkyne-allenylcyclopropanes can also undergo rhodium-catalyzed [5+2] cycloaddition to give bicyclo[5.4.0] undecatrienes or bicyclo[5.5.0]dodecatrienes (see (15)) [35]. In related work from Yu's group, they found that the alkene-VCP with an internal cyclopropane with the *cis*-substitution can react well by following the [5+2] pathway (see (16)) while the *trans* isomer underwent the [3+2] process (see [3+2] section and (32)) [36].

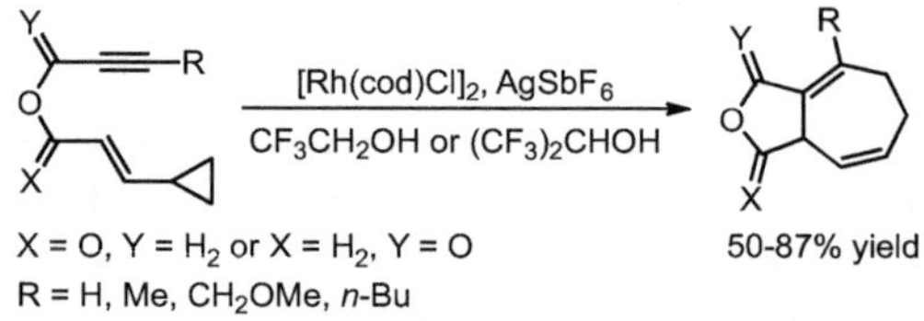


(14)

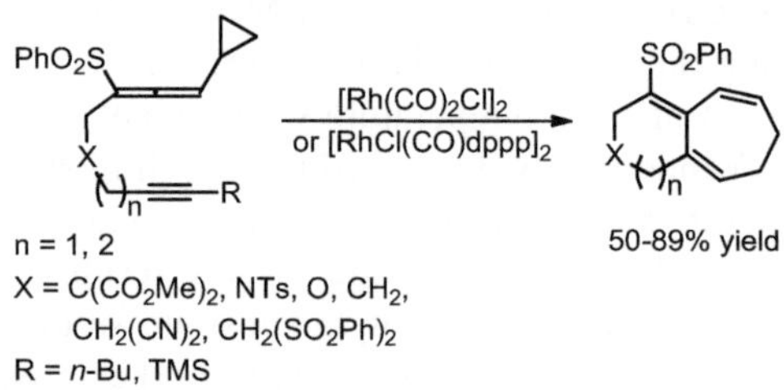


(15)

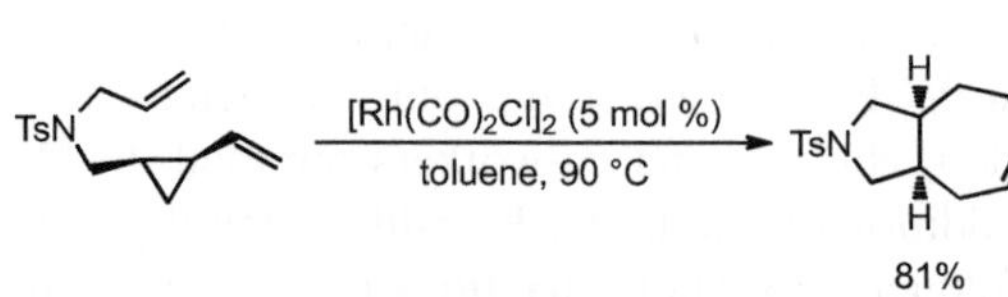


(16)

Wender's group has also developed rhodium-catalyzed intermolecular [5+2] cycloadditions. At first, they found the catalysis system of $Rh(PPh_3)_3Cl$ for the intramolecular reactions was not effective at all for the intermolecular reactions. To effect the intermolecular [5+2] cycloadditions, $[Rh(CO)_2Cl]_2$ must be used and oxygen substitution of the cyclopropane was necessary (see (17)) [37–39]. Then they successfully expanded the substrate to unactivated vinylcyclopropanes by adjusting the substituents. For monosubstituted alkynes, the substitution on the olefin terminus directs the formation of single isomer that minimized steric hindrance (see (18)) [40]. The [5+2] cycloadditions can also be applied to VCPs with allenes (see (19)) [41]. It should be noted that the alkyne substituent did not interfere with the reaction, indicating that allenes as reaction partners were superior to alkyne in the [5+2] cycloadditions. Curiously, the authors didn't report the corresponding intermolecular [5+2] cycloadditions of VCPs with alkenes.

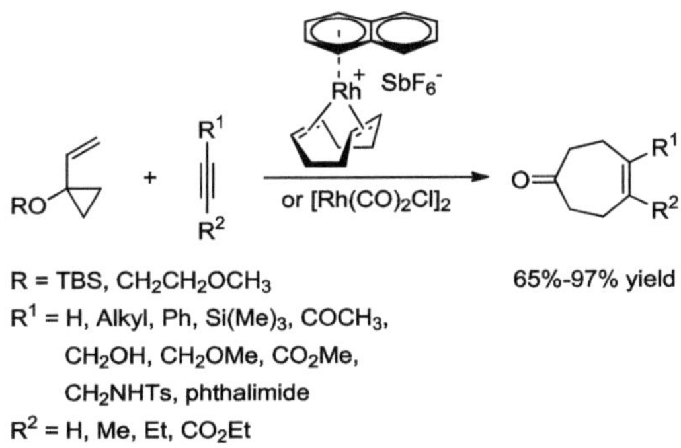


(17)

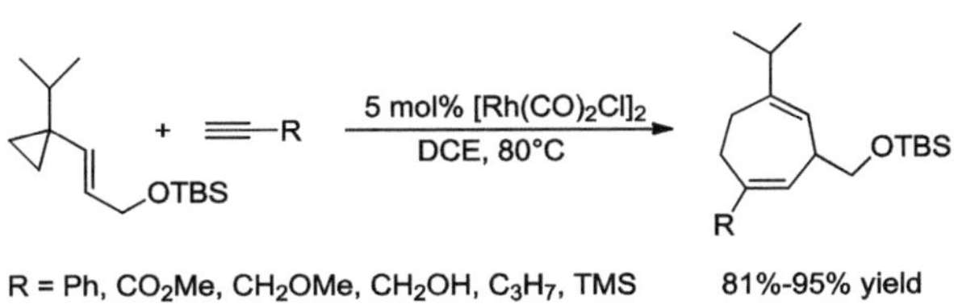


(18)

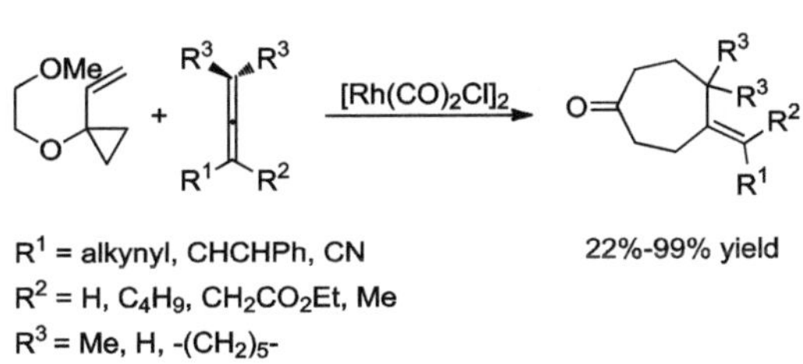


(19)

The rhodium-catalyzed [5+2] cycloadditions have been used in cascade with other processes to synthesize molecules of more complexity in a single operation. The first example was reported by Martin who developed a cascade sequence involving allylic alkylation and [5+2] cycloaddition (see (20)) [42, 43]. The catalyst $[Rh(CO)_2Cl]_2$ could be used to catalyze both the highly regioselective allylic alkylation and the following intramolecular [5+2] cycloaddition. As another example, Wender and co-workers combined intermolecular [5+2] cycloaddition with

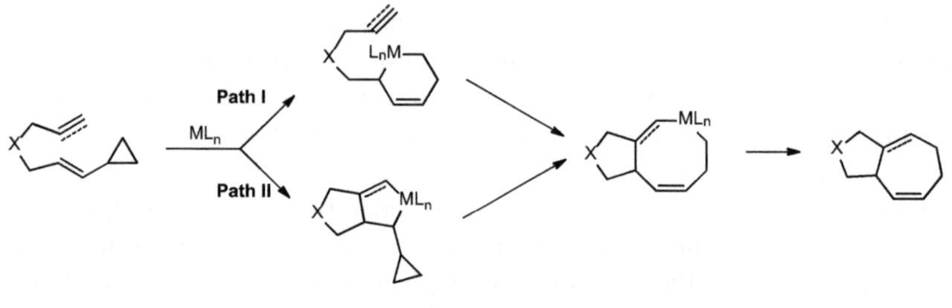


**Scheme 5** The mechanism of transition metal-catalyzed [5+2] cycloaddition

Nazarov reaction to develop a highly efficient construction of bicyclo[5.3.0]decane systems (see (21)) [44]. In addition, intermolecular [5+2] cycloadditions combined with Diels-Alder reactions have been used by Wender to construct polycyclic structures in one pot (see (22)) [45].

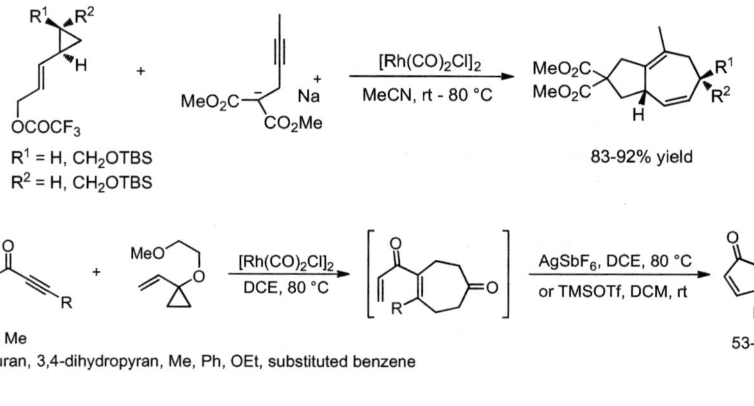


(20)

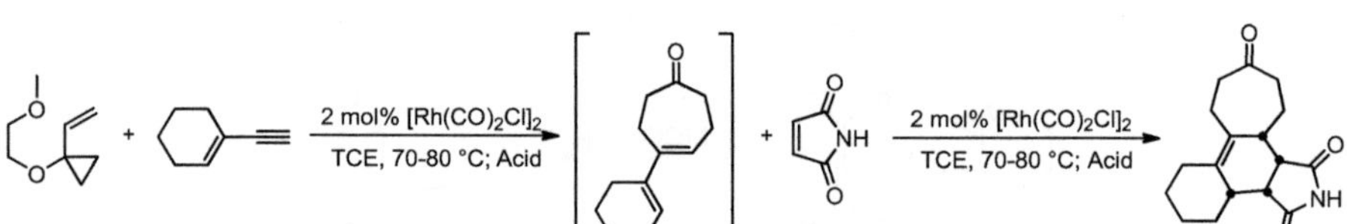


(21)

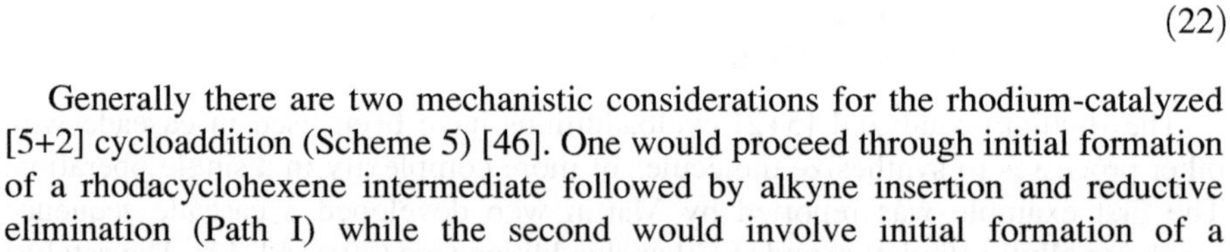


(22)

Generally there are two mechanistic considerations for the rhodium-catalyzed [5+2] cycloaddition (Scheme 5) [46]. One would proceed through initial formation of a rhodacyclohexene intermediate followed by alkyne insertion and reductive elimination (Path I) while the second would involve initial formation of a rhodacyclopentene followed by ring expansion assisted by release of the ring strain and reductive elimination (Path II). A DFT computational investigation taken up by Houk group revealed that pathway I involving a rhodacyclohexene intermediate is favored [46]. Other computational and experimental studies have disclosed other aspects of the mechanism, including the origins of different reactivities associated with different substrates, the influence of substituents on the reactivities, and the electronic and steric control of regioselectivities in the rhodium-catalyzed [5+2] cycloadditions [47–49].

## Ruthenium-Catalyzed [5+2] Cycloadditions

Trost's group successfully developed a [5+2] cycloadditions catalyzed by [CpRu $(MeCN)_3]PF_6$ in 2000 [50]. Though limited to tethered alkyne-VCPs, the [5+2] cycloadditions gave corresponding seven-membered ring products in good yields at room temperature efficiently with high regio- and stereoselectivity, and showed good compatibility with a variety of functional groups and different tethers between the alkynes and VCPs [50–53]. For substrates with 1,2-disubstituted cyclopropanes, the issue of regio- and stereoselectivity arose. With *cis* cyclopropyl substrates, steric effects appear to dominate and the less hindered bond of the cyclopropane was cleaved with high selectivity mostly (see (23)). With *trans* cyclopropyl substrates, the bond energy of the cleaving bond appears to be an important factor and the regioselectivity was usually poor as a result of both electronic and steric factor operating (see (24)). However, in both cases, the 1,2-disubstituted cyclopropyl stereochemistry was conserved and completely transferred to the product, as is same as the Wender rhodium-catalyzed [5+2] cycloaddition (see (11)). Good to excellent diastereoselectivities favoring the angular hydrogen to be anti to the homoallylic oxygen substituent were also observed for the substrates with an allylic stereogenic center (structure **8**). This methodology can also be expanded to synthesize a wide range of tricyclic systems in good yields and diastereoselectivities [54, 55] (Fig. 2). In contrast to rhodium-catalyzed [5+2] reaction, the ruthenium-catalyzed [5+2] reaction was believed to proceed via a ruthenacyclopentene (Path II, Scheme 5) intermediate as was supported by experimental facts. A DFT investigation by Houk has also been conducted to verify the mechanism and rationalize the regio- and stereoselectivities [56].

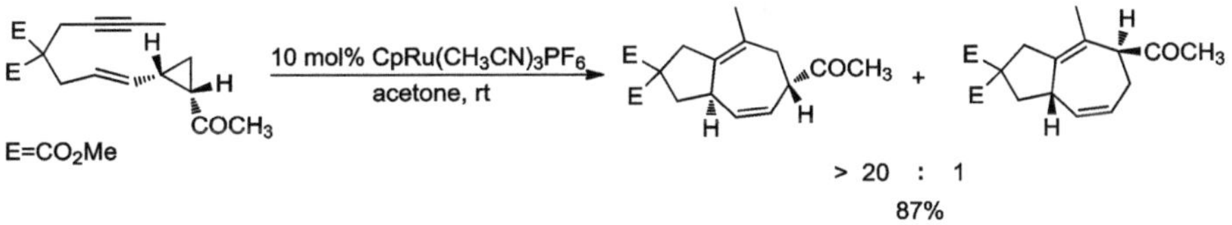


(23)

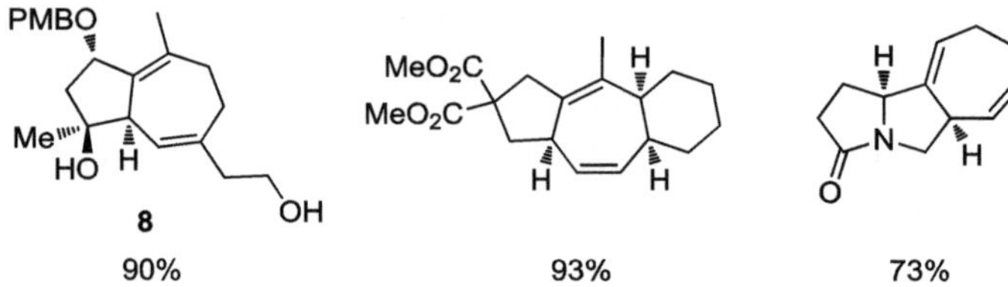


**Fig. 2** Some representative examples of Ru-catalyzed [5+2] cycloadducts

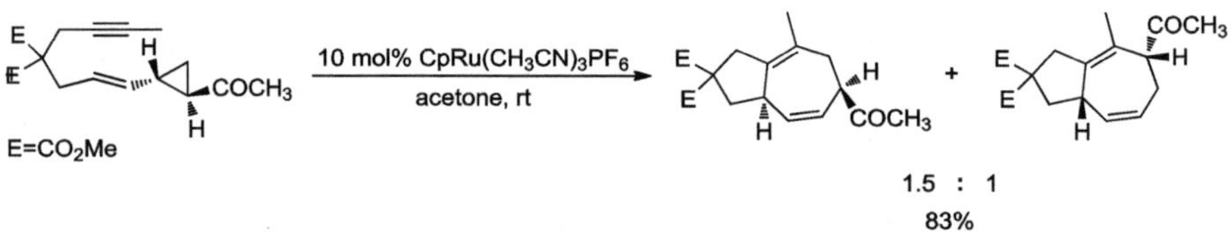


(24)

## Iron- and Nickel-Catalyzed [5+2] Cycloadditions

Iron and nickel have also been used to catalyze the intramolecular [5+2] cycloadditions. Louie's group reported a nickel-promoted intramolecular [5+2] cycloaddition when studying the nickel-catalyzed rearrangement of cyclopropylen-ynes (see (25)) [57]. The reaction course strongly depended on the substituents of the substrates and the ligands used, and has also recently been studied theoretically by Houk and co-workers [58].

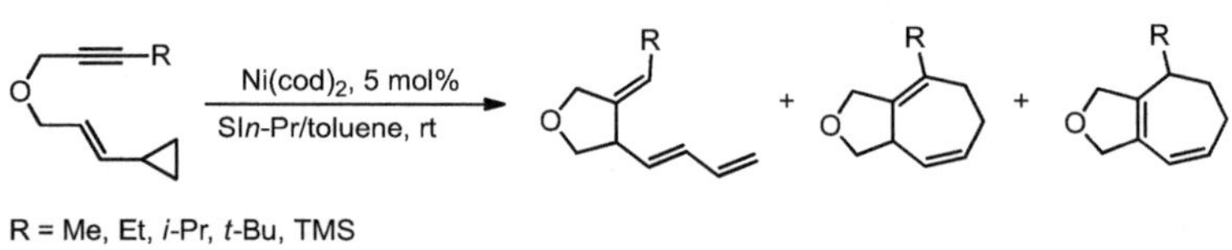


(25)

Two ferrate complexes **A** and **B** were used by Fürstner and co-workers to catalyze the intramolecular [5+2] cycloadditions of alkyne-VCPs to give the corresponding seven-membered products in good to excellent yields with a good substrate scope and diastereoselectivity (see (26)) [59].

A B

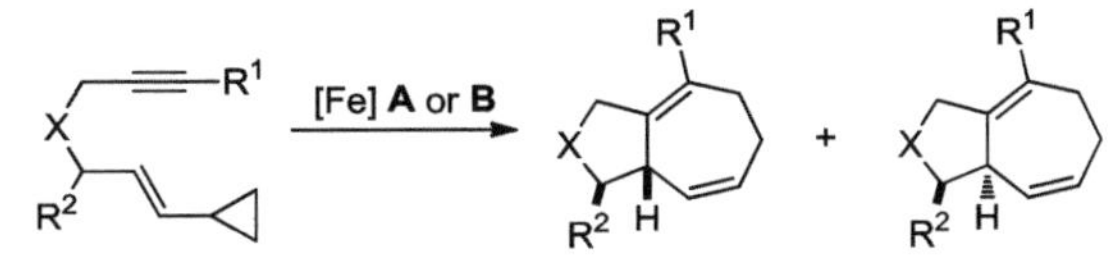


major minor

X = $C(CO_2Et)_2$, $C(CH_2OAc)_2$, $C(CH_2OTBS)_2$, $CH_2C(CO_2Et)_2$, NTs
$R^1$ = H, $SiMe_3$, $CO_2Et$, aryl
$R^2$ = H, Me

54-99% yield (26)

### 2.1.3 [5+2+1] and [5+1+2+1] Cycloadditions

In 2002, Wender's group first realized the rhodium-catalyzed three-component intermolecular [5+2+1] cycloaddition of VCPs, carbonyl-substituted alkynes, and CO [60]. Initially, they envisioned that the rhodacyclooctene intermediate of the [5+2] cycloaddition may be intercepted by CO, thus leading to eight-membered rings, but instead they got bicyclo[3.3.0]octenones arising from transannular closure of the intended eight-membered products (see (27)). The reaction proceeded in good to excellent yields and with high or complete regioselectivity for unsymmetrical alkynes. However, when the [5+2+1] conditions were applied to terminal alkynes, a four-component [5+1+2+1] pathway involving two units of CO operated, giving hydroxyindanones putatively from the initially formed nine-membered ring intermediate through tautomerization, electrocyclic closure, and elimination (see (28)) [61]. The three-component [5+2+1] reaction can also be applied to allenes (see (29)), as was demonstrated that the allene **9** without a carbonyl substitution (the allenes with carbonyl substitutions; see (19)), VCP and CO reacted well to give a mixture of cyclooctanedione and its transannular aldol product [41].

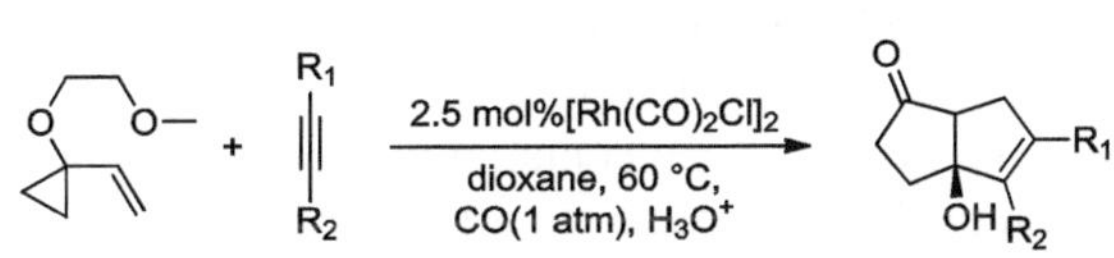


(27)

$R_1$ = $COCH_3$, $CO_2Et$, $CONH_2$, CHO, $CO_2Me$
$R_2$ = Et, TMS, Ph, Me, $CO_2Me$

48-97% yield

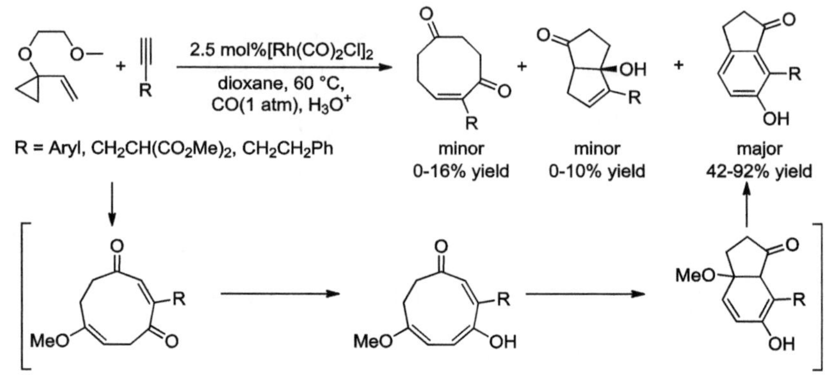


(28)

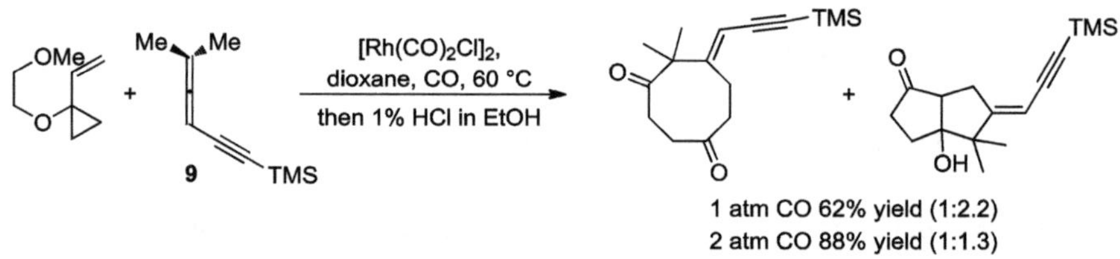


(29)

Another impressive example of rhodium-catalyzed [(5+2)+1] cycloaddition was reported by Yu's group in 2007 [49]. Their DFT calculation results indicated the activation energy of reductive elimination step in the [5+2] cycloaddition of ene-VCPs was about 25–30 kcal/mol, while the CO insertion and migratory reductive elimination had activation energies of about 13–14 and 23–24 kcal/mol, respectively. They ascribed this to the more facile reductive elimination of $(sp^2)$ C-M-$(sp^3)$C in the [(5+2)+1] process compared to the reductive elimination of $(sp^3)$ C-M-$(sp^3)$C in the [5+2] reaction. Therefore they hypothesized that ene-VCPs, which are less reactive in the $[RhCl(CO)_2]_2$-catalyzed [5+2] cycloaddition reported by Wender [21], can be made reactive either by reacting at higher temperature or by introducing CO as an additional component (see (30)). Experimental results were in agreement with their hypothesis. The intramolecular [(5+2)+1] cycloaddition of ene-VCPs and CO worked well in good to excellent yields, compatible with a variety of tether types and substitution patterns, and allowed for the preparation of 5/8- and 6/8-fused ring systems even containing quaternary centers. Thus the new computationally designed and experimentally verified Rh(I)-catalyzed [(5+2)+1] cycloaddition provided a convenient and efficient method for constructing bicyclic cyclooctenones. Furthermore, Yu's group expanded this strategy to build [5-8-5]/[6-8-5] tricyclic cyclooctenones which are important skeletons of natural products (see (31)) [62]. Recently, Yu and coworkers used both DFT calculations and experiments to study the mechanism of the [(5+2)+1] reaction and provided rationalization of the stereochemistry and reactivities of substrates with different configurations (Scheme 6) [63].

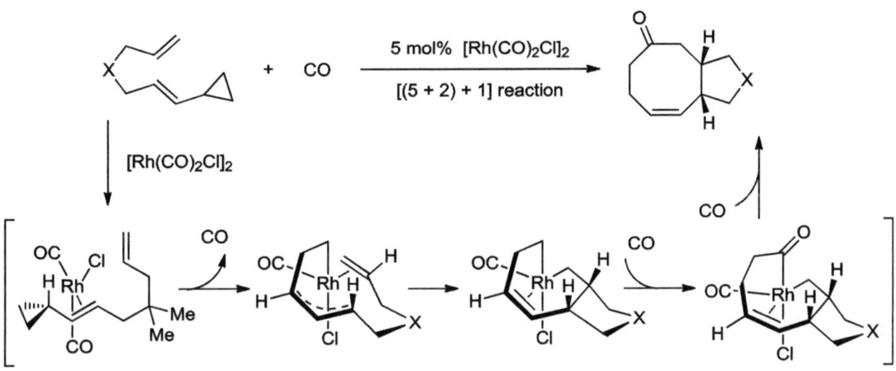


**Scheme 6** The proposed mechanism of Rh-catalyzed [(5+2)+1] cycloaddition of VCPs

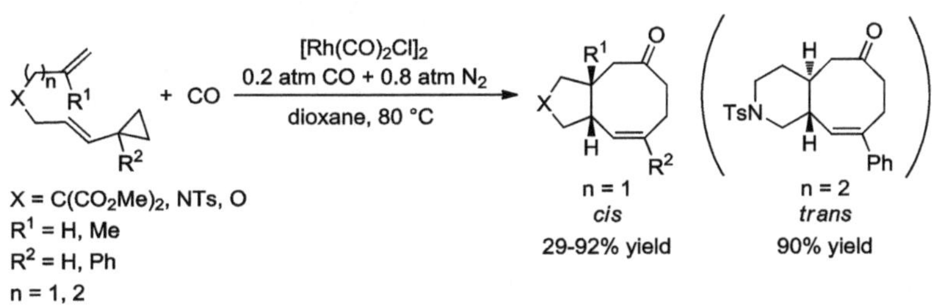


(30)

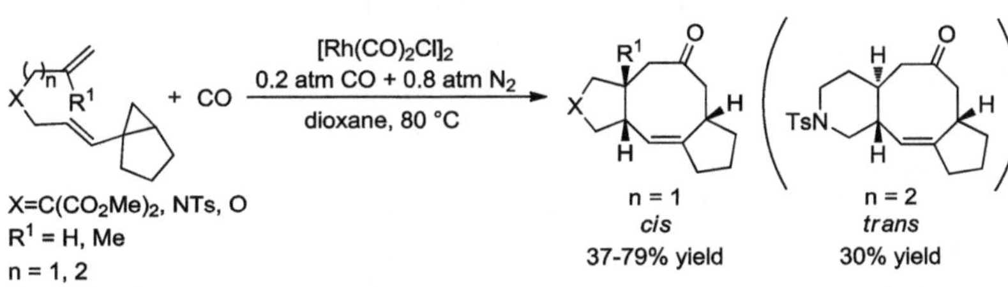


(31)

## 2.2 *VCP as a Three-Carbon Synthon in Transition Metal-Catalyzed Cycloaddition*

### 2.2.1 [3+2] Cycloadditions

VCP derivatives were first found by Tsuji and co-workers as three-carbon components to undergo a palladium-catalyzed intermolecular formal [3+2] cycloaddition with α,β-unsaturated esters and ketones in 1985, leading to vinyl-substituted cyclopentane derivatives in good yields [64]. This cycloaddition was proposed to proceed through a stepwise ionic mechanism in which two zwitterionic intermediates **10** and **11** of π-allyl palladium complex were involved (Scheme 7). Plietker's group recently reported that nucleophilic ferrate $Bu_4N[Fe(CO)_3(NO)]$ with ligand **L3** can act as an alternative catalyst for this transformation [65]. Since the discovery of the Tsuji's [3+2] cycloaddition, much effort has been devoted to developing

Scheme 7 The mechanism of Pd-catalyzed [3+2] cycloaddition of VCPs

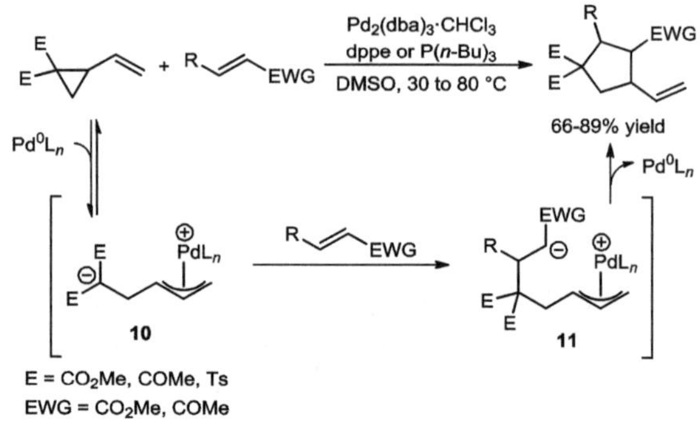


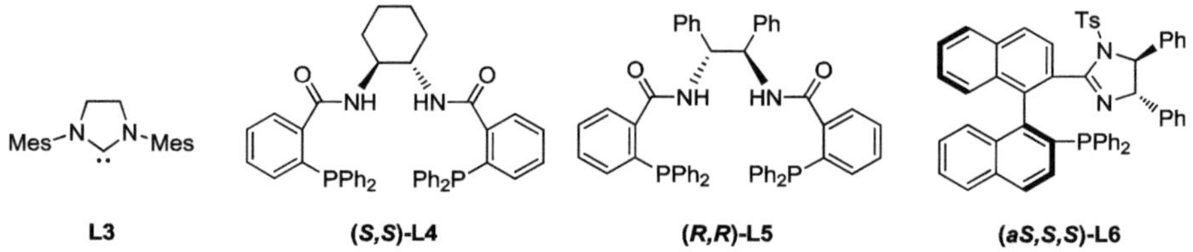


**Fig. 3** Ligands of Pd catalyst

asymmetric variants. For example, Trost's group developed a high diastereo- and enantioselective dynamic kinetic asymmetric palladium-catalyzed [3+2] cycloaddition of VCPs and azlactone- or Meldrum's acid alkylidenes using the chiral ligand **L4** and **L5** [66, 67]. Shi also reported a palladium-catalyzed asymmetric formal [3+2] cycloaddition of VCPs and β,γ-unsaturated α-keto esters employing the chiral imidazoline-phosphine ligand **L6** [68, 69] (Fig. 3).

The Tsuji's [3+2] cycloaddition virtually proceeds via dipolar intermediates stabilized both by palladium and the activating groups on the cyclopropane. There had not been an example of unactivated VCPs acting as three-carbon synthons until Yu's group first reported a rhodium-catalyzed intramolecular [3+2] cycloaddition of *trans*-vinylcyclopropane-enes in 2008 [36]. In this reaction, *trans*-2-ene-VCPs underwent an unexpected intramolecular [3+2] cycloaddition (Scheme 8a) rather than a [5+2] cycloaddition (see (15)). This strategy provided an efficient and diastereoselective approach to *cis*-fused bicyclic cyclopentanes (Scheme 8a). Interestingly, the reaction with *cis*-2-ene-VCP substrate gave a [5+2] cycloadduct. The different reaction course of a [3+2] or [5+2] cycloaddition was ascribed to proximity of different carbons in the reductive elimination step for different substrates. α-Ene-vinylcyclopropanes also underwent the similar [3+2] cycloaddition to provide a new approach to bicyclo[4.3.0]nonane and bicyclo[5.3.0]decane skeletons as reported by Yu's group (Scheme 8b) [70].

The 1-substituted-VCPs were also tested to undergo [3+2] cycloaddition by Yu's group, and it worked well with Rh(I)-phosphine complex (Scheme 9a) [71]. Impressively, this strategy was suitable for 1-ene-VCPs, 1-yne-VCPs, as well as 1-allene-VCPs, thus providing access to a wide range of cyclopentane- and cyclopentene-

**Scheme 8** Rh-catalyzed [3+2] cycloadditions of ene-VCPs

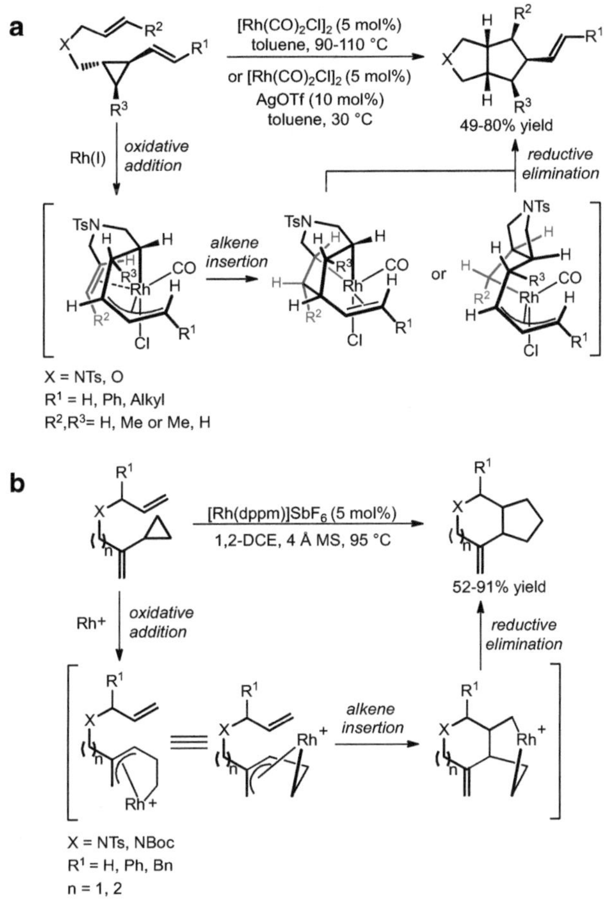


embedded bicyclic structures with a vinyl-substituted quaternary stereocenter. During the reaction course, the vinyl substituent plays an important role as a "spectator" binding group in facilitating the ring-opening of the cyclopropane through coordination with Rh. DFT study has been conducted to investigate the reaction mechanism [72]. An asymmetric rhodium-catalyzed [3+2] cycloaddition of 1-yne-VCPs, but not 1-ene-VCPs, has also been developed by Yu's group by use of (*R*)-$H_8$-BINAP as a chiral ligand (Scheme 9b) [73].

### 2.2.2 Higher Order Cycloadditions

Introducing CO as a reactant to the system of rhodium-catalyzed [3+2] cycloaddition of 1-ene-VCPs and 1-yne-VCPs has led to the corresponding carbonylative [(3+2)+1] cycloaddition, a homologous Pauson–Khand reaction (Scheme 10a)

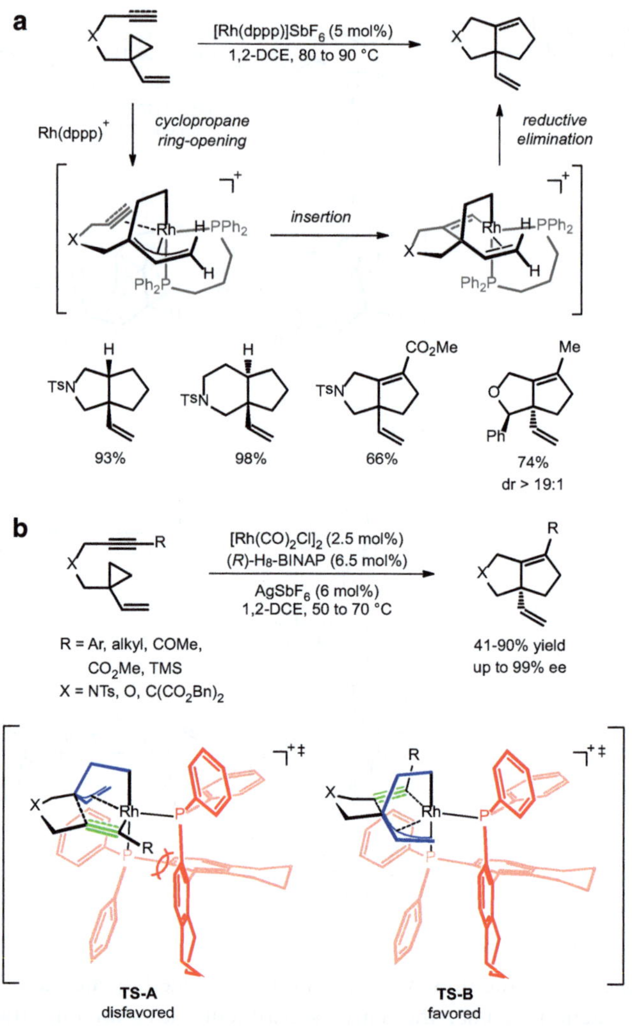


**Scheme 9** Rh(I)-catalyzed [3+2] cycloaddition of 1-substituted-VCPs

[74]. Multifunctional bicyclic cyclohexenone and cyclohexanone products were obtained in good yields with good tolerance of various functional groups and different tethers. Interestingly, for 1-yne-VCPs having an internal alkyne, a competing formal [5+1]/[2+2+1] process arose under some conditions (Scheme 10b) [75]. In the reaction, two carbonyl groups and the entire VCP moiety were incorporated into the products, rendering an unprecedented bicarbonylative cycloaddition mode of 1-yne-VCP. As indicated by its name, the reaction can be viewed as a formal cascade reaction consisting of a [5+1] cycloaddition (between the VCP and CO) and a subsequent intramolecular Pauson–Khand [(2+2)+1] cycloaddition

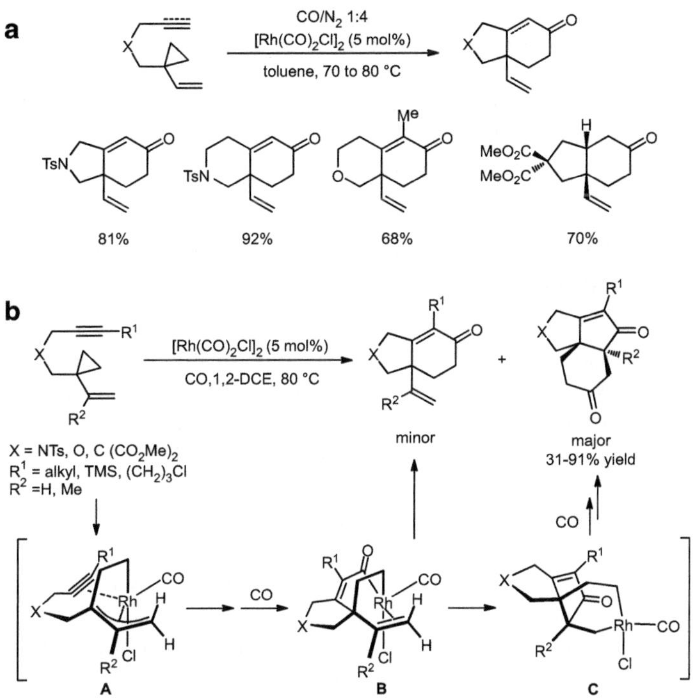


**Scheme 10** Rh(I)-catalyzed (**a**) [(3+2)+1] cycloaddition and (**b**) [5+1]/[(2+2)+1] cycloaddition

(between alkyne, in situ generated cyclohexenone, and CO), though both experimental and DFT computational studies suggested an alternative mechanism (Scheme 10b). Optimization of the reaction conditions made the reaction a practical method to synthesize functionalized angular 5/5/6-diones in generally moderate to excellent yields.

In addition, a rhodium-catalyzed [(3+3)+1] cycloaddition of biscyclopropanes with a vinyl substituent was reported by Chung's group [76]. Seven-membered ring products can be obtained from the two different types of vinyl-substituted biscyclopropanes in moderate to good yields (see (32) and (33)).

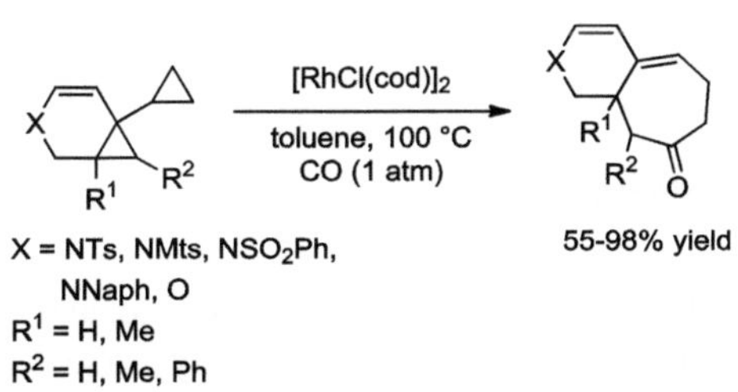


(32)

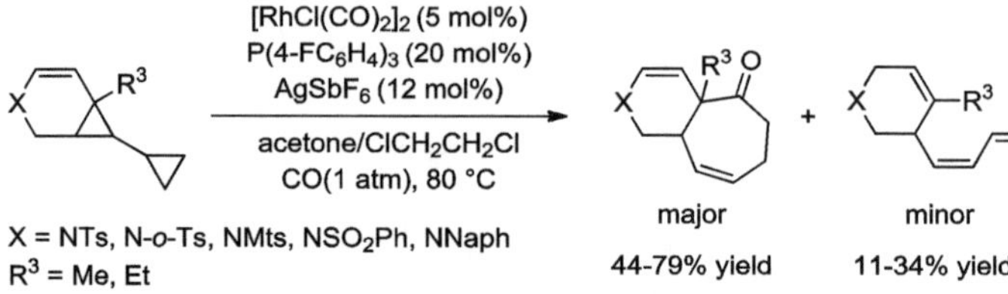


(33)

# 3 MCP Chemistry in Transition Metal-Catalyzed Cycloadditions

Methylenecyclopropanes (MCPs) are another type of important versatile building blocks bearing cyclopropane rings in organic synthesis. The unique structure confers rich reactivities to MCPs and the chemistry of MCPs has been profoundly investigated by numerous groups since the 1970s [77, 78]. The chemistry of MCPs includes reactions originating from reactivities of the double bond alone, reactions involving the cyclopropane ring opening leading to open chain compounds, as well as the cycloadditions of MCPs as three-carbon synthons. In this review we will focus on transition metal-catalyzed cycloadditions of MCPs as three-carbon synthons via carbon–carbon bond activation mechanism for synthesis of carbocycles. Generally, transition metal-catalyzed cycloadditions of MCPs and other unsaturated components can proceed with cyclopropyl ring cleavage via three different reaction pathways leading to regioisomeric products (Scheme 11, as illustrated in the case of intramolecular [3+2] cycloadditions). The oxidative addition of the transition metal into the distal C–C bond leads to the formation of the metallacyclobutane species followed by insertion of the multiple bond (pathway **a**). On the other hand, a proximal bond cleavage proceeds through either oxidative addition of the transition metal into the proximal C–C bond or formation of a metallacyclopentane with the multiple bond followed by cyclopropylmethyl-butenyl rearrangement (pathway **b**). In addition, MCPs could go through the formation of reactive metal-TMM intermediate (pathway **c**). The different reaction course is influenced by many factors, including different metals (Ni or Pd mostly), the type and number of the ligands bonded to the metal, the type, number, and position of substituents on the methylenecyclopropanes, and the electronic properties of the olefin subunit. Due to the multiple factors that may operate, it is usually difficult to predict the accurate reaction course.

## 3.1 [3+2] Cycloadditions

Since Noyori reported the first example of intermolecular [3+2] cycloaddition between MCP and olefins using a nickel catalyst (see (34)) [79, 80], the intermolecular [3+2] cycloaddition of MCP with different reaction partners has evolved into a powerful method to synthesize cyclopentanoid skeletons due to the

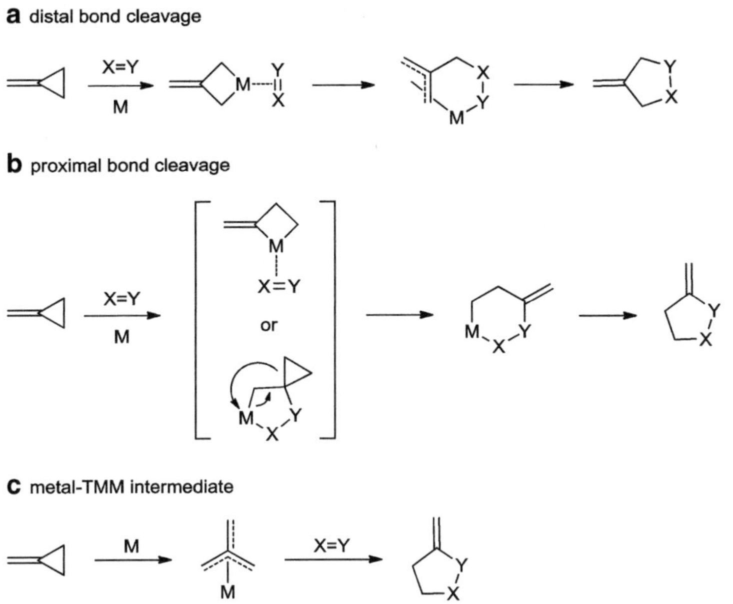


**Scheme 11** Pathways of transition metal-catalyzed cycloadditions of MCPs

extensive efforts from the groups of Noyori ([79, 80] and see reviews: [77, 78, 84]), Binger (see reviews: [77, 78, 84]), Trost (see reviews: [77, 78, 84]), and others (see reviews: [77, 78, 84]).

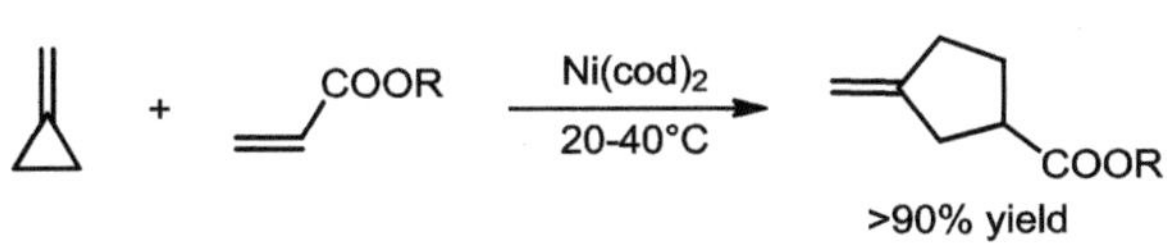


(34)

The [3+2] cycloaddition is mostly catalyzed by Ni or Pd catalysts. The MCPs can have substituents on the olefin or cyclopropane, and the two-atom partners can be electron neutral or deficient alkenes, alkynes, and carbon–heteroatom multiple bonds. Since there are different reaction courses of MCP in cycloaddition, introducing substituents on either MCP or the two-atom reaction partners complicates the reaction even more, considering the associated selectivity issues. Consequently, cycloaddition with multi-substituted substrates often gives mixtures though sometimes good selectivity can be achieved by adjusting reaction conditions and substituents. The intramolecular version of [3+2] cycloaddition, mainly developed by Motherwell [81], Nakamura [82], Lautens [83], and Mascareñas [87–91], may address the issues of selectivity to some extent, and leads to polycyclic structure meanwhile. All these have been summarized by several excellent reviews [1, 84–86] and herein we will only update the [3+2] cycloaddition of MCP for synthesis of carbocycles with some recent representative examples.

After Mascareñas reported the first intramolecular [3+2] MCP-alkyne cycloaddition catalyzed by either $Pd_2(dba)_3$ [87] or the first-generation Grubbs ruthenium carbene complex [88], Mascareñas' group found that the protocol can be applied to cycloaddition of MCPs with alkenes (see (35)) [89] as well as MCPs with allenes (see (36)) [90]. The authors also conducted DFT study about the mechanism of these transformations [89, 91]. Buono reported an impressive intermolecular [2+1]/[3+2] cycloaddition sequence of norbornadiene with several alkynes catalyzed by secondary phosphine oxide-based platinum complexes (see (37)) [92]. MCPs were demonstrated to be the intermediate participating in the [3+2] cycloaddition which, however, is limited to alkynes bearing a heteroatom substituent on the propargylic carbon atom. Zhang's group disclosed a Ni-catalyzed intramolecular [3+2] cycloaddition of MCPs to arylalkynes via proximal bond cleavage to prepare cyclopenta[*a*]indene derivatives (see (38)) [93].

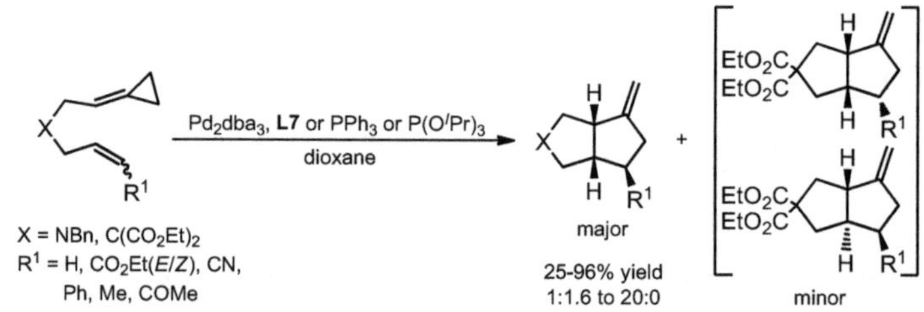


(35)

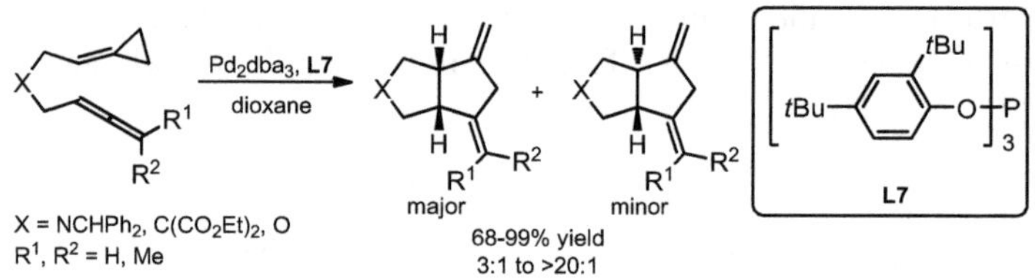


(36)

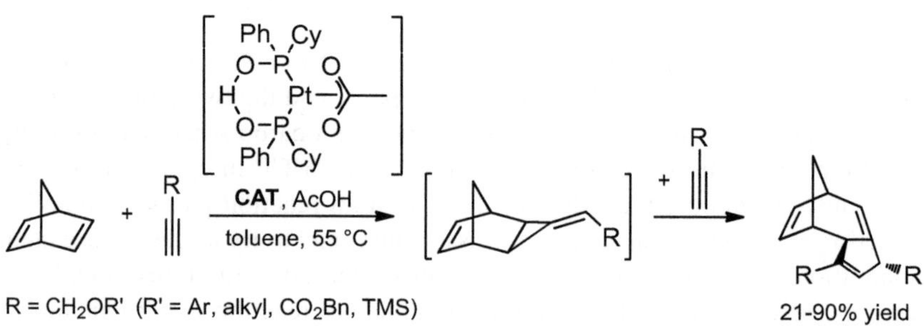


(37)

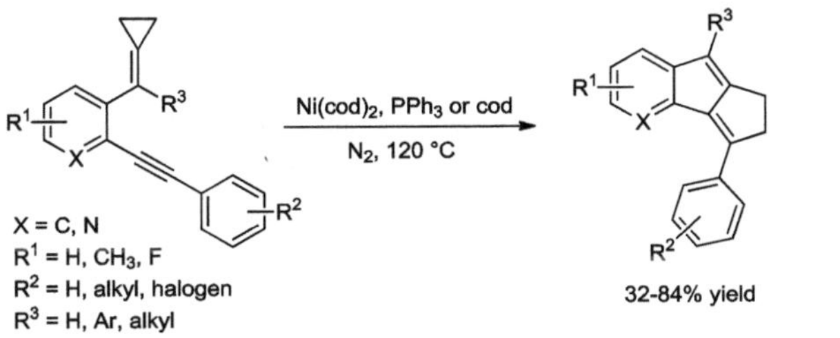


(38)

## 3.2 Other Types of Cycloadditions

de Meijere's group reported an example of cobalt-catalyzed [3+1] cycloaddition of MCPs. The reaction afforded cyclobutanones arising from proximal CO insertion of the cyclopropane ring in moderate to high yields (see (39)) [94, 95].

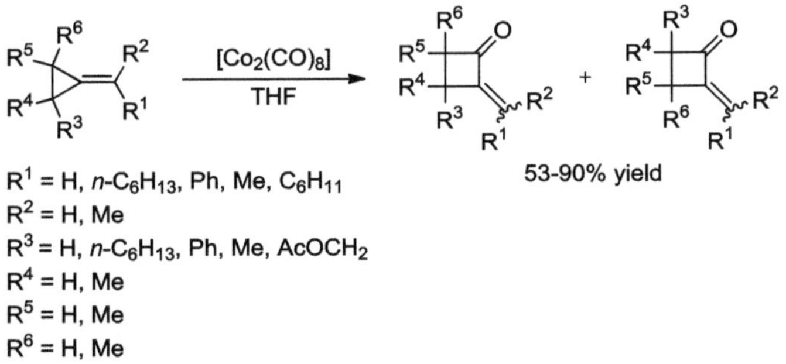


(39)

de Meijere's group also reported a novel [4+1] cycloaddition of Fischer carbenes and MCPs in which MCPs acted as a four-carbon synthon [95]. The process of the reaction was believed to involve a [2+2] cycloaddition of the MCP to the Fischer carbene, followed by a cyclopropylmethylmetal to homoallylmetal rearrangement, CO insertion, and reductive elimination (Scheme 12).

Kamikawa developed a nickel-catalyzed [3+1+1] cycloaddition of MCPs with Fischer carbenes to produce methylenecyclopentanones (see (40)) [96].

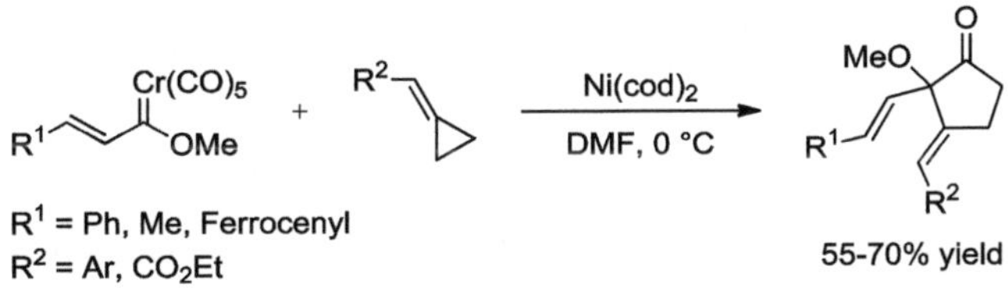


(40)

A combined theoretical and experimental development of a rhodium-catalyzed [(3+2)+1] cycloaddition of ene-MCPs and CO was described by Evans' group in 2012 [97]. The reaction proceeded well in good to excellent yields with high diastereoselectivity to give *cis*-fused bicyclohexenones (see (41)). The first highly enantioselective reaction, according to the authors, involving an MCP employing a chiral P,N-ligand was also described (see (42)).

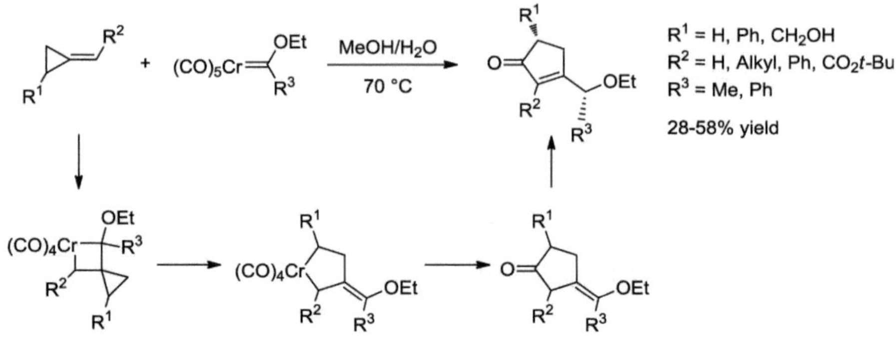


**Scheme 12** [4+1] Cycloaddition of Fisher carbenes

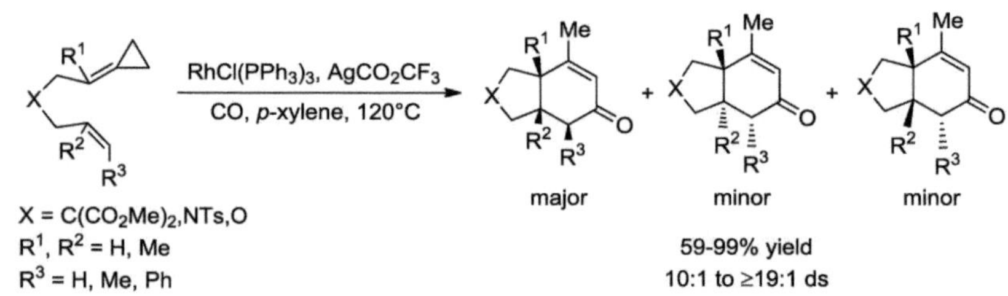


(41)

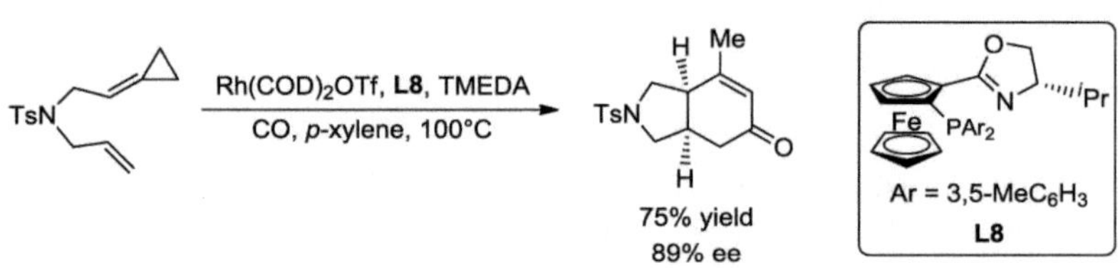


(42)

In 2010, Ohashi and Ogoshi found an interesting nickel-catalyzed [3+3] cyclodimerization of ester-substituted MCPs giving six- and five-membered ring derivatives (see (43)) [98].

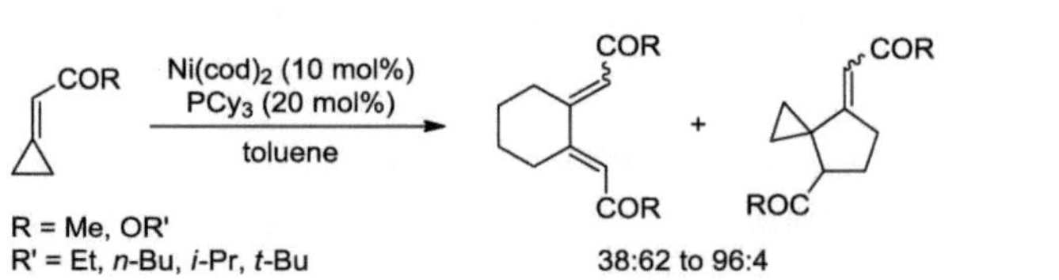


(43)

Moreover, MCPs have been used as three-carbon components participating in [3+2+2] or [4+3] cycloadditions to build seven-membered rings. Early in the 1980s, Binger reported some examples of [3+2+2] and [4+3] cycloadditions: a multi-component [3+2+2] cycloaddition of MCP and two units of allenes catalyzed by Pd(0) [99], a mono-component [3+2+2] cycloaddition of three units of MCPs catalyzed by Ni(0) [100], and a [4+3] cycloaddition of MCPs and dienes catalyzed by Pd(0) [78] (Scheme 13). Saito and co-workers have made some impressive contributions in this area. They developed a series of Ni(0)-catalyzed [3+2+2] and

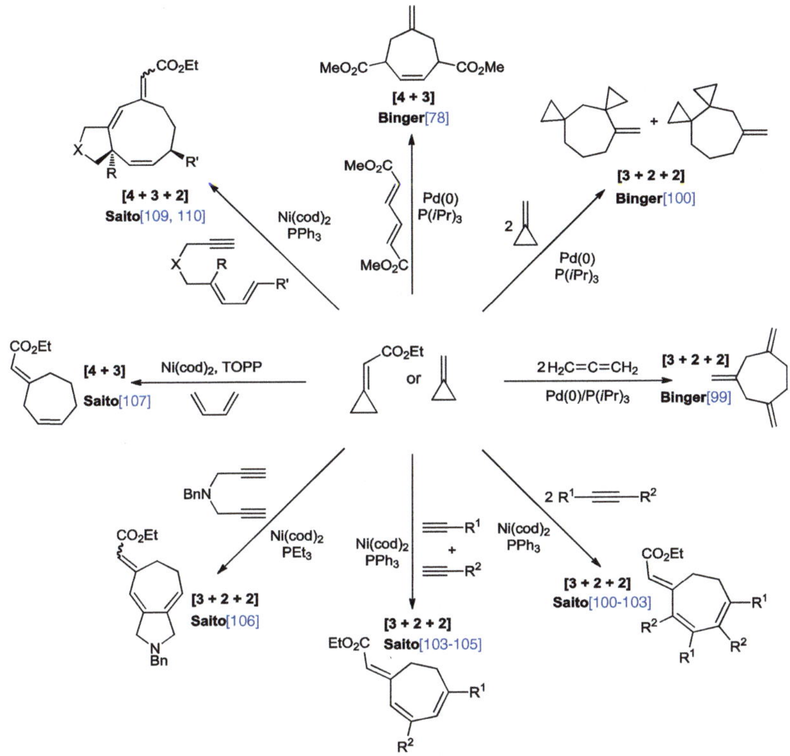


**Scheme 13** Multicomponent intermolecular cycloadditions of MCPs by Binger or Saito

[4+3] cycloadditions, including [3+2+2] cycloadditions of MCPs with two units of alkynes [101–106], and [4+3] cycloadditions of MCPs with dienes [107] (Scheme 13). A DFT study of the mechanism of nickel-catalyzed [3+2+2] cycloaddition has been carried out recently [108]. They also developed a Ni(0)-catalyzed [4+3+2] cycloaddition of MCPs and diene-ynes to synthesize nine-membered carbocycles [109, 110] (Scheme 13). Evans also reported an example of Rh(I)-catalyzed [3+2+2] cycloaddition (see (44)) [111]. In this reaction, ene-MCPs and electron-deficient alkynes were chosen to build bicyclo[5.3.0]decane compounds at high temperature. Mascareñas reported the Ni(0)-catalyzed intermolecular [3+2+2] cycloaddition of yne-MCPs and alkenes to build bicyclo[5.4.0]undecane systems (see (45)) [112], the Pd(0)-catalyzed intramolecular [3+2+2] cycloaddition of ene-yne-MCPs to produce the tricyclic products (see (46)) [113], and the Pd(0)-catalyzed intramolecular [4+3] cycloaddition of diene-MCPs to build bicyclo[5.3.0] decane systems (see (47)) [114].

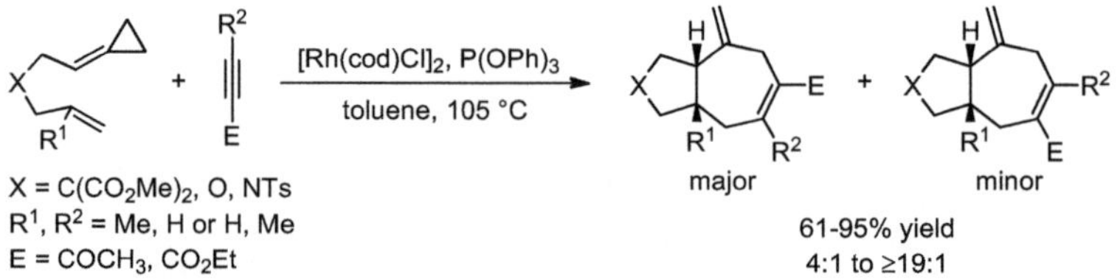


(44)

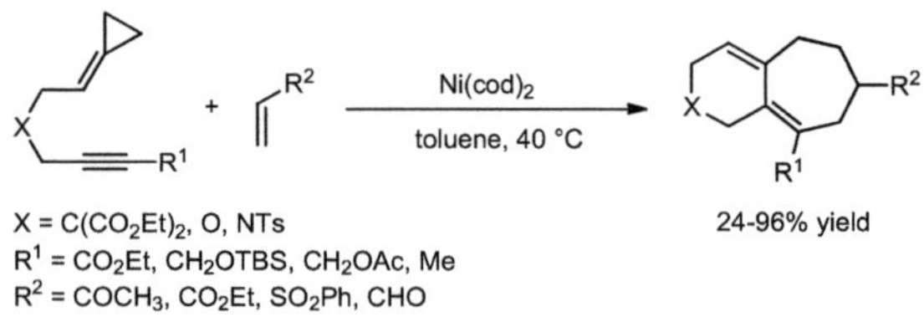


(45)

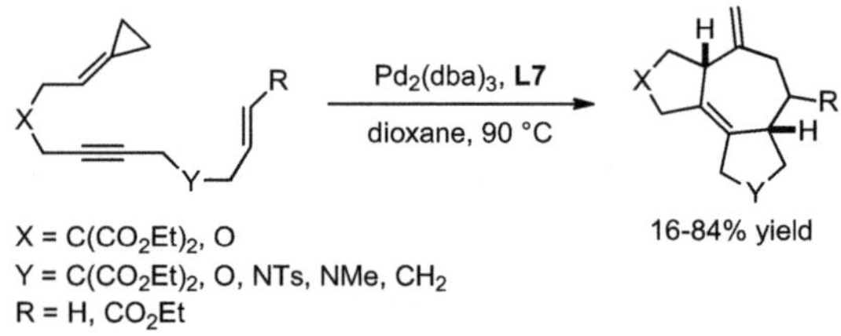


(46)

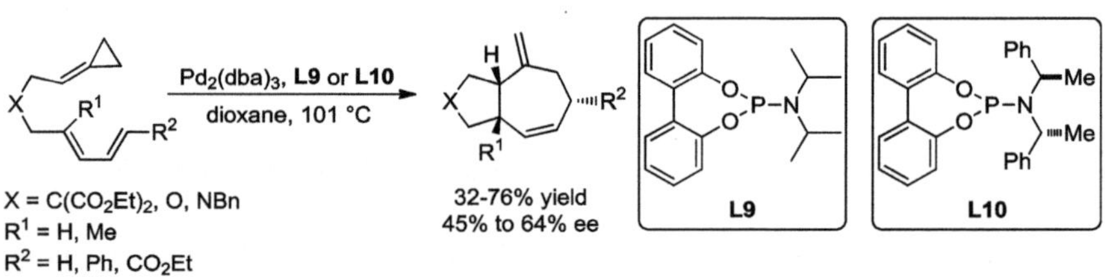


(47)

# 4 Other Types of Cyclopropanes

Carbonyl-activated or imino-activated cyclopropanes can undergo Ni-catalyzed dimerization, or [3+2] cycloaddition with α,β-unsaturated ketones, generally at a much slower rate, to give five-membered carbocycles. This topic was tactfully investigated by Ogoshi [115–117] and Montgomery [118, 119] (Scheme 14).

In 1999, Narasaka reported an example of intramolecular [(3+2)+1] cycloaddition of unactivated yne-CPs with carbon monoxide catalyzed by Rh(I) at 160°C to give the corresponding bicyclic products in low to moderate yields (Scheme 15) [120].

A novel rhodium-catalyzed carbonylation of spiropentanes leading to cyclopentenones was uncovered by Murakami in 2007 (Scheme 16) [121]. The reaction was proposed to proceed by the following pathway: oxidative addition of C–C σ-bond of the unsubstituted cyclopropane affords a spirocyclic rhodacyclobutane

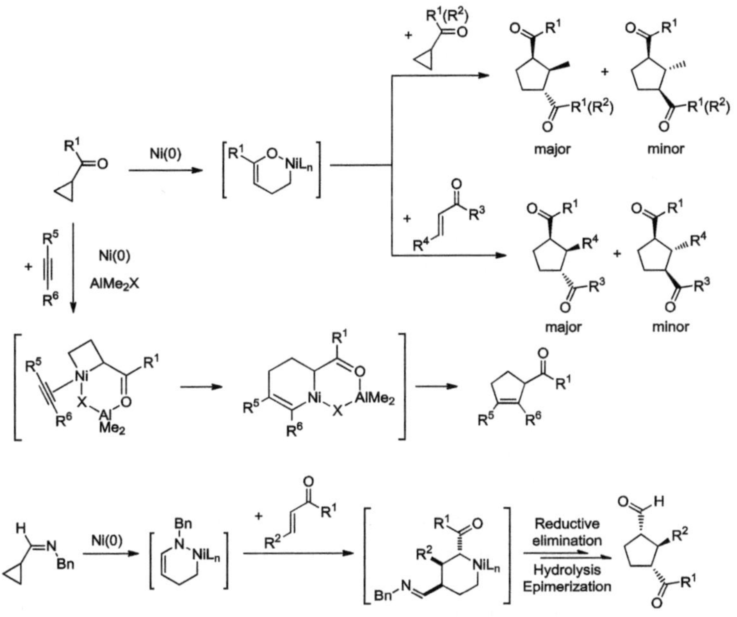


**Scheme 14** Ni-catalyzed [3+2] cycloaddition of activated CPs

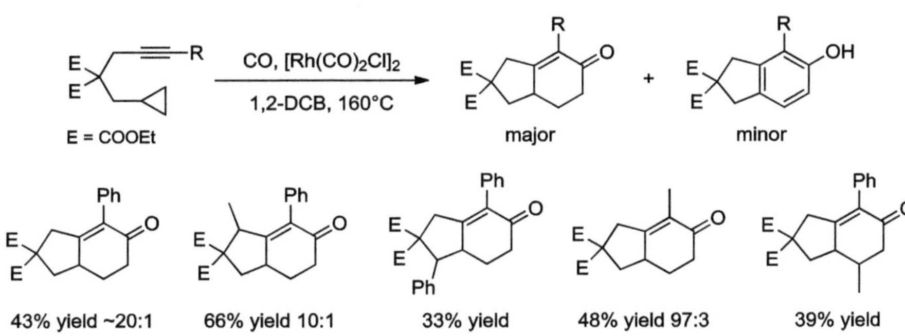


**Scheme 15** Rh-catalyzed [(3+2)+1] cycloaddition of unactivated yne-CPs

species, which undergoes CO insertion to give a rhodacyclopentanone intermediate. β-Carbon elimination converts the rhodacyclopentanone into a six-membered rhodacycle. Finally, reductive elimination followed by alkene isomerization leads to the cyclopentenone product.

Shi and co-workers reported a Pauson–Khand type [3+2+1] cycloaddition reaction of ene-vinylidenecyclopropanes and CO in the presence of rhodium(I) catalyst [122]. Both cyclic and noncyclic substituent could be tolerated (see (48)). Shi also reported a tandem process that converted 1,4-enynes tethered by cyclopropyl group

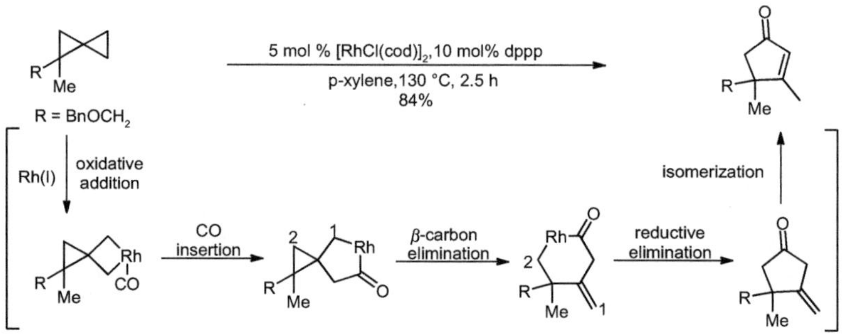


**Scheme 16** Rhodium-catalyzed carbonylation of spiropentanes

to phenol bicyclic derivatives with rhodium catalyst (see (49)) [123]. A plausible mechanism was proposed by the authors (Scheme 17). Initially, a Pauson–Khand cycloaddition induced by Rh(I) involving the alkyl group, alkenyl group, and a unit of CO gave the tricyclic spiro intermediate. The following step was analogous to the transformation reported by Murakami [121]. The ring expansion of the [2.2]spiro intermediate was promoted by Rh(I) followed by carbon monoxide insertion. Then reductive elimination and isomerization completed the catalytic cycle and afforded the desired product.

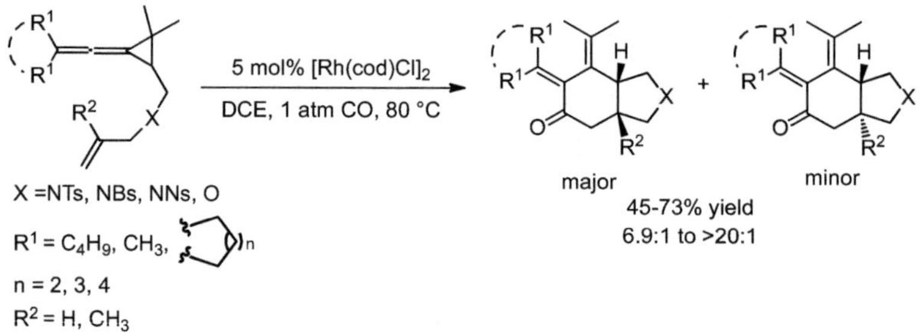


(48)

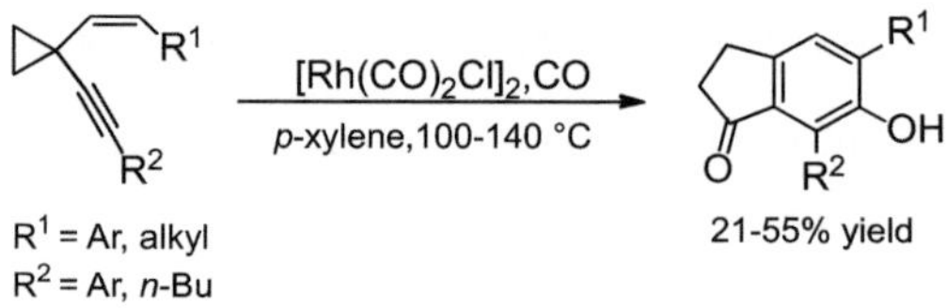


(49)

Bower reported a directing group enhanced rhodium-catalyzed [(3+2)+1] cycloaddition reaction of nitrogen-tethered yne-CPs and CO (see (50)) [124]. The authors used the urea as the directing group to induce the oxidative addition at the proximal bond of the amino-cyclopropanes.

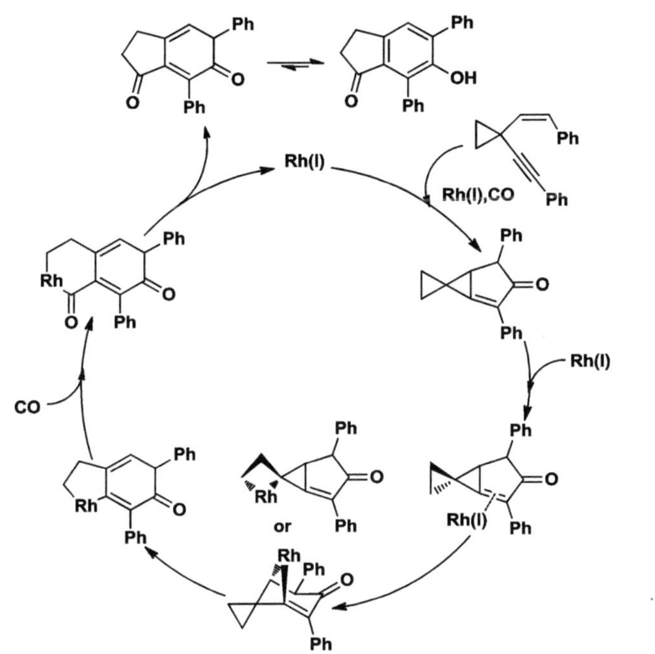

**Scheme 17** Mechanism of Rh(I)-catalyzed [3+2+1] cycloaddition

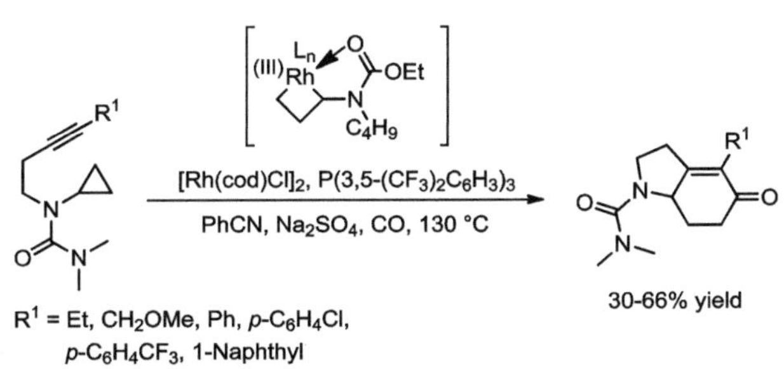


(50)

Yu's group reported the first Rh-catalyzed [7+1] cycloaddition of buta-1,3-dienylcyclopropanes and CO for the synthesis of cyclooctadienones (see (51)) [125]. This reaction showed good compatibility with different functional groups, providing an efficient entry to the eight-membered carbocyclic rings.

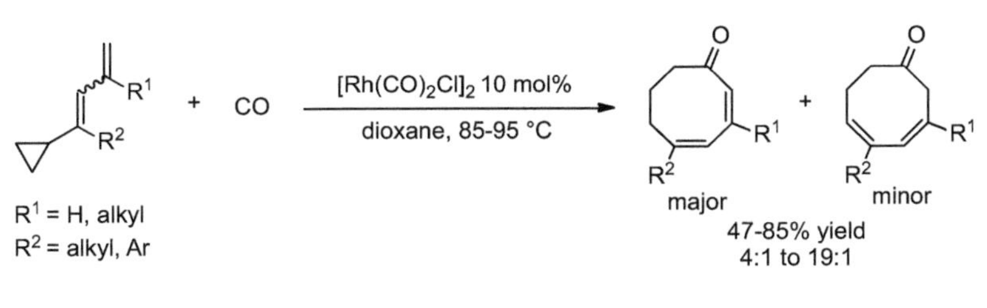


(51)

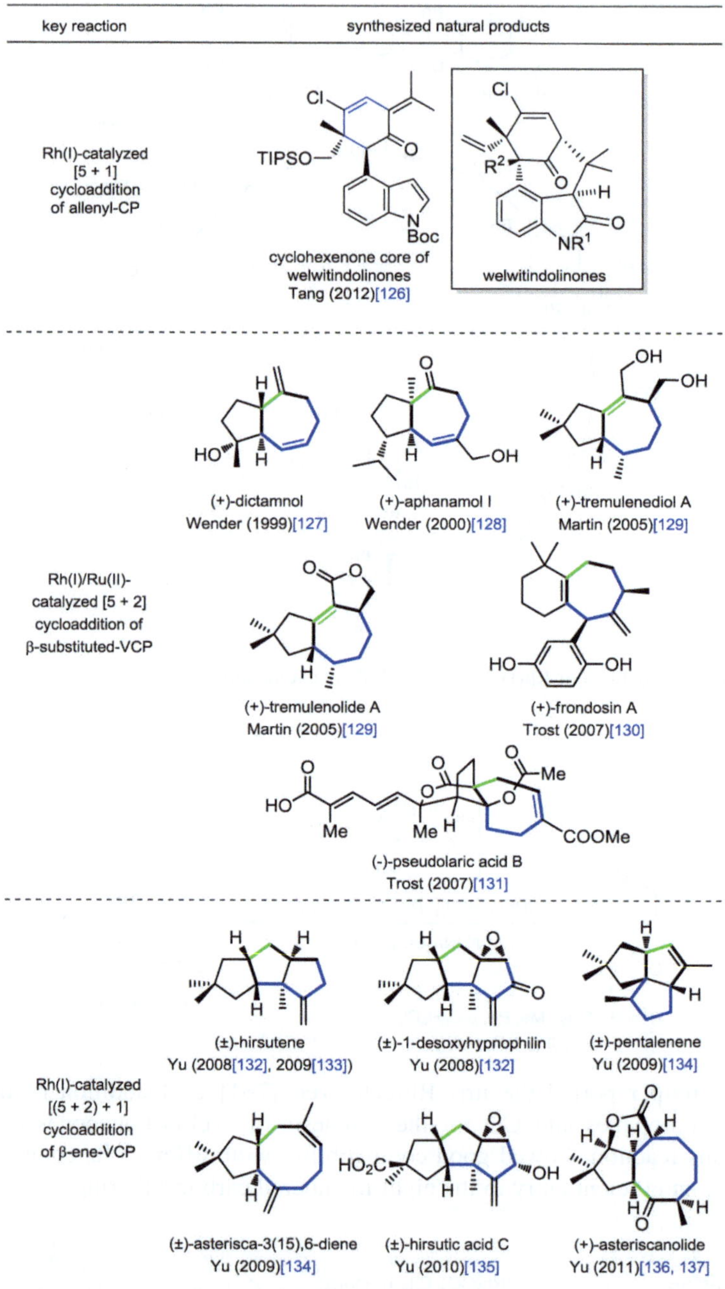


**Fig. 4** (continued)

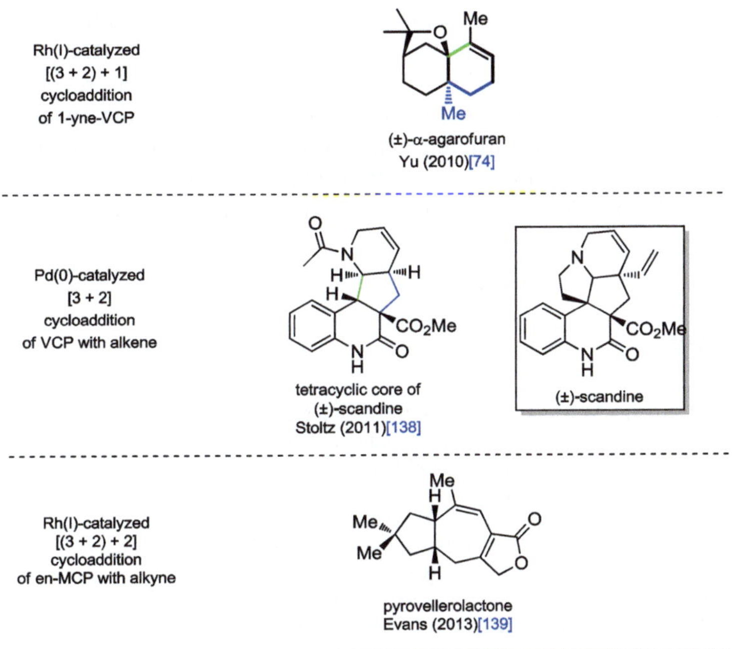


**Fig. 4** Applications in natural products synthesis

# 5 Applications in Natural Products Synthesis

Transition metal-catalyzed cyclopropane-based cycloadditions provide efficient strategies for the construction of (poly)cyclic structures. More importantly, the cycloadditions feature atom- and step-economy, good regio- and stereoselectivity, and excellent functional group tolerance. More and more total syntheses of natural products benefit from the rich cycloaddition chemistry of VCPs and MCPs. To date, numerous total syntheses of natural products have been accomplished utilizing these methodologies as key steps. Listed in Fig. 4 are some selected examples.

# 6 Conclusion

During the past several decades, with the aid of transition metals, cyclopropanes have been incorporated into cycloadditions via C–C bond cleavage and have led to many desirable types of transformations, mainly involving vinylcyclopropanes (VCPs) and methylenecyclopropanes (MCPs) [140]. These methods are complementary to traditional cycloaddition reactions. In particular, the VCP- or MCP-involved cycloadditions can provide efficient accesses to different sized ring structures of considerable complexity which are otherwise difficult to prepare by traditional approaches. The methodologies of VCPs and MCPs cycloadditions have also found many applications in total synthesis of natural products. Further advancement can be expected that include developing new synthons of cyclopropanes and combinations of different reaction partners leading to potentially new types of cycloadditions. More mechanistic studies and applications in total synthesis of the VCP- and MCP-cycloadditions are also highly desirable.

## References


1. Rubin M, Rubina M, Gevorgyan V (2007) Chem Rev 107:3117
2. Ben-Shoshan R, Sarel S (1969) Chem Commun 883
3. Victor R, Ben-Shoshan R, Sarel S (1970) Tetrahedron Lett 49:4253
4. Sarel S (1978) Acc Chem Res 11:204
5. Aumann R (1974) J Am Chem Soc 96:2631
6. Schulze MM, Gockel U (1996) Tetrahedron Lett 37:357
7. Schulze MM, Gockel U (1996) J Organomet Chem 525:155
8. Taber DF, Kanai K, Jing Q, Bui G (2000) J Am Chem Soc 122:6807
9. Taber DF, Joshi PV, Kanai K (2004) J Org Chem 69:2268
10. Kurahashi T, de Meijere A (2005) Synlett 2619
11. Jiang GJ, Fu XF, Li Q, Yu ZX (2012) Org Lett 14:692
12. Iwasuwa N, Owada Y, Matsuo T (1995) Chem Lett 2:115
13. Owada Y, Matsuo T, Iwasuwa N (1997) Tetrahedron 53:11069
14. Murakami M, Itami K, Ubukata M, Tsuji I, Ito Y (1998) J Org Chem 63:4
15. Shu D, Li X, Zhang M, Robichaux PJ, Tang WP (2011) Angew Chem Int Ed 50:1346
16. Shu D, Li X, Zhang M, Robichaux PJ, Guzei IA, Tang WP (2012) J Org Chem 77:6463
17. Wender PA, Takahashi H, Witulski B (1995) J Am Chem Soc 117:4720
18. Wender PA, Husfeld CO, Langkopf E, Love JA (1998) J Am Chem Soc 120:1940
19. Wender PA, Husfeld CO, Langkopf E, Love JA, Plleuss N (1998) Tetrahedron 54:7203
20. Wender PA, Glorius F, Husfeld CO, Langkopf E, Love JA (1999) J Am Chem Soc 121:5348
21. Wender PA, Sperandio D (1998) J Org Chem 63:4164
22. Wender PA, Dykman AJ, Husfeld CO, Kadereit D, Love JA, Rieck H (1999) J Am Chem Soc 121:10442
23. Wender PA, Dykman AJ (1999) Org Lett 1:2089
24. Gilbertson SR, Hoge GS (1998) Tetrahedron Lett 39:2075
25. Wang B, Cao P, Zhang XM (2000) Tetrahedron Lett 41:8041
26. Lee SI, Park SY, Park JH, Jung IG, Choi SY, Chung YK, Lee BY (2006) J Org Chem 71:91
27. Saito A, Ono T, Hanzawa Y (2006) J Org Chem 71:6437
28. Wender PA, Williams TJ (2002) Angew Chem Int Ed 41:4550
29. Wender PA, Love JA, Williams TJ (2003) Synlett 1295
30. Gómez FJ, Kamber NE, Deschamps NM, Cole AP, Wender PA, Waymouth RM (2007) Organometallics 26:4541
31. Wender PA, Lesser AB, Sirois LE (2012) Angew Chem Int Ed 51:2736
32. Xu X, Liu P, Lesser A, Sirois LE, Wender PA, Houk KN (2012) J Am Chem Soc 134:11012
33. Wender PA, Haustedt LO, Lim J, Love JA, Willams TJ, Yoon JY (2006) J Am Chem Soc 128:6302
34. Shintani R, Nakatsu H, Takatsu K, Hayashi T (2009) Chem Eur J 15:8692
35. Inagaki F, Sugikubo K, Miyashita Y, Mukai C (2010) Angew Chem Int Ed 49:2206
36. Jiao L, Ye SY, Yu ZX (2008) J Am Chem Soc 130:7178

37. Wender PA, Rieck H, Fuji M (1998) J Am Chem Soc 120:10976
38. Wender PA, Dyckman AJ, Husfeld CO, Scanio MJC (2000) Org Lett 2:1609
39. Wender PA, Sirois LE, Stemmler RT, Williams TJ (2010) Org Lett 12:1604
40. Wender PA, Barzilay CM, Dyckman AJ (2001) J Am Chem Soc 123:179
41. Wegner HA, de Meijere A, Wender PA (2005) J Am Chem Soc 127:6530
42. Ashfeld BL, Miller KA, Smith AJ, Tran K, Martin SF (2005) Org Lett 7:1661
43. Ashfeld BL, Miller KA, Smith AJ, Tran K, Martin SF (2007) J Org Chem 72:9018
44. Wender PA, Stemmler RT, Sirois LE (2010) J Am Chem Soc 132:2532
45. Wender PA, Gamber GG, Scanio MJC (2001) Angew Chem Int Ed 40:3895
46. Yu ZX, Wender PA, Houk KN (2004) J Am Chem Soc 126:9154
47. Liu P, Sirois LE, Cheong PHY, Yu ZX, Hartung IV, Rieck H, Wender PA, Houk KN (2010) J Am Chem Soc 132:10127
48. Yu ZX, Cheong PHY, Liu P, Legault CY, Wender PA, Houk KN (2008) J Am Chem Soc 130:2378
49. Wang Y, Wang J, Su J, Huang F, Jiao L, Liang Y, Yang D, Zhang S, Wender PA, Yu ZX (2007) J Am Chem Soc 129:10060
50. Trost BM, Toste FD, Shen H (2000) J Am Chem Soc 122:2379
51. Trost BM, Shen HC (2000) Org Lett 2:2523
52. Trost BM, Shen HC (2001) Angew Chem Int Ed 40:1114
53. Trost BM, Shen HC, Schulz T, Koradin C, Schirok H (2003) Org Lett 5:4149
54. Trost BM, Toste FD (2001) Angew Chem Int Ed 40:2313
55. Trost BM, Shen HC, Horne DB, Toste FD, Steinmetz BG, Koradin C (2005) Chem Eur J 11:2577
56. Hong X, Trost NM, Houk KN (2013) J Am Chem Soc 135:6588
57. Zuo G, Louie J (2005) J Am Chem Soc 127:5798
58. Hong X, Liu P, Houk KN (2013) J Am Chem Soc 135:1456
59. Fürstner A, Majima K, Martin R, Krause H, Kattnig E, Goddard R, Lehmann CW (2008) J Am Chem Soc 130:1992
60. Wender PA, Gamber GG, Hubbard RD, Zhang L (2002) J Am Chem Soc 124:2876
61. Wender PA, Gamber GG, Hubbard RD, Pham SM, Zhang L (2005) J Am Chem Soc 127:2836
62. Huang F, Yao ZK, Wang Y, Wang YY, Zhang JL, Yu ZX (2010) Chem Asian J 5:1555
63. Liao W, Wang YY, Jiao L, Yu ZX (2014) unpublished result
64. Shimizu I, Ohashi Y, Tsuji J (1985) Tetrahedron Lett 26:3825
65. Dieskau AP, Holzwarth MS, Plietker B (2012) J Am Chem Soc 134:5048
66. Trost BM, Morris PJ (2011) Angew Chem Int Ed 50:6167
67. Trost BM, Morris PJ, Sprague SJ (2012) J Am Chem Soc 134:17823
68. Mei LY, Wei Y, Xu Q, Shi M (2012) Organometallics 31:7591
69. Mei LY, Wei Y, Xu Q, Shi M (2013) Organometallics 32:3544
70. Li Q, Jiang GJ, Jiao L, Yu ZX (2010) Org Lett 12:1332
71. Jiao L, Lin M, Yu ZX (2010) Chem Commun 46:1059
72. Jiao L, Lin M, Yu ZX (2011) J Am Chem Soc 133:447
73. Lin M, Kang GY, Guo YA, Yu ZX (2012) J Am Chem Soc 134:398
74. Jiao L, Lin M, Zhuo LG, Yu ZX (2010) Org Lett 12:2528
75. Lin M, Li F, Jiao L, Yu ZX (2011) J Am Chem Soc 133:1690
76. Kim SY, Lee SI, Choi SY, Chung YK (2008) Angew Chem Int Ed 47:4914
77. Trost BM (1986) Angew Chem Int Ed Engl 25:1
78. Binger P, Büch HM (1987) Top Curr Chem 135:77
79. Noyori R, Odagi T, Takaya H (1970) J Am Chem Soc 92:5780
80. Noyori R, Kumagai Y, Umeda I, Takaya H (1972) J Am Chem Soc 94:4018
81. Corlay H, Lewis RT, Motherwell WB, Shipman M (1995) Tetrahedron 51:3303
82. Yamago S, Nakamura E (1988) J Chem Soc Chem Commun 1112
83. Lautens M, Ren Y, Delanghe PHM (1994) J Am Chem Soc 116:8821
84. Lautens M, Klute W, Tam W (1996) Chem Rev 96:49

85. Nakamura I, Yamamoto Y (2002) Adv Synth Catal 344:111
86. Yu L, Guo R (2011) Org Prep Proced Int 43:209
87. Delgado A, Rodríguez JR, Castedo L, Mascareñas JL (2003) J Am Chem Soc 125:9282
88. López F, Delgado A, Rodríguez JR, Castedo L, Mascareñas JL (2004) J Am Chem Soc 126:10262
89. García-Fandiño R, Gulías M, Mascareñas JL, Cárdenas DJ (2012) Dalton Trans 41:9468
90. Trillo B, Gulías M, López F, Castedo L, Mascareñas JL (2006) Adv Synth Catal 348:2381
91. García-Fandiño R, Gulías M, Castedo L, Granja JR, Mascareñas JL, Cárdenas DJ (2008) Chem Eur J 14:272
92. Achard T, Lepronier A, Gimbert Y, Clavier H, Giordano L, Tenaglia A, Buono G (2011) Angew Chem Int Ed 50:3552
93. Yao B, Li Y, Liang ZJ, Zhang YH (2011) Org Lett 13:640
94. Kurahashi T, de Meijere A (2005) Angew Chem Int Ed 44:7881
95. Kurahashi T, Wu YT, Meindl K, Rühl S, de Meijere A (2005) Synlett 805
96. Kamikawa K, Shimizu Y, Takemoto S, Matsuzaka H (2006) Org Lett 8:4011
97. Mazumder S, Shang D, Negru DE, Baik MH, Evans PA (2012) J Am Chem Soc 134:20569
98. Ohashi M, Taniguchi T, Ogoshi S (2010) Organometallics 29:2386
99. Binger P, Schuchardt U (1980) Chem Ber 113:1063
100. Binger P, Brinkman A, McMeeking J (1977) Justus Liebigs Ann Chem 1065
101. Saito S, Masuda M, Komagawa S (2004) J Am Chem Soc 126:10540
102. Saito S, Komagawa S, Azumaya I, Masuda M (2007) J Org Chem 72:9114
103. Komagawa S, Takeuchi K, Sotome I, Azumaya I, Masu H, Yamasaki R, Saito S (2009) J Org Chem 74:3323
104. Komagawa S, Saito S (2006) Angew Chem Int Ed 45:2446
105. Yamasaki R, Terashima N, Sotome I, Komagawa S, Saito S (2010) J Org Chem 75:480
106. Maeda K, Saito S (2007) Tetrahedron Lett 48:3173
107. Saito S, Takeuchi K (2007) Tetrahedron Lett 48:595
108. Komagawa S, Wang C, Morokuma K, Saito S, Uchiyama M (2013) J Am Chem Soc 135:14508
109. Saito S, Maeda K, Yamasaki R, Kitamura T, Nakagawa M, Kato K, Azumaya I, Masu H (2010) Angew Chem Int Ed 49:1830
110. Yamasaki R, Ohashi M, Maeda K, Kitamura T, Nakagawa M, Kato K, Fujita T, Kamura R, Kinoshita K, Masu H, Azumaya I, Ogoshi S, Saito S (2013) Chem Eur J 19:3415
111. Evans PA, Inglesby PA (2008) J Am Chem Soc 130:12838
112. Saya L, Bhargava G, Navarro MA, Gulías M, López F, Fernández I, Castedo L, Mascareñas JL (2010) Angew Chem Int Ed 49:9886
113. Bhargava G, Trillo B, Araya M, López F, Castedo L, Mascareñas JL (2010) Chem Commun 46:270
114. Gulías M, Durán J, López F, Castedo L, Mascareñas JL (2007) J Am Chem Soc 129:11026
115. Ogoshi S, Nagata M, Kurosawa H (2006) J Am Chem Soc 128:5350
116. Tamaki T, Nagata M, Ohashi M, Ogoshi S (2009) Chem Eur J 15:10083
117. Tamaki T, Ohashi M, Ogoshi S (2011) Angew Chem Int Ed 50:12067
118. Liu L, Montgomery J (2006) J Am Chem Soc 128:5348
119. Liu L, Montgomery J (2007) Org Lett 9:3885
120. Koga Y, Narasaka K (1999) Chem Lett 7:705
121. Matsuda T, Tsuboi T, Murakami M (2007) J Am Chem Soc 129:12596
122. Lu BL, Wei Y, Shi M (2012) Organometallics 31:4601
123. Chen GQ, Shi M (2013) Chem Commun 49:698
124. Shaw MH, Melikhova EY, Kloer DP, Whittingham WG, Bower JF (2013) J Am Chem Soc 135:4992
125. Yao ZK, Li JJ, Yu ZX (2010) Org Lett 13:134
126. Zhang M, Tang WP (2012) Org Lett 14:3756
127. Wender PA, Fuji M, Husfeld CO, Love JA (1999) Org Lett 1:137

128. Wender PA, Zhang L (2000) Org Lett 2:2323
129. Ashfeld BL, Martin SF (2005) Org Lett 7:4535
130. Trost BM, Hu Y, Horne DB (2007) J Am Chem Soc 129:11781
131. Trost BM, Waser J, Meyer A (2007) J Am Chem Soc 129:14556
132. Jiao L, Yuan C, Yu ZX (2008) J Am Chem Soc 130:4421
133. Fan X, Tang MX, Zhuo LG, Tu YQ, Yu ZX (2009) Tetrahedron Lett 50:155
134. Fan X, Zhuo LG, Tu YQ, Yu ZX (2009) Tetrahedron 65:4709
135. Yuan C, Jiao L, Yu ZX (2010) Tetrahedron Lett 51:5674
136. Liang Y, Jiang X, Yu ZX (2011) Chem Commun 47:6659
137. Liang Y, Jiang X, Fu XF, Ye S, Wang T, Yuan J, Wang Y, Yu ZX (2012) Chem Asian J 7:593
138. Goldberg AFG, Stoltz BM (2011) Org Lett 13:4474
139. Evans PA, Inglesby PA, Kilbride K (2013) Org Lett 15:1798
140. Jiao L, Yu ZX (2013) J Org Chem 78:6842

Top Curr Chem (2014) 346: 233–258
DOI: 10.1007/128_2014_545
© Springer-Verlag Berlin Heidelberg 2014
Published online: 23 April 2014

# Transition Metal-Catalyzed C–C Bond Activation of Four-Membered Cyclic Ketones

**Tao Xu, Alpay Dermenci, and Guangbin Dong**

**Abstract** With the advent of new synthetic methodologies, carbon–carbon bond (C–C) activation strategies have uncovered not only new fundamental reactivity but also the potential for use as a highly efficient synthetic protocol. This chapter specifically discusses the use of four-membered ketone-based starting materials for C–C activation initiated transformations using a variety of transition metals. The two major modes of activation, oxidative addition and β-C elimination, are presented as each pathway shows different mechanistic details and the ability to effect several types of reactions. Applications to the synthesis of complex molecules are presented and perspectives on future applications are considered.

**Keywords** β-Carbon elimination · C–C activation · Cycloaddition · Oxidative addition

## Contents



T. Xu and G. Dong (✉)
Department of Chemistry and Biochemistry, University of Texas at Austin, 1 University Station A5300, Austin, TX 78712-0165, USA
e-mail: gbdong@cm.utexas.edu

A. Dermenci
Department of Chemistry and Biochemistry, University of Texas at Austin, 1 University Station A5300, Austin, TX 78712-0165, USA

Pfizer Inc., Worldwide Medicinal Chemistry, 445 Eastern Point Road, Groton, CT 06340, USA

# 1 Introduction

Given the ubiquity of carbon–carbon (C–C) and carbon–hydrogen (C–H) bonds, the ability to disconnect and/or functionalize either selectively would provide synthetic chemists with an atom-economic [1] and straightforward method to construct biologically interesting or complex molecules [2–5]. In contrast to the developing and fruitful area of C–H functionalization ([6] and references cited therein), C–C activation/functionalization is less developed, emerging as a new area in the synthetic community [7–9]. In general, there are two primary modes of C–C single bond cleavage: direct oxidative addition (Scheme 1A), and β-carbon elimination (Scheme 1B).

The challenges associated with oxidative addition of a C–C bond onto a transition metal are twofold. First, the reductive elimination (reverse reaction of oxidative addition) is usually an exergonic reaction and thus thermodynamically favored, which makes the oxidative addition of C–C bonds disfavored and therefore a sluggish process. More often than not, oxidative additions take place at high temperature or need other driving forces such as strain release, forming aromatic compounds, and/or chelation-derived assistance [10, 11]. Second, C–C bonds typically have neighboring C–H bonds that are more "exposed" which causes kinetic competition to C–C bond activation [12–14]. In other words, during interaction with a transition metal, C–H activation is often more favorable due to the statistical abundance and favorable orbital trajectory of C–H bonds.

Regarding the second mode of C–C activation, β-carbon elimination poses similar challenges, though as a primarily intramolecular process it does not involve the same kinetic barriers with a transition metal. Furthermore, when acyclic substrates are employed, a byproduct is generated alongside the β-C elimination reaction. In this case, the β-C elimination process generates an entropy increase, lowering the activation barrier. However, generally, transition metal-mediated β-C elimination reactions are still thermodynamically challenging due to formation of weak metal–carbon bonds, and often less competitive compared to the more common β-H elimination.

Due to the above-mentioned thermodynamic and kinetic challenges to cleave C–C σ bonds, strain-release provided by small-sized rings serves as one of the most important driving forces for C–C activation. A large number of novel and synthetically useful transformations based on this mode of reactivity have been realized, particularly during the past two decades. For example, reactions with

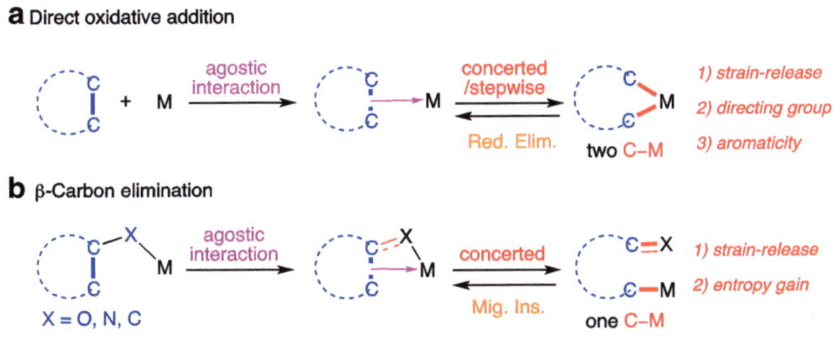


**Scheme 1** Two methods to activate C–C bonds

cyclopropanes are of high synthetic value, and have been extensively developed [15] and reviewed. However, activation of the related four-membered ring compounds has received much less attention [16]. In particular, the four-membered ring compounds containing a ketone moiety are unique substrates, because the carbonyl can serve as a reacting group or a convenient handle to control site-selectivity (see below). On the other hand, given the possibility for decarbonylation (see below), these compounds can behave as either a four-carbon or a three-carbon synthon, leading to distinct transformations. Herein, while a number of excellent reviews on C–C activation have been reported previously [9–16], this review specially focuses on C–C bond cleavage and further transformations of four-membered ring ketones, including cyclobutanones, cyclobutenone/benzocyclobutenones, and cyclobutenediones/benzocyclobutenediones.

## 2 C–C Bond Activation of Cyclobutanones

Seminal studies of C–C bond activation [17–25] demonstrated that C–C bonds adjacent to a carbonyl group are subject to C–C bond cleavage when treated with late transition metals. The pioneering work by Murakami and Ito showed that cyclobutanones are suitable substrates for catalytic C–C bond activation transformations.

### 2.1 *Rh-Catalyzed C–C Bond Activation via Oxidative Addition*

In 1994, Murakami and co-workers [26] found that when cyclobutanone (**1**) was treated with an equimolar amount of $(PPh_3)_3RhCl$ in refluxing toluene, decarbonylation took place to produce cyclopropane (**4**) in quantitative yield along with the unreactive complex *trans*-$[Rh(CO)Cl(PPh_3)_2]$ (1):

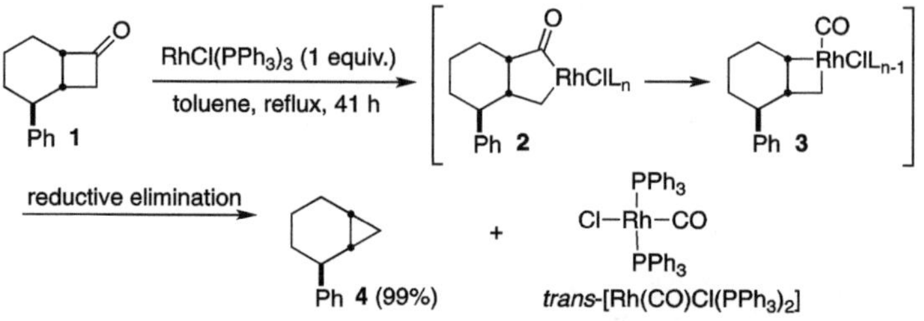


(1)

This decarbonylation reaction was initially believed to proceed through direct oxidative addition of Rh onto the less hindered C–C bond adjacent to the carbonyl group to give five-membered rhodacycle (**2**), followed by carbon monoxide extrusion to yield the four-membered rhodacycle (**3**), which then undergoes reductive elimination to furnish the observed product (**4**). These results are in firm agreement with a stoichiometric decarbonylation reaction previously reported by Rusina [27]. Formation of the thermodynamically stable but catalytically inert *trans*-[Rh (CO)Cl($PPh_3$)$_2$] and release of the ring strain are two major driving forces for this reaction.

A more detailed study [28] found that *trans*-[Rh(CO)Cl($PPh_3$)$_2$] could catalyze this decarbonylation reaction, although requiring higher temperatures and with lower efficiency (Fig. 1). CO Extrusion from the *trans*-[Rh(CO)Cl($PPh_3$)$_2$] at higher temperature (137–144°C) was believed to regenerate the active catalyst ($PPh_3$)$_2$RhCl. Further screening revealed that 5 mol% [Rh(cod)Cl]$_2$ and 10 mol% $AsPh_3$ would afford **6a** in 68% isolated yield and only trace amounts (2%) of side product **7a**. The electron-deficient nature of $AsPh_3$ may promote reductive elimination, thus explaining the increased selectivity for **6a**. Figure 1 also shows that **4** and **6b** were isolated in 80% and 70% yield, respectively. To obtain **6c** and **6d**, 5 mol% [Rh(cod)dppb]$BF_4$ was employed as catalyst and provided yields of 99% and 77%, respectively. This decarbonylation reaction constitutes one of the first examples of $Rh^I$-catalyzed C–C bond activation of cyclobutanones.

Additionally, Murakami and co-workers coupled hydrogenation with C–C bond insertion. Under a hydrogen atmosphere (50 atm), $Rh^I$-catalyzed C–C bond activation/hydrogenation produced 2-methyl-1,4-butandiol derivatives [28]. Figure 2 summarizes the substrate scope for this reaction. The reaction is compatible with esters and arylhalides, but typically high-pressure hydrogen (50 atm) gas is required for reduction. The yields are generally high except for the substrate with an α-substitution (**9j**).

The reaction was further extended to a cascade sequence by enabling a double C–C bond cleavage reaction (Scheme 2) [29]. Cyclohexenone (**14a**), for example, was successfully isolated in 28% yield when spirocyclobutanone (**10a**) was treated with 5 mol% cationic [Rh(cod)dppe]$BF_4$ (condition A). The yield was improved to 80% and 89% when using catalysts such as Rh(dppe)$_2$Cl and Rh(dppp)$_2$Cl, respectively (conditions B and C).

Mechanistically, Murakami and co-workers suggest that $Rh^I$ first inserts into the α C–C bond of the cyclobutanone ring in **10a** via oxidative addition to generate the

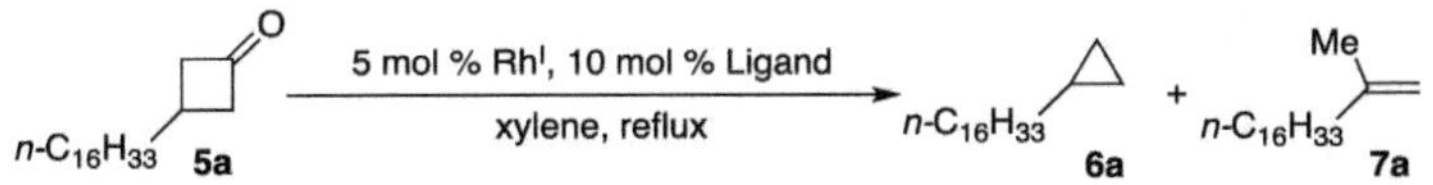


| Entry | Catalyst | Ligand | Time | Yield[a,b] | |
|---|---|---|---|---|---|
| 1 | *trans*-[Rh(CO)Cl($PPh_3$)$_2$] | None | 23 h | **6a**[a] 17% | **7a**[a] 6% |
| 2 | *trans*-[Rh(CO)Cl($PPh_3$)$_2$] | None | 5 d | **6a**[a] 40% | **7a**[a] 22% |
| 3 | [Rh(cod)dppb]$BF_4$ | None | 1 d | **6a**[a] 36% | **7a**[a] 32% |
| 4 | [Rh(cod)Cl]$_2$ | $AsPh_3$ | 7 d | **6a**[b] 68% | **7a**[b] 2% |

[a] GC yield. [b] Isolated yield.

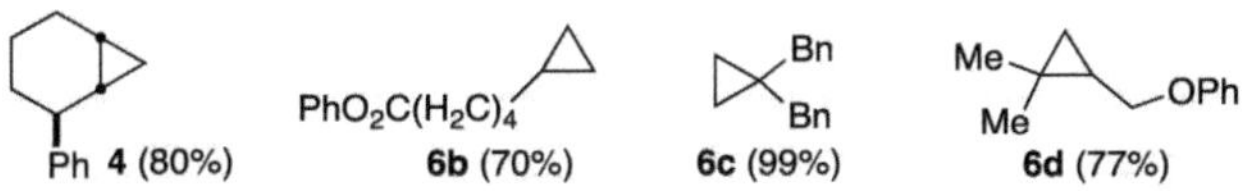


**Fig. 1** Catalytic decarbonylation of cyclobutanones

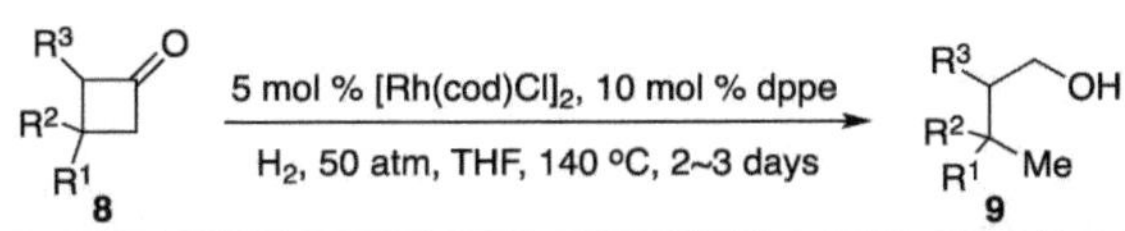


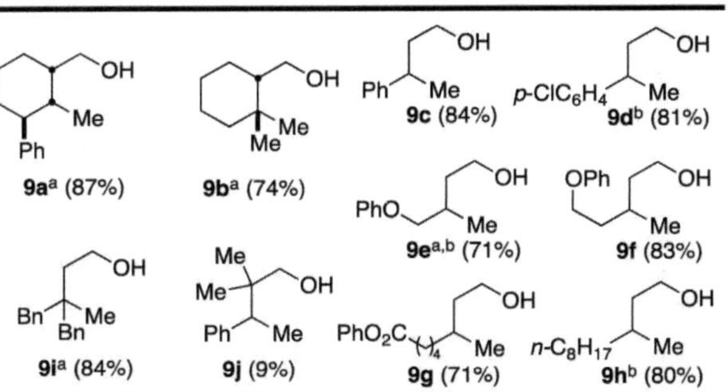


[a] Toluene was used as solvent. [b] After hydrogenolysis, reaction mixture was treated with $NaBH_4$.

**Fig. 2** Rh-catalyzed C–C activation/hydrogenation cascade

five-membered rhodacycle (**11**), which subsequently proceeds via β-C elimination to afford seven-membered rhodacycle **12**. Reductive elimination and double bond isomerization furnished cyclohexenone (**14a**). The authors believe strain release of the spiro four-membered ring provide a significant driving force for the reaction and is applicable to a variety of other three or four-membered spirocyclic cyclobutanones (Fig. 3). In substrates with nonequivalent C–C bonds, β-C elimination at the less sterically hindered C–C bond is favored, as demonstrated by substrates **10d** and **10e** in Fig. 3.

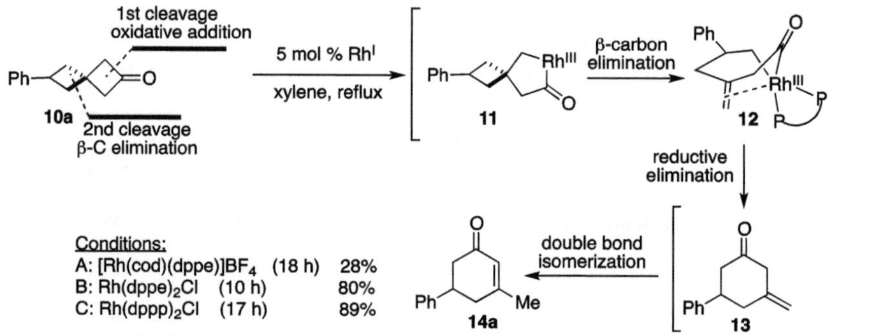


**Scheme 2** Rh-catalyzed successive C–C bond cleavage

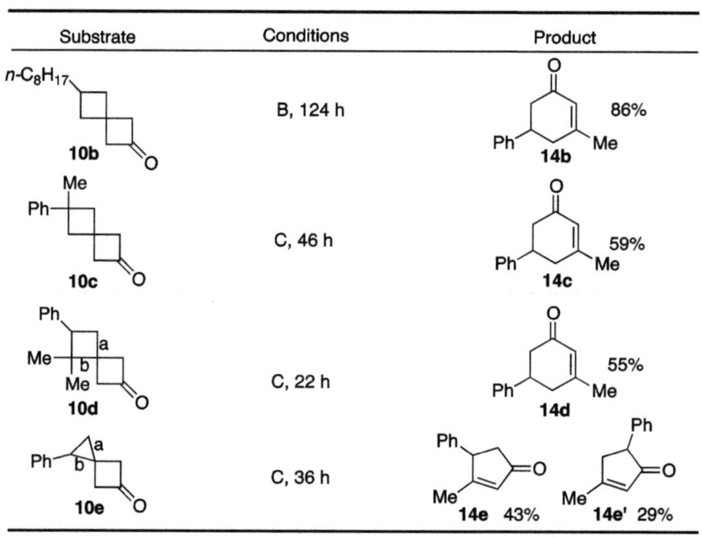


**Fig. 3** Substrate scope for successive C–C bond cleavage

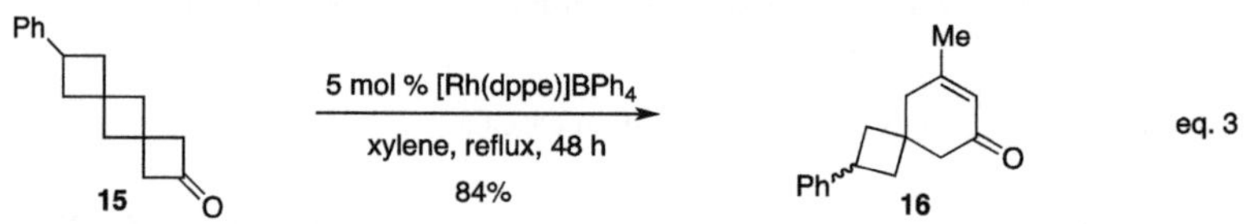


**Scheme 3** Double C–C bond cleavage

A triple C–C bond cleavage cascade was also attempted (Scheme 3) under the optimized conditions described below with compound **15**; however, only double C–C activation was observed, giving cyclohexenone **16** as the sole product.

A sequential C–C bond activation/C–O bond cleavage reaction was subsequently reported by the same group [30]. It was discovered that alternative reaction

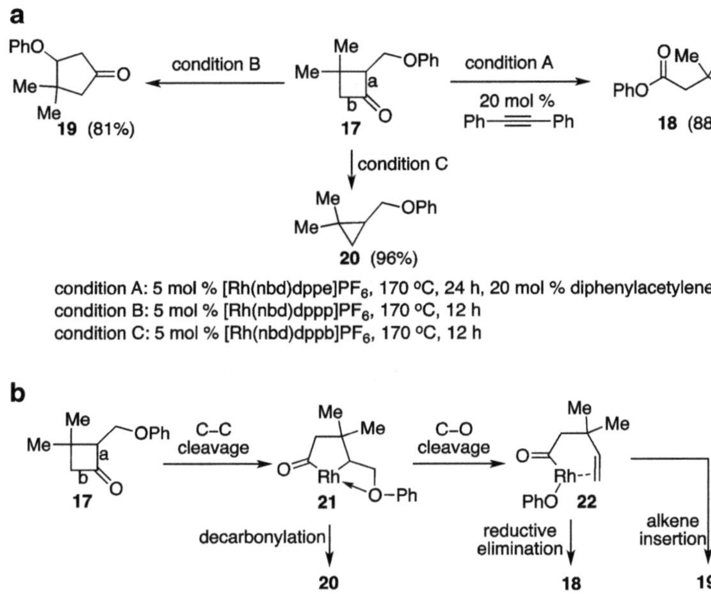


**Scheme 4** Rh-catalyzed C–C bond activation/C–O bond cleavage cascade

pathways are possible with different bidentate ligands (Scheme 4a). While two possible C–C bonds in cyclobutanone **17** can be activated, Murakami and co-workers suggested bond "a" would preferably undergo C–C bond activation resulting from the directing ability of the benzylic ether. Alternatively cleavage of bond "b" followed by decarbonylation and CO reinsertion can give the same intermediate **21**. The ether (–OPh) directing effects apparently do not govern the reaction, as cyclopentanone **19** was the predominant product (condition B in Scheme 4a). Cleavage of bond "a" can be induced by addition of diphenylacetylene to produce ester **18** (condition A in Scheme 4a). Presumably coordination of the diphenylacetylene competes with the olefin in intermediate **22** to prevent OPh reinsertion, thus favoring reductive elimination to produce **18**. Also, the decarbonylation product cyclopropane **20** could be achieved if a ligand with a large bite angle, such as dppb, was employed (condition C in Scheme 4a).

In 2000, inspired by the work of Liebeskind [31], Wender and co-workers reported [32] an intramolecular Rh-catalyzed [6+2] cycloaddition reaction between vinylcyclobutanone and terminal alkenes (Scheme 5). In this transformation 5 mol% $[Rh(CO)_2Cl]_2$, 10 mol% $PPh_3$, and 10 mol% AgOTf were employed and cyclooctenone **24** was afforded in 92% yield as a single diastereoisomer. Besides sulfonamides, other linkers such as ether and geminal diesters were also found to be compatible with this reaction condition using specified catalyst precursors.

In 2002, Murakami and co-workers reported [33] that they successfully trapped the five-membered rhodacycle **26** (Scheme 6a) intramolecularly with an alkene to afford benzocyclo[3.2.1]octanone **27**. The $^{13}$C-labeled substrate **25** strongly

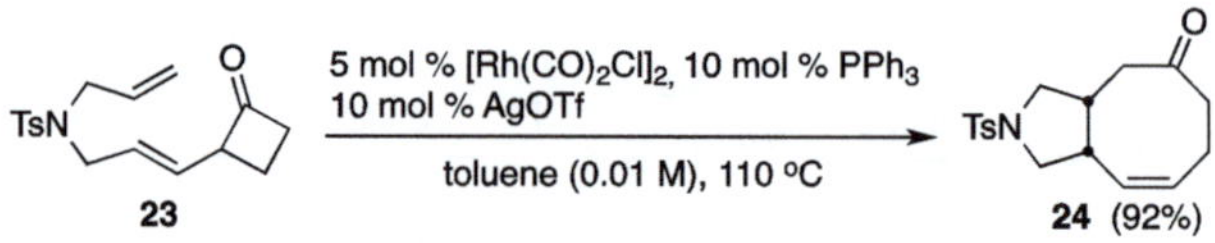


**Scheme 5** Rh-catalyzed [6+2] cycloaddition via C–C bond activation

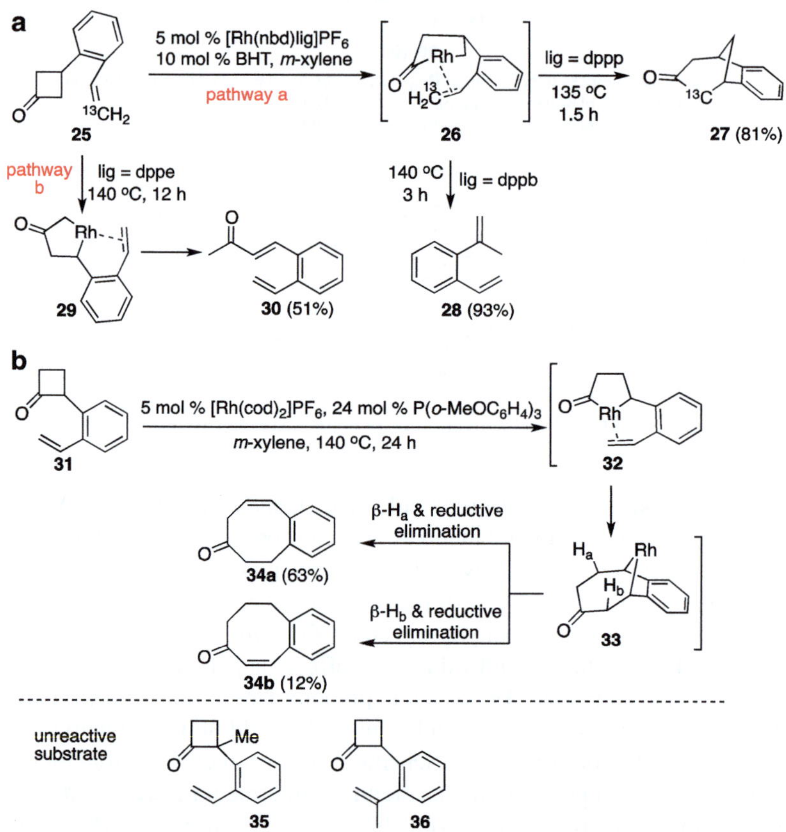


**Scheme 6** Rh-catalyzed intramolecular cyclobutanone-alkene coupling

supports "pathway a" where the alkene inserts into the α C–C bond of the cyclobutanone with $[Rh(nbd)dppp]PF_6$ as catalyst. The reaction outcome significantly depends on the ligand used, as switching from dppp to dppb or dppf gave completely different products. Decarbonylation was observed with dppb to give **28**. It is proposed that the wider bite angle with cationic Rh favors a four-membered rhodacycle as a result of steric repulsion, thus promoting a decarbonylation pathway. With dppe as ligand, the β C–C bond is likely cleaved (**29**) (pathway b) which after β-H elimination and reductive elimination yields **30** in 51% yield. In this scenario, the alkene presumably serves as a directing group to initiate the C–C cleavage.

The substitution of the cyclobutanone plays an important role in the outcome of the reaction as shown in a later report by Murakami [34]. 2-Substituted cyclobutanone (**31,** Scheme 6b) afforded benzocyclooctenones **34** under the reaction conditions. It was proposed that $Rh^{I}$ is directed by the terminal olefin to insert into the more hindered α C–C bond, followed by migratory insertion to form intermediate **33**. Non-selective β-H elimination of either $H_a$ or $H_b$ afforded the isomeric mixture of olefins **34a** and **34b**. Further exploration of the substrate scope found that additional steric bulk inhibits the reaction as neither substrate **35** nor **36** reacted.

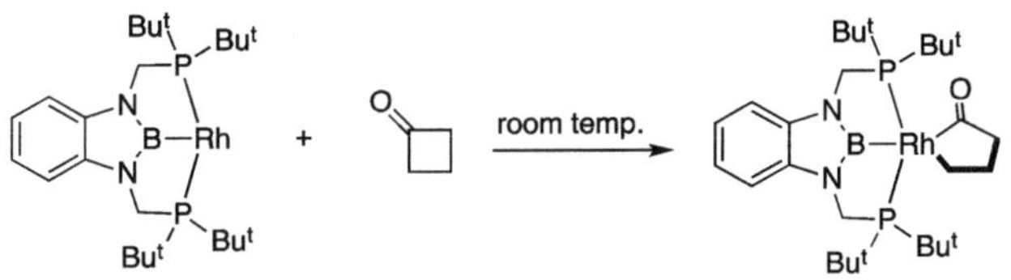


(2)

Very recently, Matsuda et al. reported [35] a pincer-$Rh^{I}$ complex that can cleave the α C–C bond of cyclobutanone at room temperature (2). The reactivity is attributed to the highly electron-donating nature of the boron ligand as well as the unsaturated coordination on the rhodium center.

## 2.2 *Rh-Catalyzed C–C Bond Activation via β-C Elimination*

In addition to direct oxidative addition, β-C elimination is another commonly used strategy to activate C–C bonds (Fig. 4). Utilizing this strategy, Murakami and co-workers reported both racemic and enantioselective C–C activation/C–O forming reactions with phenol-substituted cyclobutanone (**37**) in 2000 [36] and 2007 [37], respectively (Scheme 7a). They propose that the reaction follows a four-step sequence[1] in the catalytic cycle: (1) generation of the rhodium aryloxide **39** (Scheme 7b (i)), (2) nucleophilic addition to the carbonyl group to form rhodium cyclobutanolate **40** (Scheme 7b (ii)), (3) enantioselective β-C elimination to generate **41** (Scheme 7b (iii)), and (4) protonolysis affording **38a** and regenerated catalyst (Scheme 7b (iv)). In the last step, β-H elimination could also take place, thus providing **38b** as the product after isomerization. To further explore the transformation and the proposed mechanism, 2-substituted cyclobutanones **42** and **44** were subjected to the standard conditions (Scheme 7c) to provide seven-membered lactone **43** (β-C elimination with bond "a") and γ-lactone **45** (β-C elimination with bond "b"), respectively.

[1] Murakami suggested an oxidative addition mechanism in [35].

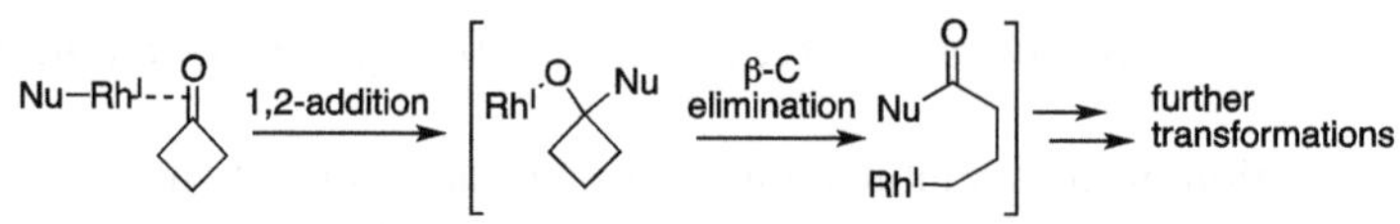


**Fig. 4** General C–C activation strategy via β-C elimination

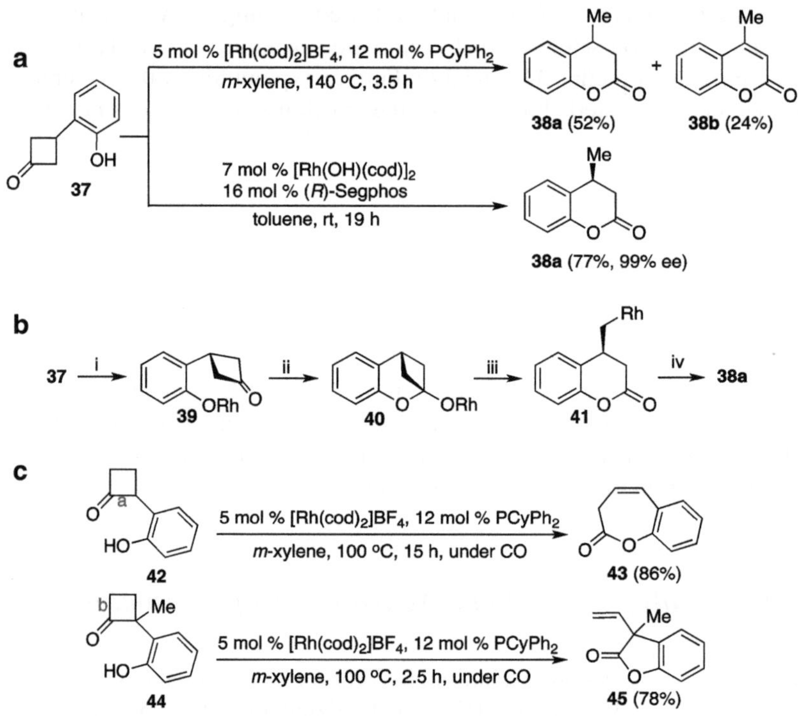


**Scheme 7** Rh-catalyzed addition/β-C elimination of cyclobutanones

Besides using Rh-alkoxide as the nucleophile, aryl-Rh species [38–40] have also been demonstrated to add onto cyclobutanone carbonyl groups, following a similar addition/β-C elimination sequence to afford either ring-opening or ring-expansion products (Scheme 8). For example, arylboronic acid/esters undergo transmetallation with $Rh^{I}$, forming aryl-Rh intermediates, which readily undergo nucleophilic addition into the cyclobutanone moiety, ultimately leading to the products shown. Murakami and co-workers later reported [41, 42] that $Pd^{II}$ could also catalyze this reaction by the analogous mechanistic pathway.

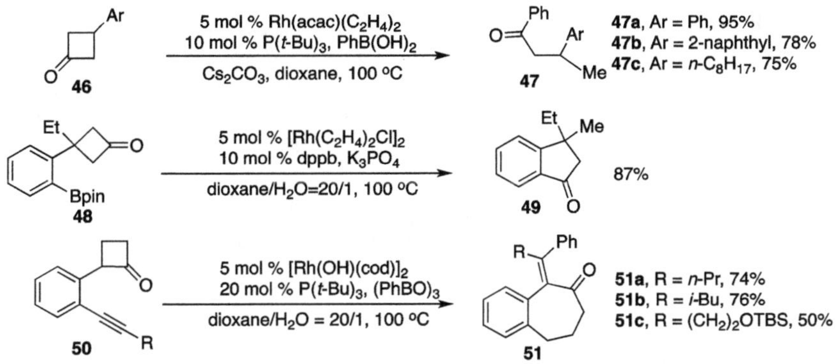


**Scheme 8** Aryl-Rh-catalyzed addition/β-C elimination of cyclobutanones

## *2.3 Ni-Catalyzed C–C Bond Activation via β-C Elimination*

In the arena of C–C bond activation via β-C elimination, Ni shows complementary reactivity to Rh and in fact has unique characteristics: (1) as a first row transition metal, Ni is usually more reactive than its second and/or third row counterparts when cyclometalation [43] is involved; (2) $Ni^0$-catalyzed aldehyde and alkyne/alkene coupling reactions have been developed [44].

With $Ni^0$ as a catalyst, an intermolecular [4+2] cycloaddition [45] reaction with cyclobutanone **52** and 4-octyne **53a** produced cyclohexenone **54a** in 95% yield. The proposed reaction mechanism is illustrated in Scheme 9. Presumably the reaction of **52** and **53a** with $Ni^0$ would proceed through oxidative cyclization to give oxanickelacyclopentene (**55**). β-C elimination cleaves the cyclobutane ring to generate **56** and leads to formation of product **54a** after reductive elimination. Overall, a formal [4+2] cycloaddition was accomplished with $Ni^0$ via β-C elimination. In contrast, Rh was not an effective catalyst for this transformation.

Murakami and co-workers further developed this reaction to a [4+2+2] cycloaddition [46, 47]. Cyclobutanone (**52**) can be effectively coupled with 1,6-diyne (**53b**) to afford bicyclo[6,3,0]undecadienone **54b** in excellent yield (91%). Two possible mechanisms were proposed for this transformation (Scheme 10). Either pathway leads to intermediate **57c**, which upon β-C elimination and reductive elimination of the four-membered ring furnishes the final product.

In addition, the Louie and Aissa groups reported similar transformations by activation of Boc-protected azetidinone (a recent DFT calculation suggests an alternative oxidative addition mechanism for the alkyne insertion into azetidinones [48]) and/or 3-oxetane as the coupling partner [49, 50]. A [4+2] coupling between protected azetidinones and internal alkynes was independently reported by the Louie [51] and Murakami groups [52]. As shown in Scheme 11, protected azetidinone (**58**) and internal alkynes (**59**) can undergo oxidative metallocyclization to afford the sterically more favored intermediate **61b**, which will afford the

**Scheme 9** Formal intermolecular [4+2] cycloaddition catalyzed by Ni$^0$

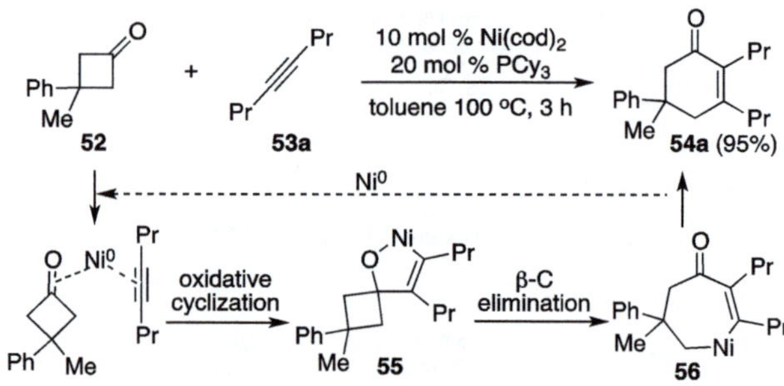


**Scheme 10** Formal [4+2+2] cycloaddition catalyzed by Ni$^0$

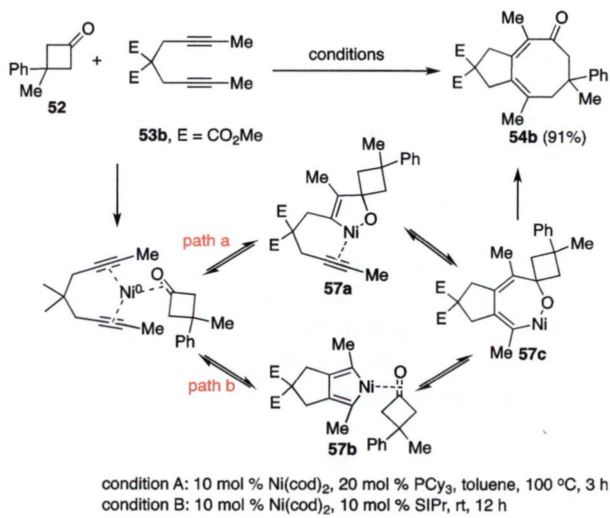


**Scheme 11** Ni-catalyzed 3-piperidone synthesis

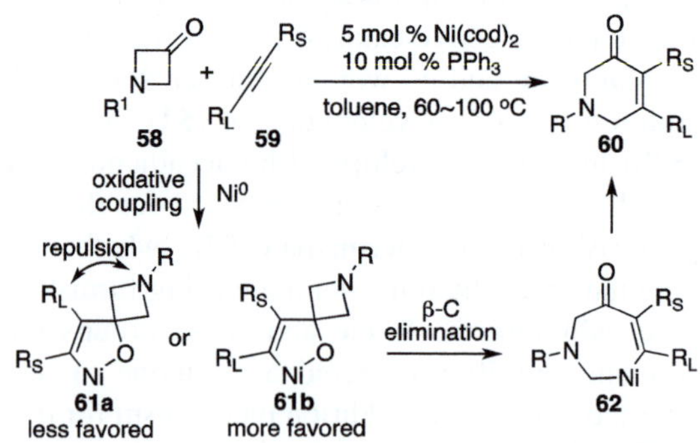


piperidone (**60**) after β-C elimination and reductive elimination. The yield of this reaction ranges from 56% to quantitative.

Besides coupling with alkynes, in 2006 Murakami and co-workers reported [53] a Ni-catalyzed intramolecular coupling of cyclobutanones with alkenes. An asymmetric version of this reaction was reported [54] by the same group in 2012

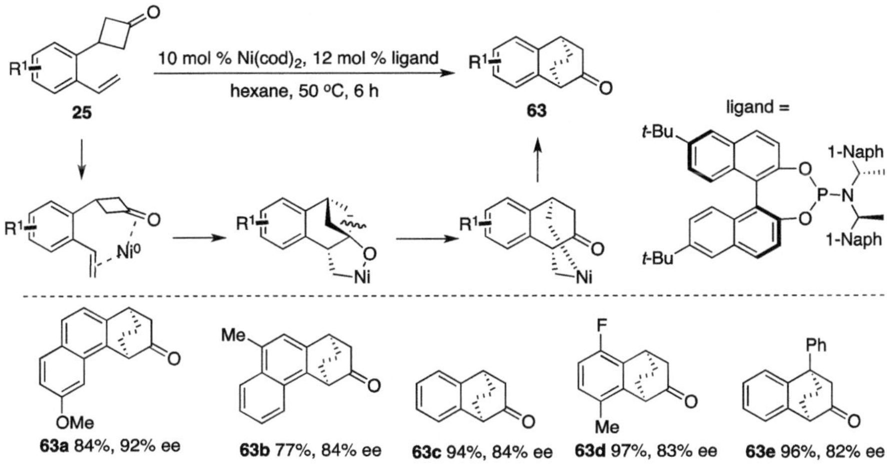


**Scheme 12** Asymmetric carboacylation of olefins catalyzed by $Ni^0$

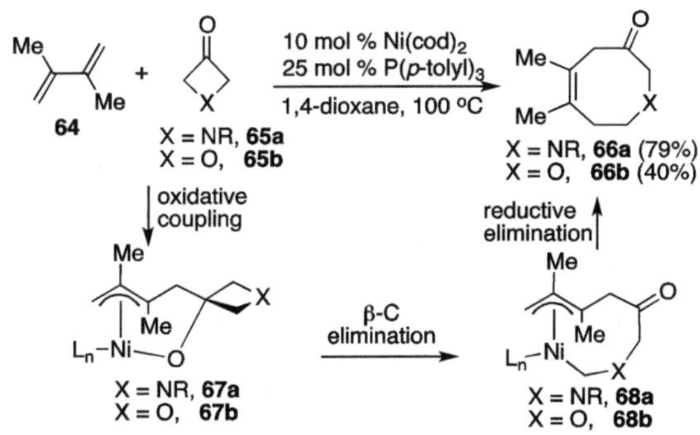


**Scheme 13** Ni-catalyzed cycloaddition of 1,3-dienes with heterocyclic four-membered ketones

(Scheme 12) (during the preparation of this manuscript, a new asymmetric reaction was reported [55]) [56]. A similar mechanism was proposed and benzobicyclo [2,2,2]octenone **63** was isolated in high yield (77–97%) and ee (80–93%). Thus far, this is a unique example of an intermolecular carboacylation of alkenes via C–C bond activation. One year later, Louie and co-workers reported a nickel-catalyzed cycloaddition of 1,3-dienes with 3-azetidinones and 3-oxetanones [57]. In their report, the combination of $Ni(cod)_2$ and monodentate phosphine $P(p\text{-tolyl})_3$ was found to successfully couple 1,3-dienes and 3-azetidinones/3-oxetanones and afford eight-membered heterocycles in medium to good yield. It is interesting to note that only 2,3-substituted dienes were suitable substrates, primarily because of sterics (Scheme 13).

# 3 C–C Bond Activation of Cyclobutenones

In addition to cyclobutanone-based substrates, their unsaturated counterparts, cyclobutenones also participate in C–C activation transformations (for thermal opening of cyclobutenones, see [58]). Although activation of cyclobutenones follows the same guiding principles (Scheme 1), they can proceed under alternate mechanistic pathways that allow for distinct outcomes and products. Due to their unsymmetrical nature, C–C cleavage reactions with cyclobutenones often have interesting site-selectivity challenges (Fig. 5). Cyclobutenones are considered as vinyl ketene equivalents [58]; thus, cleavage of the C1–C4 bond is generally kinetically favored. However, $sp^2$ carbon–metal bonds are known to be stronger than $sp^3$ carbon–metal bonds; thus cleavage of the C1–C2 bond can be thermodynamically preferred. Besides thorough studies of ring openings with stoichiometric transition metals, to date a number of catalytic and synthetically useful transformations have been developed.

## *3.1 Stoichiometric C–C Bond Activation*

Studies towards C–C bond activation of cyclobutenones predated cyclobutanones research with the pioneering work by Liebeskind and co-workers. They found [59, 60] that when cyclobutenone **69** was treated with an equimolar amount of $Rh(PPh_3)_3Cl$, rhodacyclopentenone **70** precipitated from the reaction via cleavage of the C1–C4 bond (Scheme 14). Cyclobutenones containing electron-deficient substituents were more reactive. A single-crystal X-ray structure was obtained for **70d**, which supported the molecular structure of the Rh-complex. The same transformation can also be performed on benzocyclobutenones, e.g., **71**. A mixture of products was observed when the reaction was stopped after 5 h. However, when the reaction was heated for 5 days, activation of the C1–C2 bond (bond "a") was observed as the major product affording a 30/1 ratio of **72b**/**72a**. It was found that **72a** can isomerize to **72b** upon heating at high temperature, suggesting **72a** is the kinetic product (130°C, 6 h). The authors speculated that the methylenedioxyl group in **71** may coordinate to the rhodium, leading to the more thermodynamically preferred product.

While these rhodacycles were found to be inert with alkynes [60], the concept of single C–C bond activation of cyclobutenones and/or benzocyclobutenones using late transition metals was still established. With an attempt to discover more reactive metallacycles, cobalt complex **73** was prepared and used in the stoichiometric C–C bond activation of cyclobutenones (**74**, Scheme 15).

Cobaltacyclopentenone **75a** was successfully obtained when **73** reacted with cyclobutenone **74a**, albeit in low yield, most likely because **74a** is less electrophilic. When a Lewis acid ($ZnCl_2$) was employed to enhance reactivity, a different regioisomeric product (**75b**) was observed. The proposed rationale involves a

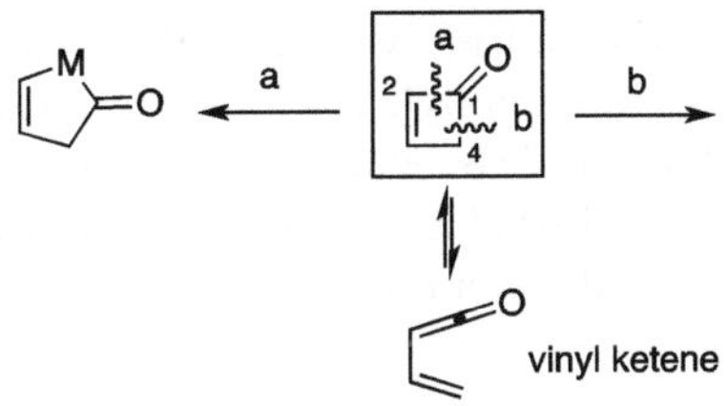


**Fig. 5** General strategy of C–C activation using cyclobutenone

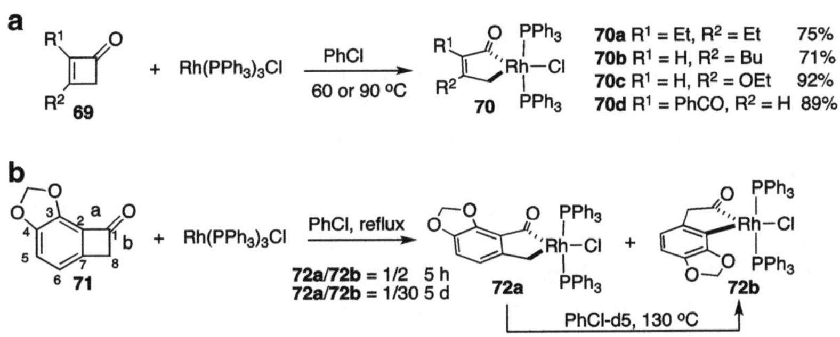


**Scheme 14** Rh-mediated stoichiometric C–C bond activation of cyclobutenones/ benzocyclobutenones

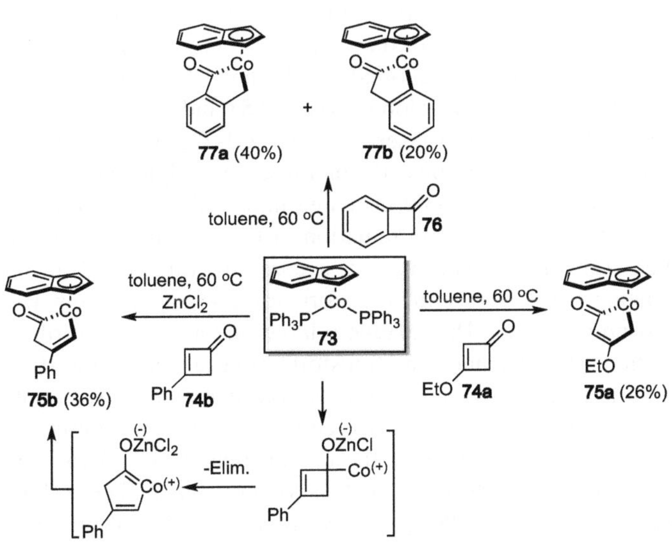


**Scheme 15** $Co^I$ complex-mediated C–C bond activation of cyclobutenone

stepwise C–C bond cleavage mechanism, wherein $ZnCl_2$ activation of **73** leads to cobalt nucleophilic attack at the carbonyl, followed by an α-C elimination and isomerization to afford metallacycle **75b**. When benzocyclobutenones were used as substrates both C–C bond cleaved products were observed; however, no isomerization of **77a** to **77b** was observed, even under harsher conditions.

## 3.2 Rh-Catalyzed C–C Bond Activation of Cyclobutenones

Liebeskind and co-workers showed that stoichiometric cobaltacyclopentenone species [61] could react with alkynes to furnish phenol derivatives, and later a Ni-catalyzed transformation was developed [62] (Fig. 6). The reaction shows less regioselectivity in alkyne insertion for internal unsymmetrical alkyne substrates.

C–C bond activation of cyclobutenones followed by ring expansion via β-C elimination cascade serves as a unique strategy to form medium-sized rings. Liebeskind and co-workers designed [31] a double C–C bond cleavage reaction of cyclopropyl-substituted cyclobutenones to product seven-membered rings. When substrate **78** was treated with 5 mol% $Rh(PPh_3)_3Cl$, cycloheptadienone **79** and its isomer **79′** were isolated in satisfactory yields (Scheme 16a). This method was also extended (Scheme 16b) to cyclobutyl-substituted substrate **82** where cyclooctadienone **83** was obtained in 90% yield.

Other catalytic transformations involving C–C bond activation of cyclobutenones have also been developed. For example, Kondo and co-workers reported [63, 64] an Rh-catalyzed dimerization of cyclobutenone **84** to form pyranones **85** (Scheme 17). Furthermore, they demonstrated that the rhodacyclopentenone intermediate can be trapped with reactive alkenes, such as norbornene, to give decarbonylation product **86** or direct insertion product **87**.

In 2012, Xu and Dong reported [65] the Rh-catalyzed intramolecular carboacylation between benzocyclobutenones (**88**) and olefins (Scheme 18). One unique feature is that the olefin inserts into the C1–C2 bond instead of the more reactive C1–C8 bond. They propose the olefin serves as both a directing group and trapping reagent for the C–C bond cleavage intermediate **89**. Through migratory insertion followed by reductive elimination, a tricyclic fused-ring compound **90** was furnished from this transformation, a core structure found in many natural products (Scheme 18). This racemic transformation was optimized with dppb as the bidentate phosphine ligand. The relatively large bite angle was attributed to facilitate this reaction. This "cut and sew" transformation can enable insertion of various olefins, including mono-, di-, and even tri-substituted alkenes with both alkyl and aromatic substituents. They also discovered that addition of a Lewis acid, such as $ZnCl_2$, as a co-catalyst can enhance the overall reactivity and can enable one to include more challenging substrates, such as tri-substituted alkenes (**90g**) and those that form hydropyran rings (**90e**). The asymmetric version of this transformation was developed later by the same group using (*R*)-dtbm-segphos as the chiral ligand and produced tricyclic ring scaffold **90** in 92–99% ee [66].

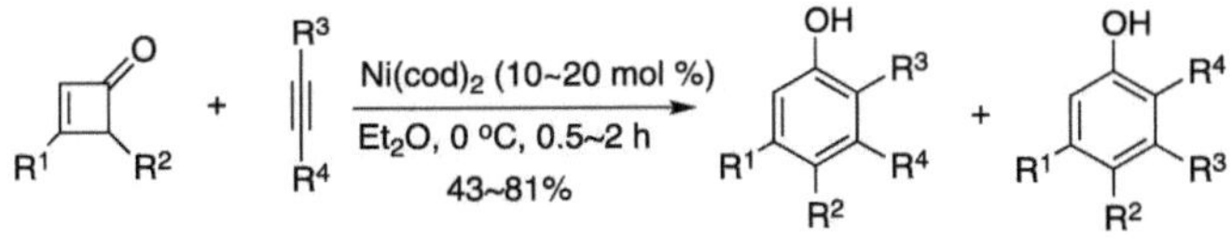


**Fig. 6** Ni-catalyzed cyclobutenone-alkyne couplings

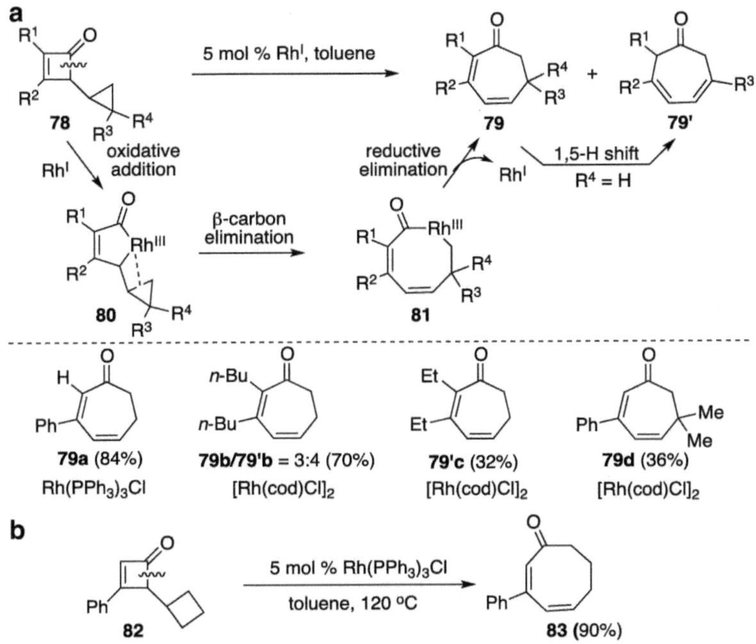


**Scheme 16** Rh-catalyzed C–C bond activation to make medium-sized rings

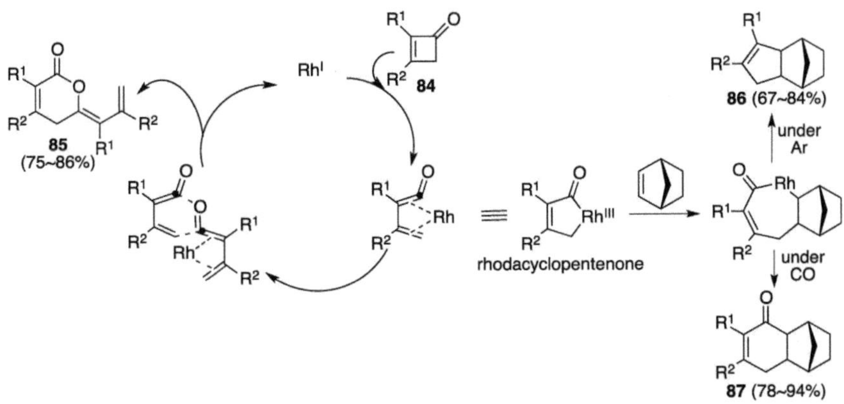


**Scheme 17** Intermolecular norbornene insertion via C–C bond activation

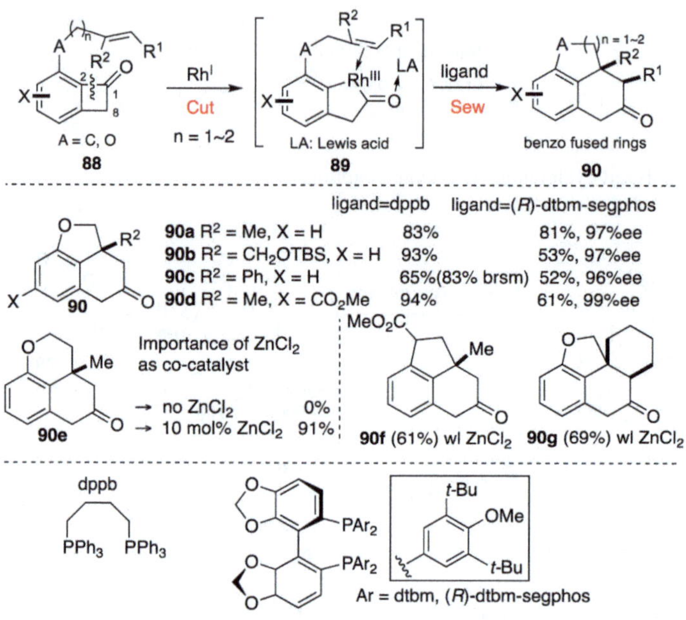


**Scheme 18** Rh-catalyzed carboacylation of olefins via C–C bond activation

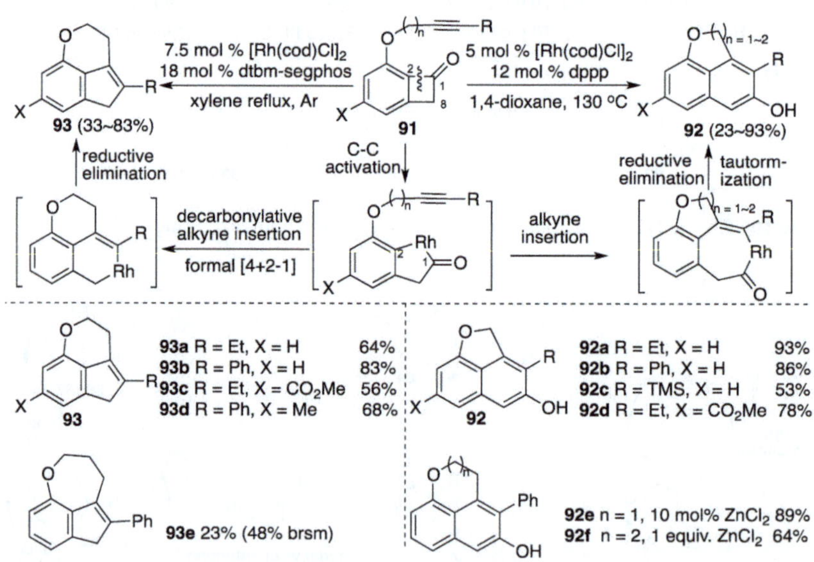


**Scheme 19** Rh-catalyzed intramolecular alkyne insertion via C–C bond activation

Catalytic intramolecular alkyne insertions into benzocyclobutenones were also recently developed [67] by Dong and co-workers (Scheme 19). Besides selectively forming the normal "cut and sew" product **92** (β-naphthols), the decarbonylative

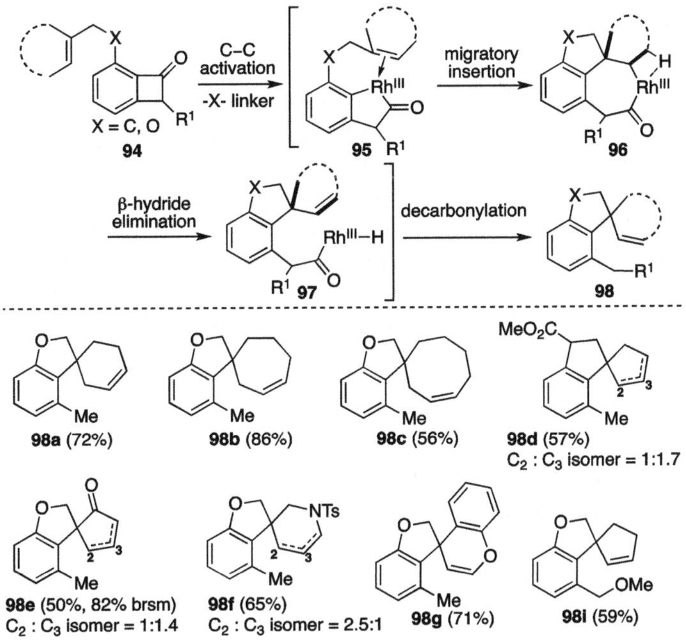


**Scheme 20** Rh-catalyzed multi-substituted olefin insertion featuring C–C activation/β-H elimination sequence to produce spirocyclic rings

insertion product (**93**) became the dominating product by switching to slightly different reaction conditions. While the reason is still unclear, dtbm-segphos ligand gave the highest selectivity for decarbonylation products. With this divergent strategy, a variety of fused β-naphthol and indene scaffolds could be obtained in good yields with high functional group tolerance.

Spirocyclic rings are commonly found in a variety of natural products, yet efficient methods to build these structural motifs with high functional group compatibility are limited. Very recently, Xu, Savage, and Dong reported [68], a Rh-catalyzed spirocyclization, via C–C bond activation of benzocyclobutenone **94** that contains a tri-substituted cyclic olefin. $[Rh(CO)_2Cl]_2$ (5 mol%) with tris (pentafluorophenyl)phosphine $[P(C_6F_5)_3]$ (10 mol%) was identified as an excellent catalytic system to carry out this transformation (Scheme 20). The electron-deficient nature of $P(C_6F_5)_3$ is the key to assisting rhodacycle **95** to insert into the sterically hindered poly-substituted olefins, which, upon β-H elimination and decarbonylation, leads to spirocycle products. Selective olefin chain walk was observed for a number of substrates (e.g. **98a–c**) whereas the cause for such selectivity is unclear. Substrates containing various ring sizes can undergo decarbonylative spirocyclization. In addition, many sensitive functional groups, such as dienes, ketones, enamides, esters, benzyl and vinyl ethers, and unprotected tertiary alcohols, are all compatible.

# 4 C–C Bond Activation of Cyclobutenediones

While several different substrates have been presented thus far, cyclobutenediones were among the first four-membered cyclic ketone substrates studied for C–C bond activation, primarily due to the combination of high strain energy, relative stability, and ready availability. The first report of cyclobutenedione C–C bond cleavage was published in 1973, when Kemmitt and co-workers found [69, 70] that benzocyclobutenedione (**99**) could react with $Pt(PPh_3)_4$ at even ambient temperature to afford platinumcyclopentadione (**100**) as red crystals (Scheme 21). Cyclobutenediones are also suitable substrates. This stoichiometric study was at the forefront of C–C bond activation of cyclobutenediones and benzocyclobutenediones.

## *4.1 Stoichiometric C–C Bond Activation of Cyclobutenediones*

Kemmitt and co-workers' pioneering work using $Pt(PPh_3)_4$ to effect the C–C bond cleavage of cyclobutenediones led to further developments with other transition metals [71–81]. In the 1980s, Liebeskind and co-workers found [71, 72] that when benzocyclobutenedione **99** was treated with $Rh(PPh_3)_3Cl$, $Co(PPh_3)_3Cl$ or $Fe(CO)_5$, metallacyclopentadiones **102** could be isolated in satisfactory yields (Scheme 22). In the case of rhodium, a kinetic product similar to **101** was detected initially, which most likely isomerizes slowly to the thermodynamically favored product **102**.

The phthaloylmetal (**102**) species can serve as a reactive four-atom precursor for the synthesis of 1,4-quinones [73–81]. Liebeskind and co-workers extensively investigated these intermediates, especially a phthaloylcobalt complex (**102a**). They observed that the reactions of **102a** with alkynes were extremely sluggish; however, addition of 2 equiv. of silver salt boosted the reactivity to provide 1,4-benzoquinone products in moderate to high yields [78]. The development of the methodology led to a total synthesis of nanaomycin A [77] (Scheme 23). Furthermore, mechanistic studies revealed that additional $PPh_3$ ligand decreased the reaction rate; dimethylglyoxime was found to be a more suitable ligand, which stabilized the phthaloylcobalt species while maintaining the reactivity with alkynes.

## *4.2 Catalytic C–C Bond Activation of Cyclobutenediones*

In 2000, Mitsudo and co-workers reported the first example of catalytic C–C bond activation/olefin insertion of cyclobutenediones [82, 83] using $Ru_3(CO)_{12}$ as the catalyst (Scheme 24). The authors proposed that $Ru_3(CO)_{12}$ inserts into bond "b" similar to $Pt(PPh_3)_4$ insertion (Scheme 21), after which decarbonylation and insertion into norbornene yields cyclopentenone **107**. When $^{13}CO$ was used, the

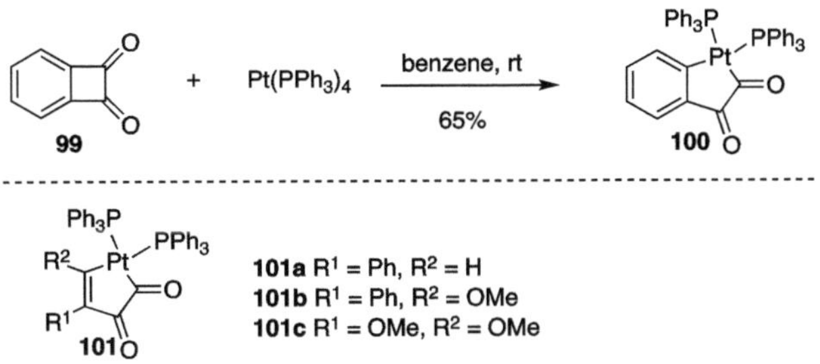


**Scheme 21** Pt-mediated stoichiometric C–C bond activation

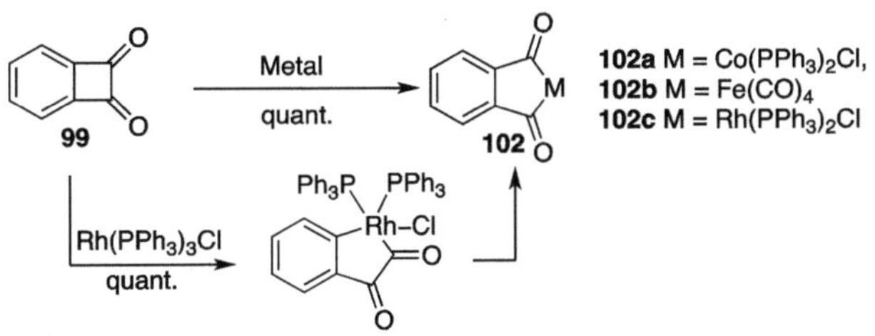


**Scheme 22** Activation of benzocyclobutenediones with Fe, Co, and Rh

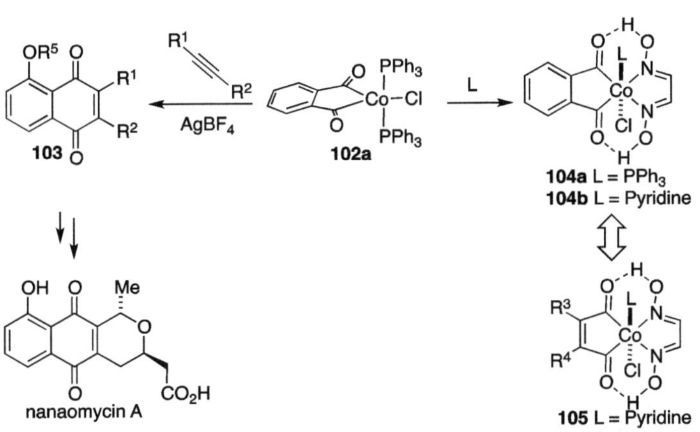


**Scheme 23** Total synthesis of nanaomycin A featuring C–C bond activation

$^{13}$C-labeled **107** was observed in 70% yield. The authors suggested an equilibration between the decarbonylated rhodacycle **109** and its disassembled form **110**, although CO exchange with complex **111a** is also possible. In addition, under high CO pressure (50 atm), the decarbonylation was suppressed and the direct norbornene-insertion product, hydroquinones, was isolated as the major product.

An intramolecular decarbonylative alkene insertion into cyclobutenediones to give azabicycloalkenones **113** was reported by Yamamoto [84]. The authors found

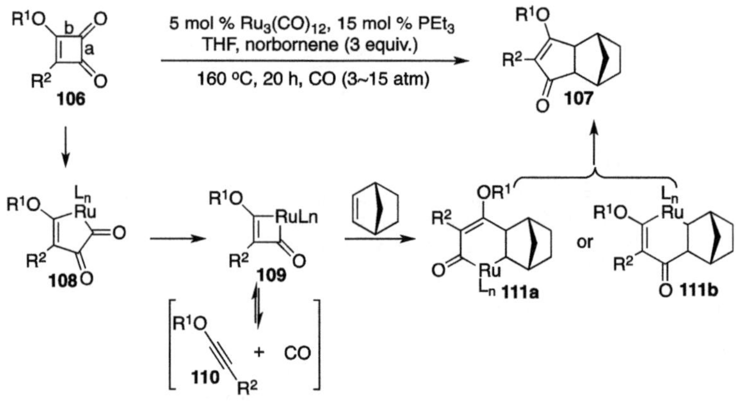


**Scheme 24** Intermolecular decarbonylative carboacylation of norbornene

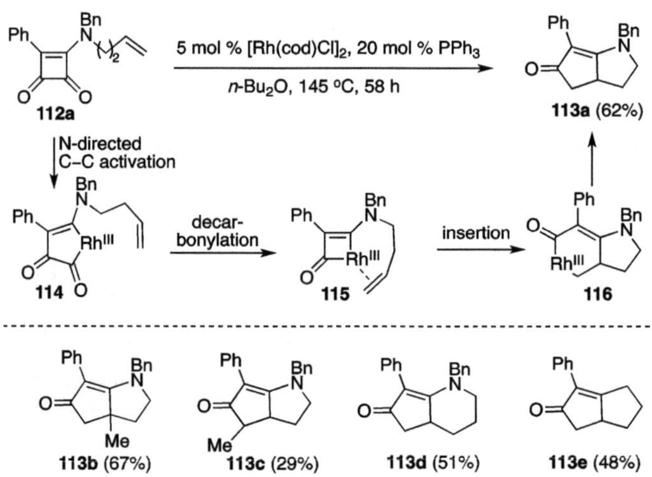


**Scheme 25** Rh-catalyzed intramolecular decarbonylative carboacylation

that the in situ generated Wilkinson's catalyst provided optimal results. An analogous mechanism involving C–C bond cleavage, decarbonylation, alkene migratory insertion, and reductive elimination, was proposed (Scheme 25). The nitrogen-linkage was not necessary as the methylene-mediated substrate also provided the desired product (**113e**).

## 5 Conclusion

While still in a developing stage, transition metal-catalyzed C–C bond activation of four-membered ring ketones has emerged as a useful synthetic methodology to enable transformations that are difficult via conventional approaches. Given that most of these four-membered ring ketones can be readily accessed, these methods provide new strategies to prepare various ring systems from relatively simple starting materials. Thus, these advancements allow for C–C bonds to be treated as a useful functional group rather than an inert bond with little synthetic value. Clearly, limitations still exist with most of these methods, such as needs of high reaction temperature, high catalyst loading and substrate restraints; consequently, few practical applications have been demonstrated to date. We expect future research in this field will likely focus on development of more efficient catalytic systems and reliable transformations with broad substrate scope and functional group tolerance while making their applications in complex molecule synthesis more practical.

**Acknowledgement** We thank UT Austin and CPRIT for a startup fund, NIGMS (R01GM109054-01) and the Welch Foundation (F 1781) for research grants. We thank Prof. Yoshiaki Nakao for proofreading this review chapter and thoughtful suggestions, and we also thank Dr. Jotham W. Coe for his generous efforts in editing the manuscript.

## References

1. Trost BM (1991) Science 254:1471–1477
2. Corey EJ, Cheng X-M (2009) The logic of chemical synthesis. Wiley-VCH, Weinheim
3. Nicolaou KC, Sorensen EJ (1996) Classics in total synthesis: targets, strategies, methods. Wiley-VCH, Weinheim
4. Nicolaou KC, Snyder SA (2003) Classics in total synthesis II: more targets, strategies, methods. Wiley-VCH, Weinheim
5. Nicolaou KC, Chen J (2011) Classics in total synthesis III. Wiley-VCH, Weinheim
6. Yu J-Q, Shi Z-J (eds) (2010) C–H activation. Springer, Berlin, Heidelberg
7. Winter C, Krause N (2009) Angew Chem Int Ed 48:2460–2462
8. Najera C, Snasano JM (2009) Angew Chem Int Ed 48:2452–2456
9. Seiser T, Cramer N (2009) Org Biomol Chem 7:2835–2840
10. Rybichinski B, Milstein D (1999) Angew Chem Int Ed 38:870–883
11. Jun C-H (2004) Chem Soc Rev 33:610–618
12. Murakami M, Ito Y (1999) Top Organomet Chem 3:97–129
13. Perthuisot C, Edelbach BJ, Zubirs DL, Simhai H, Iverson CN, Muller C, Satoh T, Jones WD (2002) J Mol Catal A 189:157–168
14. Ruhland K (2012) Eur J Org Chem 14:2683–2706
15. Ruben M, Rubina M, Gevorgyan V (2007) Chem Rev 107:3117–3179
16. Seiser T, Saget T, Tran DN, Cramer N (2011) Angew Chem Int Ed 50:7740–7752
17. Tipper CFH (1955) J Chem Soc 2045–2046
18. Cassar L, Eaton PE, Halper J (1970) J Am Chem Soc 92:6366–6368
19. Bishop KC (1976) Chem Rev 76:461–486

20. Crabtree RH, Dion RP (1984) J Chem Soc Chem Commun 1260–1261
21. Suggs JW, Jun CH (1984) J Am Chem Soc 106:3054–3056
22. Liebeskind LS, Baysdon SL, South MS, Lyer S (1985) Tetrahedron 41:5839–5853
23. Periana RA, Bergman RG (1986) J Am Chem Soc 108:1346–7355
24. Hartwig JF, Andersen RA, Bergman RG (1989) J Am Chem Soc 111:2717–2719
25. Gozin M, Wesman A, Ben-David Y, Milstein D (1993) Nature 364:699–701
26. Murakami M, Amii H, Ito Y (1994) Nature 370:540–541
27. Rusina A, Vlcek A (1965) Nature 206:295–296
28. Murakami M, Amii H, Shigeto K, Ito Y (1996) J Am Chem Soc 118:8285–8290
29. Murakami M, Takahashi K, Amii H, Ito Y (1997) J Am Chem Soc 119:9307–9308
30. Murakami M, Takahashi T, Amii H, Ito Y (1998) J Am Chem Soc 120:9949–9950
31. Huffman MA, Liebeskind LS (1993) J Am Chem Soc 115:4895–4896
32. Wender PA, Correa AG, Sato Y, Sun R (2000) J Am Chem Soc 122:7815–7816
33. Murakami A, Itahashi T, Ito Y (2002) J Am Chem Soc 124:13976–13977
34. Matsuda T, Fujimoto A, Ishibashi M, Murakami M (2004) Chem Lett 33:876–877
35. Masuda Y, Hasegawa M, Yamashita M, Nozaki K, Ishida N, Murakami M (2013) J Am Chem Soc 135:7142–7145
36. Murakami M, Tsuruta T, Ito Y (2000) Angew Chem Int Ed 39:2484–2486
37. Matsuda T, Shigeno M, Murakami M (2007) J Am Chem Soc 129:12086–12087
38. Matsuda T, Makino M, Murakami M (2004) Org Lett 6:1257–1259
39. Matsuda T, Shigeno M, Makino M, Murakami M (2006) Org Lett 8:3379–3381
40. Matsuda T, Makino M, Murakami M (2004) Angew Chem Int Ed 44:4608–4611
41. Matsuda T, Shigeno M, Murakami M (2008) Org Lett 10:5219–5221
42. Ishida N, Ikemoto W, Murakami M (2012) Org Lett 14:3230–3232
43. Montgomery J (2013) Organonickel chemistry. In: Organometallics in synthesis: fourth manual. Wiley, Hoboken, pp 319–428
44. Oblinger E, Montgomery J (1997) J Am Chem Soc 119:9065–9066
45. Murakami M, Ashida S, Matsuda T (2005) J Am Chem Soc 127:6932–6933
46. Murakami M, Ashida S, Matsuda T (2006) J Am Chem Soc 128:2166–2167
47. Ashida S, Murakami M (2008) Bull Chem Soc Jpn 81:885–893
48. Li Y, Lin Z (2013) Organometallics 32:3003–3011
49. Kumar P, Zhang K, Louie J (2012) Angew Chem Int Ed 51:8602–8606
50. Ho KYT, Aissa C (2012) Chem Eur J 18:3486–3489
51. Kumar P, Louie J (2012) Org Lett 14:2026–2029
52. Ishida N, Yuhki T, Murakami M (2012) Org Lett 14:3898–3901
53. Murakami M, Ashida S (2006) Chem Commun 4599–4601
54. Liu L, Ishida N, Murakami M (2012) Angew Chem Int Ed 51:2485–2488
55. Parker E, Cramer N (2014) Organometallics 33:780–787
56. Souillart L, Parker E, Cramer N (2014) Angew Chem Int Ed 53:3001–3005
57. Thankur A, Facer ME, Louie J (2013) Angew Chem Int Ed 52:12161–12165
58. Danheiser RL, Gee SK (1984) J Org Chem 49:1672–1674
59. Huffman MA, Liebeskind LS (1990) Organometallics 9:2194–2196
60. Huffman MA, Liebeskind LS (1992) Organometallics 11:255–266
61. Huffman MA, Liebeskind LS (1990) J Am Chem Soc 112:8617–8618
62. Huffman MA, Liebeskind LS (1991) J Am Chem Soc 113:2771–2772
63. Kondo T, Tagushi Y, Kaneko Y, Niimi M, Mitsudo T (2004) Angew Chem Int Ed 43:5369–5372
64. Kondo T, Miimi M, Nomura M, Wada K, Mitsudo T (2007) Tetrahedron Lett 48:2837–2839
65. Xu T, Dong G (2012) Angew Chem Int Ed 51:7567–7571
66. Xu T, Ko HM, Savage NA, Dong G (2012) J Am Chem Soc 134:20005–20008
67. Chen P, Xu T, Dong G (2013) Angew Chem Int Ed 53:1674–1678
68. Xu T, Savage NA, Dong G (2013) Angew Chem Int Ed 53:1891–1895

69. Evans JA, Everitt GF, Kemmitt RDW, Russell DR (1973) J Chem Soc Chem Commun 158–159
70. Hamner ER, Kemmitt RDW, Smith MA (1974) J Chem Soc Chem Commun 841–842
71. Liebeskind LS, Baysdon SL, South MS, Blount JF (1980) J Organomet Chem 202:C73–C76
72. Liebeskind LS, Baysdon SL, South MS (1980) J Am Chem Soc 102:7398–7400
73. Baysdon SL, Liebeskind LS (1982) Organometallics 1:771–775
74. Liebeskind LS, Leeds JP, Baysdon SL, Iyer S (1984) J Am Chem Soc 106:6451–6453
75. Jewell CF Jr, Liebeskind LS, Williamson M (1985) J Am Chem Soc 107:6715–6716
76. Iyer S, Liebeskind LS (1987) J Am Chem Soc 109:2759–2770
77. South MS, Liebeskind LS (1984) J Am Chem Soc 106:4181–4185
78. Liebeskind LS, Baysdon SL, Goedken V, Chidambaram R (1986) Organometallics 5:1086–1092
79. Cho SH, Wirtz KR, Liebeskind LS (1990) Organometallics 9:3067–3072
80. Liebeskind LS, Chidambaram R (1987) J Am Chem Soc 109:5025–5026
81. Hoberg H, Herrera A (1981) Angew Chem Int Ed 20:876–877
82. Kondo T, Nakamura A, Okada T, Suzuki N, Wada K, Mitsudo T (2000) J Am Chem Soc 122:6319–6320
83. Mitsudo T, Kondo T (2001) Synlett 309–321
84. Yamamoto Y, Kuwabara S, Hayashi H, Nishiyama H (2006) Adv Synth Catal 348:2493–2500

# Index

MIX
Papier aus verantwortungsvollen Quellen
Paper from responsible sources
FSC® C105338

If you have any concerns about our products,
you can contact us on
**ProductSafety@springernature.com**

In case Publisher is established outside the EU,
the EU authorized representative is:
**Springer Nature Customer Service Center GmbH**
**Europaplatz 3, 69115 Heidelberg, Germany**

Printed by Libri Plureos GmbH
in Hamburg, Germany